"This book is unique in that it prov
analysis that moves beyond comn
able victims to show there are no universal
Gender is highly relevant but in complex ways." – *Annica Kronsell, Professor, Political Science, Lund University, Sweden*

"These are elegantly written essays that urgently address the dearth of information about the implications of climate change by gender in the rich countries. This volume takes stock of the current global order and sets a compelling research and political agenda for tackling the systemic changes needed for progress whilst remaining sensitive to the intersectionality of experiences brought to life through this important book." – *Isabella Bakker, FRSC, Distinguished Research Professor, York University, Canada*

"By putting gender at the centre of its analysis and policy discussions this path-breaking book highlights that climate change poses inescapable challenges for all of us. Crucially it also points to the need for activism and contestation in order to forge a sustainable and fairer world for future generations." – *Rhonda Sharp, Professor of Economics, University of South Australia, Australia and former President of the International Association for Feminist Economics*

"This is a timely volume that breaks a strange silence: it provides critical analysis and compelling evidence that gender inequality shapes experiences of and responses to climate change as much in rich countries as it does in poor countries, albeit in different ways. It is an invaluable resource to all of us who are committed to understanding climate change as a feminist and social justice issue." – *Sherilyn MacGregor, Reader in Environmental Politics, The University of Manchester, UK*

"Putting a gender lens on climate change is putting a gender lens on Aboriginal issues, forestry, natural disasters, just transition, agriculture, water, energy, jobs, health, resource extraction, government policies, food security, mitigation and adaptation, housing, and transportation. Reading this book exposes the injustices and offers concrete solutions." – *Donald Lafleur, Vice Président Exécutif, Canadian Labour Congress, Canada*

"This excellent, wide-ranging, multi-disciplinary collection makes a welcome and valuable contribution to redressing the gender balance in rich countries, and to cultivating broader gender and climate change scholarship." – *Karen Morrow, College of Law and Criminology, Swansea University, Wales, UK*

"An exciting collection of top scholars comes together in this path-breaking book to decipher the collision of two of today's hottest political topics: gender and climate change. It reveals how this massive problem of climate change is better tackled when gender forms the centre of policy solutions." – *Kennedy Stewart, MP, Opposition Science Critic, New Democratic Party, Parliament of Canada*

CLIMATE CHANGE AND GENDER IN RICH COUNTRIES

Climate change is at the forefront of ideas about public policy, the economy and labour issues. However, the gendered dimensions of climate change and the public policy issues associated with it in wealthy nations are much less understood.

Climate Change and Gender in Rich Countries covers a wide range of issues dealing with work and working life. The book demonstrates the gendered distinctions in both experiences of climate change and the ways that public policy deals with it. The book draws on case studies from the UK, Sweden, Australia, Canada, Spain and the US to address key issues such as: how gendered distinctions affect the most vulnerable; paid and unpaid work; and activism on climate change. It is argued that including gender as part of the analysis will lead to more equitable and stronger societies as solutions to climate change advance.

This volume will be of great relevance to students, scholars, trade unionists and international organisations with an interest in climate change, gender, public policy and environmental studies.

Marjorie Griffin Cohen is an economist and a Professor Emeritus of Political Science and Gender, Sexuality and Women's Studies at Simon Fraser University, Canada.

Routledge Studies in Climate, Work and Society

In *The Routledge Studies in Climate, Work and Society* series, scholars and other thinkers at the forefront of constructing a strategic link between work and climate change contribute to identifying the issues, evaluating policies and silences, tracking change, and stimulating international exchange of ideas and experience. Collectively, the books in this series will emphasise fresh thinking, strategic creativity, international and inter-sectoral comparisons and contribute to the further development of the role of work in societal responses to global warming.

Workers and Trade Unions for Climate Solidarity
Tackling climate change in a neoliberal world
Paul Hampton

Climate Change and Gender in Rich Countries
Work, public policy and action
Edited by Marjorie Griffin Cohen

CLIMATE CHANGE AND GENDER IN RICH COUNTRIES

Work, Public Policy and Action

Edited by Marjorie Griffin Cohen

LONDON AND NEW YORK

First published 2017
by Routledge
2 Park Square, Milton Park, Abingdon, Oxon OX14 4RN

and by Routledge
711 Third Avenue, New York, NY 10017

Routledge is an imprint of the Taylor & Francis Group, an informa business

British Library Cataloguing-in-Publication Data
A catalogue record for this book is available from the British Library

Library of Congress Cataloging-in-Publication Data
Names: Cohen, Marjorie Griffin, 1944–editor.
Title: Climate change and gender in rich countries : work, public policy and action / edited by Marjorie Griffin Cohen.
Description: Abingdon, Oxon ; New York, NY : Routledge, 2017. |
Series: Routledge studies in climate, work and society
Identifiers: LCCN 2016055884| ISBN 9781138222397 (hb) | ISBN 9781138222403 (pb) | ISBN 9781315407906 (ebk)
Subjects: LCSH: Women and the environment—Developed countries. | Climatic changes—Economic aspects—Developed countries. | Climatic changes—Social aspects—Developed countries. | Labor—Environmental aspects—Developed countries. | Environmental policy—Developed countries.
Classification: LCC GE195.9 .C55 2017 | DDC 363.738/74561081—dc23
LC record available at https://lccn.loc.gov/2016055884

ISBN: 978-1-138-22239-7 (hbk)
ISBN: 978-1-138-22240-3 (pbk)
ISBN: 978-1-315-40790-6 (ebk)

Typeset in Bembo
by Swales & Willis Ltd, Exeter, Devon, UK

Printed and bound by CPI Group (UK) Ltd, Croydon, CR0 4YY

This book is dedicated to Devlin, Sukey, Zoe, Talia and Clara, and the world of their future.

CONTENTS

FIGURES

TABLES

CONTRIBUTORS

Margaret Alston is Professor of Social Work and Head of Department at Monash University, Australia, where she has established the Gender, Leadership and Social Sustainability (GLASS) research unit. In 2010 she was awarded an Order of Australia for her services to social work and to rural women. She has published widely in the field of gender and climate change.

Leonora C. Angeles is cross-appointed faculty member at the School of Community and Regional Planning and Institute for Gender, Race, Sexuality and Social Justice at the University of British Columbia, Vancouver, Canada. She is also faculty research associate at the UBC Centre for Human Settlements and the Institute for Asian Research.

Jonas Anshelm is a professor of Tema Technology and Social Change, Linköping University, Sweden. He is mainly occupied with the political history of ideas, and has published frequently in the area of climate discourses, geoengineering, mining and energy politics. His latest book, co-authored with Martin Hultman, is *Discourses of Global Climate Change.*

Bipasha Baruah is the Canada Research Chair in Global Women's Issues, and associate professor of women's studies and feminist research at Western University, Canada. She conducts research on women and work, and social, political and economic inequality.

Susan Buckingham has thirty years' experience of academic research on gender and environmental issues, and fifteen years' experience of volunteering in women's organizations. She now works independently on projects that advance environmental and gender justice, such as being a gender advisor on a number of European Union projects, including urban waste management in tourist areas, and marine research. Susan continues to write extensively on how gender is related to environmental problems, and is the series editor for the new Routledge book series on Gender and Environment.

Chris Buse is the Project Lead for the Cumulative Impacts Research Consortium at the University of Northern British Columbia (UNBC), Canada, which is dedicated to promoting an integrative understanding of the cumulative environmental, community and health impacts of resource development across northern British Columbia. He is also a Michael Smith Foundation for Health Research Postdoctoral Fellow and Adjunct Professor with the School of Health Sciences at UNBC. He received his PhD from the University of Toronto's Dalla Lana School of Public Health.

Nathalie J. Chalifour is an associate professor at the Centre for Environmental Law and Global Sustainability, University of Ottawa, Canada, Nathalie's main area of research is environmental law and policy, with a focus on the intersection between the environment, the economy and social justice. Her publications examine a variety of topics, including climate change, carbon taxation, environmental justice, the green economy, sustainable food policy and environmental human rights. She obtained her Doctorate of Law at Stanford University, US, and holds a Masters of Juridical Sciences, which she obtained as a Stanford Fellow and Fulbright Scholar.

Linda Clarke is Professor of European Industrial Relations at the University of Westminster, UK, Co-director of the Centre for the Study of the Production of the Built Environment (ProBE) and on the European Institute for Construction Labour Research (CLR) board. She has extensive experience of comparative research on labour, equality and diversity, vocational education and training (VET), low energy construction, and wage relations in the European construction sector.

Marjorie Griffin Cohen is an economist who is Professor Emeritus of Political Science and Gender, Sexuality and Women's Studies at Simon Fraser University, Canada. She has written extensively in the areas of political economy and public policy with special emphasis on issues concerning the economy, women, labour, electricity deregulation, energy, climate change and international trade agreements. She was recently awarded the John Kenneth Galbraith prize in Economics and Social Justice.

Kendra Coulter is an associate professor in the Centre for Labour Studies at Brock University, Canada, and the author of *Animals, Work, and the Promise of Interspecies Solidarity*. She was awarded the Canadian Association for Work and Labour Studies 2015 Book Prize for *Revolutionizing Retail: Workers, Political Action, and Social Change*. She is leading a multi-year project on humane jobs, and more information is available at http://brocku.ca/kcoulter and http://humanejobs.org.

Shayna Dolan is a graduate student in the Master of Science Community Health programme at the University of Northern British Columbia, Canada. Her thesis explores the health and well-being of women living in a rural and remote Northern community dependent on natural resource extraction.

Amber J. Fletcher is Assistant Professor of Sociology and Social Studies at the University of Regina, Canada. Her research examines the social and gender dimensions of climate change and public policy, with particular focus on how climate hazards and major changes in agricultural policy affect the lives of farm women.

Maya K. Gislason is an assistant professor in the Faculty of Health Sciences at Simon Fraser University, Canada, as well as the faculty lead for the Social Inequities and Health concentration within the Masters in Public Health Professional Program. Maya is actively engaged in building the field of ecosystem approaches to health both in Canada and internationally. Her research centres around the study of social-ecological "game changers" such as intensive resource extraction, climate change and infectious diseases and their impacts on human, animal and ecosystem health, with a particular focus on the health of Indigenous and non-Indigenous women and children.

Colin Patrick Gleeson is a reader in Energy and Buildings at the University of Westminster, UK, a chartered building services engineer and deputy director of the Centre for the Study of the Production of the Built Environment (ProBE). His research is focused on the design, construction and performance of low energy buildings and renewable energy technologies.

Penny Gurstein is Professor and Director of the School of Community and Regional Planning and the Centre for Human Settlements at the University of British Columbia, Canada. She specializes in the socio-cultural aspects of community planning with particular emphasis on those who are the most marginalized in planning processes. Her research focuses on developing strategies and interventions that encourage diversity, equity and urban sustainability in the planning and design of communities.

Martin Hultman is an associate professor at Linköping University, Sweden. His publications include *The Making of an Environmental Hero: A History of Ecomodern Masculinity, Fuel Cells and Arnold Schwarzenegger*, *Discourses of Global Climate Change* and *A Green Fatwā? Climate Change as a Threat to the Masculinity of Industrial Modernity*. His current research deals with post-humanities ethics, ecofeminism, environmental utopias and ecopreneurship. He is currently co-writing a book entitled *Ecological Masculinities*.

Michael Kim is an independent researcher in Vancouver, British Columbia, Canada. He has recently completed a Master's Degree in political science at Simon Fraser University, Canada, with a concentration in development and sustainability.

Rakibe Külcür is a regulatory environmental and health and safety consultant. Her PhD at Brunel University, London, UK, explored gender structures and working practices of ENGOs in the UK and Turkey and the implications of this for their environmental campaigns. Her academic interests include gender, environment, organizations, politics, climate change and environmental justice.

Carl Mandy is in the MA programme in Gender, Sexuality and Women's Studies at Simon Fraser University, Canada, and was previously in the MA programme in Philosophy at Birkbeck, University of London, UK. His research interests include political activism, feminist theory, political and moral philosophy, and gender and film, and he has been active in trade unions in the UK and Canada.

Sara Ortiz Escalante is a PhD candidate at the School of Community and Regional Planning, University of British Columbia, Canada. She is researching the ways urban planning impacts the lives of women nightshift workers. She is also an urban planner at Col·lectiu Punt 6, a cooperative of feminist planners and architects in Barcelona that works to rethink domestic, community and public spaces from an intersectional gender perspective.

Margot W. Parkes is a Canada Research Chair in Health, Ecosystems and Society, at the University of Northern British Columbia, Canada. Her work probes our understanding of the environment as a context for health, and seeks to integrate social and ecological determinants of health, focusing on interrelated issues spanning land and water governance and climate change. Recent research in northern British Columbia focuses on intersectoral action to improve health, and on the cumulative health, environment and community impacts of resource development.

Patricia E. Perkins is a professor in the Faculty of Environmental Studies, York University, Toronto, Canada, where she teaches ecological economics, community organizing and critical interdisciplinary research design. Her research focuses on feminist ecological economics, climate justice and participatory governance. She is the editor of *Water and Climate Change in Africa: Challenges and Community Initiatives in Durban, Maputo and Nairobi* and an editor of the journal *Ecological Economics*.

Maureen G. Reed is Professor and Assistant Director, Academic at the School of Environment and Sustainability at the University of Saskatchewan, Canada. She works with organizations and communities that seek to model sustainability, including biosphere reserves, model forests, community forests and Indigenous communities. She is now working with some of these communities to better understand how gender and culture affect their capacity to adapt to climate hazards.

Annie Rochette is the Deputy Director of the Professional Legal Training Course for the Law Society of British Columbia. Before taking this position in 2016, she was an associate professor at the Université du Québec à Montréal (UQAM), Canada, where she taught environmental law, international environmental law and methodology courses. Her research interests include feminist and ecofeminist approaches to environmental law, gender and climate change, as well as legal education, at both the national and international levels.

Jemma Tosh is a researcher at the Faculty of Health Sciences at Simon Fraser University, Canada. She is also the Director of Psygentra Consulting Inc., a company that specializes in support, consulting and research related to psychology, gender and trauma.

Christine Wall is a reader in Architectural and Construction History in the Faculty of the Built Environment and co-director of the Centre for Research into the Production of the Built Environment at the University of Westminster, UK. She has researched and published widely on the history of the built environment and recently led the Leverhulme Trust funded, oral history project Constructing Post-War Britain: Building workers' stories 1950–1970. She is an editor of the Oral History Journal and a trustee of the Construction History Society.

Bob Woollard is Professor of Family Practice at the University of British Columbia, Canada. He has extensive national and international experience in medical education, social accountability, poverty reduction, rural health services and ecosystem approaches to health. With Maya Gislason he co-chaired the development of the global consensus for an action plan on climate change for the International Association for Ecology and Health ratified at Montreal in 2014. He continues the active practice of medicine.

ACKNOWLEDGEMENTS

I am grateful to the contributors to this volume for the research undertaken and for the work involved in publishing it here. The book is a result of an international researchers' workshop held in 2015, which that included academic researchers and practitioners from trade unions, community groups and environmental organizations. This three-day workshop involved paper presentations, but also critiques of the papers and considerable discussion in order to shape the final papers that are published here.

The international workshop received funding from a Canadian Social Sciences and Humanities Research Council Connections Grant, the *Work in a Warming World* (W3) York University Research Group, the Simon Fraser University Dean of Arts and the Simon Fraser University Vice-president Research. The book was specifically supported by a W3 grant and research funding from Simon Fraser University.

Many individuals also need to be thanked for their role in bringing this research to light: Carla Lipsig-Mummé for her role as the principle investigator of the W3 project and as the series editor of the *Routledge Studies in Climate, Work and Society*. The genesis of this book grew out of the W3 project. The workshop itself was organized with the enormous help of research assistants at Simon Fraser University and I particularly want to thank Jessica Knowler, Skyler Warren, Michael Kim and Carl Mandy. As with other editions I've published, Margaret Menton Manery compiled the index and graciously and expertly helped guide the manuscript to completion. I would also like to thank Margaret Farrelly and others at Routledge for their support and assistance.

Marjorie Griffin Cohen, Vancouver, BC, Canada

ACKNOWLEDGEMENTS

PART I

Context and overview

1

INTRODUCTION

Why gender matters when dealing with climate change

Marjorie Griffin Cohen

Climate change is the one global issue that is both ubiquitous and long-term: there is no one on earth who will not be affected. In the summer of 2016 the urgency of dealing with this monumental problem was underscored by it being the hottest summer on record. But this was not an isolated event because 2015 was the hottest year for the world up until then, and 2014 was the hottest year before that. Fifteen of the sixteen warmest years on record have happened in the twenty-first century. Just as alarming is that this warming is occurring at a more rapid rate than ever seen in the past 1,000 years (*The Climate Examiner* 2016).

While climate change deniers still exist, most governments now understand that they need to take action to slow the progression of climate change to a rate that will allow life to be tolerable in most parts of the planet (UN *Paris Agreement* 2015). The problem is that it is neither certain that the current objectives (of no more than a 2°C increase) are adequate nor that the actions and plans of governments, either singly or collectively, are sufficient to make a difference to the current trajectory. The major international agreement to reduce GHG (greenhouse gas) emissions, the *Paris Agreement*, is criticized by many environmental scientists and other groups for being too weak to compel genuine change because it consists mostly of intentions and aims, rather than real instruments to compel countries to enforce compliance (Milman 2015). Others see actual climate change denial in the actions of those governments who, as do most governments represented in this book, continue to expand the fossil fuel industry while at the same time professing support for GHG reductions (Daub & Klein 2016).

This book on *Climate Change and Gender in Rich Countries: Work, Public Policy and Action* grew out of a researchers' workshop that took place in Vancouver, BC, Canada in June 2015. The workshop, "Climate change, gender and work in rich countries", brought together researchers working on gender and climate change from developed, wealthy countries to share their research and to discuss the issues inherent to making gendered issues, particularly those regarding work, a part of the discussions and debates on climate change policies (SFU, *Climate change, gender and work in rich countries* 2015). The focus on rich countries was deliberate because of the dearth of information about the implications of climate change by gender in these countries. During the workshop researchers presented research related to Australia, Canada, Spain, Sweden, the United Kingdom and the United States of America. While this was a researchers' workshop, practitioners from trade unions, environmental organizations and community groups also presented papers and discussed the challenges for including gender issues in both analysis and action on climate change.

Gender analysis in rich countries

The workshop was specifically about "rich" nations because while governments in highly developed nations have begun to look at the impact of climate change on people, and while the inclusion of gender considerations is sometimes mentioned, it is rarely central to public policy analysis of the problem, plans to deal with it, or the implementation of climate change policy. The absence of a gendered analysis in policy discussions in decision-making venues is troubling, but is recognizable as consistent with most problems and debates on policy issues in wealthy, highly developed capitalist countries. There are many reasons for this, but an obvious one is that a gendered analysis of any new policy measure is usually fairly rudimentary and an inordinate amount of effort over a long period of time is required on the part of feminists to have any issue taken seriously (Bashevkin 1998; Cohen & Pulkingham 2009; Sainsbury 1999). The other possible reason for this policy lacuna is the under-researched nature of the subject itself in wealthy countries. While feminists have long been writing about ecological issues, this has been on the fringe of feminist demands in most countries and ecological feminism tends to be considered less central to feminist thought and action than issues dealing with more rights-based issues. This may be changing, but any analysis of the gendered implications and climate change and the political action associated with it are not developed enough to penetrate the current approaches to dealing with climate change. There is neither a well-developed body of information about the effects of climate change by gender, nor how public policy that is intended to cope with or mitigate climate change affects people differently, whether by gender or any other form of difference.

Work has been done in pointing out the problems of excluding gender in climate change analyses and ideas about what needs to be done both by researchers and international organizations (MacGregor 2010; McNutt & Haryluck 2009; Nelson 2009). In North America research interests on the gender/climate change nexus so far deals most heavily on the distinction between male and female responses to climate change initiatives. It shows that women tend to have both greater knowledge about and concern for climate change than do men, and women are more likely to engage in pro-environmental behaviours than men. This phenomenon appears to have increased over time and is related to women's tendency to be more progressive in their political outlook than men (Mohai 1997; Clancy & Roehi 2003). There is also an interest in the disproportionate way that women are likely to pay for the costs of climate change mitigation initiatives, such as a carbon tax (Chalifour 2010). The methodologies used vary, with some focusing primarily on income distribution by gender and class to show the unequal impacts of climate change mitigation policies. Other approaches use the distinction between males' and females' time-and-use patterns that influence energy use, research that is usually associated with an analysis of household work and consumption patterns. Studies on consumption indicate that gender has a significant impact on the use of resources, with women having a lower use in all income levels in both the North and the South (Johnsson-Latham 2007; Cohen 2015). At least one study in the US, however, finds little evidence to indicate differences between male and female-headed households and in countries where more equal economic opportunities exist, women tend to adopt male-type lifestyles and consumption patterns (Lambrou & Piana 2006).

As Seema Arora-Jonsson shows, however, the more general references largely tend to essentialize the issue and see women as either vulnerable or virtuous in relation to the environment, with women in the South usually depicted as more affected by climate change than men, and men in the North as polluting more than women (2011). Also as a result of the lack of hard information about gender distinctions, there is a tendency to conflate the experiences of women in developing and developed countries. The usual justification for including women in climate change analysis is based on claims that everywhere "women are more vulnerable to the effects of climate change than men – primarily as they constitute the majority of the world's poor and are more dependent for their livelihood on natural resources that are threatened by climate change" (UN Women Watch).

Despite this relative neglect of gender and climate change issues in wealthy counties, the global literature in relation to developing nations is surprisingly extensive.[1] This is undoubtedly because the impact of climate change so directly affects the work that people in developing countries do in providing food, working in forests and fisheries, and living on marginal land (Agarwal 2010; Ahmed & Fajber 2009). But it is also because international agencies have been more proactive in connecting sustainable development and its funding to gender and social development policies (UN Women n.d.). The argument for inclusion of gendered approaches to climate change rests on the idea that gender issues are integral to the whole concept of development policy in that the latter deals with both the contributors to climate change, the policy associated with solutions to climate change and how poverty can be alleviated and development can be achieved in a sustainable way (Khamati-Njenga & Clancy 2002). Most importantly the argument is made that gender issues should not be marginalized in the discussion of how to pursue a sustainable economy, a message that can and should be carried over to the developed world as well.

Another possible reason for the lack of seriousness on this issue in developed nations is the sense that it is less of a problem because there is more homogeneity in experiencing climate change than in developing areas of the world. In a global perspective the equity and justice issues are framed around the North/South divide, while in rich countries the divide is more often recognized as that between generations as well as class, with gender more or less absent from considerations (Posner & Sunstein 2008). As a result, the discussion of gender and climate change in developed nations and academic research on the topic is not well advanced and is very much less related to the general structure of sustainable development than it is in developing countries. This means that when the effects of climate change mitigation strategies are examined, such as the implementation of a carbon tax, no consideration is given for the gendered impacts, and usually if there is an examination of its effects they are based on effects on different income levels (European Environmental Agency 2011; Williams III et al. 2014).

Major themes

The major themes of this book concentrate on gender, work, public policy and action. Frequently when "gender" is used it is a signal that women will be the subject. In this book the intent is to understand when and where the different experiences of males and females matter with regard to climate change. There is not an assumption that in all cases females will be more disadvantaged than males, but rather, that their very different situations, depending on location, type of work, class position and ethnicity or race will play a big role in how they are affected by climate change and government actions to deal with it.

Gender is not a simple concept, although it is now widely accepted as preferable to understanding differences by sex because sex refers to physical differences while "gender" implies a social condition applied to the biological sex of either a male or a female: the distinctions between males and females are not strictly biological, but are cultural as well. This is an idea that in itself can be problematic, in that for many there is not a clear distinction between the physical and the psychosocial aspects of sex (Halpern 2012, 35–36). The recent use of the word "genders" tends to be more inclusive than simply males and females and indicates that genders can relate to personal identification that crosses a wide spectrum of gender divides. This book will examine ways in which the social construction of gendered systems of work and social life in rich countries mean that climate change has different implications for a wide range of human conditions. This is because men and women do very different kinds of work, tend to have very different kinds of economic and social relationships, and have different physical attributes that are affected by any change. A monumental global change, such as climate change, will have differential impacts by gender, but clearly not in all circumstances, and there are certainly times when males and females will be affected in the same way (Nagel 2016). The emphasis in this book is on the male/female gender divide only because that is the current state of research, not because there is a failure to recognize the potential distinctions for other gendered groups.

The idea of recognizing the "intersectionality" of experience in research is evident throughout the chapters in this book. This relates to the multiplicity of cultural experiences that affect people, particularly with regard to those who experience some type of social dominance or discrimination. For many, gender is not the dominant oppression, although its intersection with a dominant oppression, such as racial identity, can mean the experience is different because of gender. Intersectionality as a distinct concept was identified by Kimberlé Williams Crenshaw in the context of the issue of violence against women and greatly advanced ideas about how oppressed peoples can experience their oppressions differently (1991). Usually an intersectional analysis deals with race, class and gender. In this book the intersections of various ways of identifying people aside from gender are important. This includes differences based on a variety of issues such as ability, age, ethnicity, racial identity, location, class, education and physical location.

One group that is frequently the point of study in this book are Aboriginal peoples in Canada. The distinct historical position of these groups, who are descendants of the original inhabitants of the country, means that they are especially vulnerable.[2] Their vulnerability is also shaped by location and gender. The chapters specifically dealing with the Aboriginal experience are those by Michael Kim on the Inuit in the Arctic and Ellie Perkins on the activism of Aboriginal women on climate change. Both Maureen Reed in her chapter on forest-based communities and Maya Gislason and her research groups' examination of resource extracting communities include the particular significance of living in these types of communities and the meaning this has for Aboriginal women and children. Other forms of difference that intersect with gender in this book relate especially to geographical location and the incidence of poverty, but also significant is the type of work performed and the conditions under which it occurs. The impact of climate change on specific areas is well known and a first obvious place for research; so too is the impact on those living in poverty. Less understood is the relationship between climate change and the need for humane work. In this a gendered examination is merited and is undertaken in various ways throughout the volume. In a particularly novel approach to climate change research, Kendra Coulter raises the issue of humane work across animal species. She shows that an examination of the gendered nature of the implications of climate change is significant for animal as well as human life.

Work is a major theme in this book because of its centrality to life and the transformations required in both the economy and work that are inherent in climate change and the ways that governments deal with it. The gendered conditions of work and its value is dominant in feminist analysis of capitalist economies and how these economies develop and change. Over time feminist attention expanded from a concentration on understanding the gendered distinctions in paid work to the recognition that unpaid work is crucial to the functioning of the system as well. In this volume all the chapters deal with aspects of work, whether paid or unpaid, or they deal with aspects of life that support work and working lives. The implications of climate change for paid work receives considerable attention among various labour organizations, including the International Labour Organization (ILO) and trade unions in industrialized nations. This attention primarily is on the ways that climate degradation poses significant risks to employment in "dirty" industries where males work and how this disadvantage for workers can be avoided. The call for a "just transition" to green jobs and a green economy is the major emphasis for action and argues that climate change can be a catalyst to shift economic growth from dirty to clean industries (International Trade Union Confederation (ITUC) 2015). While labour groups appear to be open to the inclusion of women's work in policy initiatives, relatively little concrete action or analysis is evident to date. This inability to integrate a gender analysis relates to a general bias about the hierarchies of work, in that what is considered as most significant for the economy is highly gendered. But this gender-blindness also relates to the assumption that simply greening dirty sectors of the economy will be what is required. This approach is an important issue because economic policy is strongly related to ideas about what sectors of the economy drive economic growth and when, typically, male-type jobs are seen as most the productive, these areas receive the most attention. The intent of the book is to extend this analysis to examine the significance of unpaid work and to show that the kinds of work that women traditionally do is both essential and important for economic change considerations.

The themes of the gendered nature of public policy and political action are integral to dealing with climate change. Each of the articles in the book deals with these issues from a feminist perspective, although the analyses range from those who focus on specific instances where action and public policy could be more proactive, even under the current conditions, to those whose analyses deal with the more systemic aspects of the wider socio-economic circumstances of wealthy industrialized economies. All of the approaches to climate change and gender go beyond what Maria Mies and Vandana Shiva refer to as "catch-up with men in their society", by demanding simply an equal share of a diminishing access to nature (Mies & Shiva 1993, 7). Similarly, there is not a sense that advances in science and technology will be sufficient for dealing with the problems related to climate change. Rather, a much greater system reorientation needs to take place so that any solutions to climate change are significant enough to make a real difference, and that more is needed than the "green nudges" that are typical of current climate change policy (Schubert 2016).

This wider approach would of necessity involve rethinking the global order, which privileges a specific type of economic system. As many critics of the current world order understand, economic and environmental changes are inseparable and that "the rise of a capitalist world-economy and the rise of a capitalist world-ecology were two moments of the same world historical process" (Daschuk 2013, xi). The current international economic paradigm, most evident through free trade agreements, largely conditions the actions of governments to promote policies supporting export-led growth and the proliferation of production of things and services not used within nations. This leads to irrational production, at least for meeting human needs, and greater inequality both worldwide and within nations. The current prominent solutions to climate change do little to meet the requirements to reduce GHG emissions, and continue the pre-eminence of market structures that are damaging for humans. An alternate approach would construct an economy that focuses on meeting real human needs and could lead to both a more equitable society and one that provides sufficient material goods and services for all people. These goals could be reached only with strong political will and a common understanding of what constitutes a good life that does not degrade the environment. This could be approached on a rational basis with the consideration that gendered issues are a serious part of the analysis. The process of rethinking a rational economy would also necessitate re-thinking the traditional work-roles assigned by gender, particularly with the privileging of certain kinds of work, mainly that work associated with males, as more significant than the work of others.

Action on climate change is a political act, but the authors of this collection recognize that while individual responsibility is important, it is insufficient for dealing with the magnitude of the problem. This, again, returns to the systemic nature of climate change and the need for systemic changes for progress to be made. It is in this context that a gendered analysis can lead to a more inclusive approach to dealing with the issue.

The book

The major objective of the book is to expand ideas about how the future could unfold if gender is taken seriously in acting on solutions to mitigate further climate change. The collective sense is that the current gender-blind approach denies both an analysis of what is really happening to people and obscures solutions that could involve all people and radically change the nature of society. The current absence of a gendered understanding of climate change is a form of climate change denial, in that it underestimates the magnitude of changes needed and the extent to which this includes both the social and economic systems.

The book is organized in five major sections. Aside from the first section, which sets the context for the book, other sections deal with paid and unpaid work, vulnerability and insecurity, rural and resource communities, and activism and public policy.

Part I: Context and overview

The first section, which includes this introduction, sets the context for the main themes explored. In the chapter "Masculinities of global climate change: Exploring ecomodern, industrial and ecological masculinity", Martin Hultman and Jonas Anshelm, two researchers from Sweden, make the important point that the different approaches to climate change are based on masculinist views that are distinguishable. They examine the different masculinist discourses that are embodied in the knowledge about climate change and what this means for different types of actions to deal with it. The analysis shows how "ecomodern" masculinity in the form of both industrial fatalism and green Keynesianism dominate the current discourses and suggest industrial and ecological masculinity as alternative approaches. This analysis of the varieties of masculinities involved in climate change debates allows a more nuanced approach to existing public policy directions from a distinctly feminist perspective.

Susan Buckingham and Rakibe Külcür's chapter "It's not just the numbers: Challenging masculinist working practices in climate change decision-making in UK government and environmental non-governmental organizations", sets the tone of the book for the way it analyses public policy approaches to gender sensitivity. It explores the effectiveness of "gender mainstreaming" by looking specifically at UK evidence. Buckingham and Külcür are able to show that despite the persistent calls from organizations like the UN, gender mainstreaming has not been integrated into climate change related decision-making. This does not mean that "gender mainstreaming" is not practised, but rather that it has been hijacked as a pro-business argument for policy decisions, rather than used as a genuine focus on gender equality in decision-making. Buckingham and Külcür argue that simply recruiting more women into senior posts will have little effect in the context of existing organizational structures and more fundamental changes are necessary in how workplaces are organized.

Part II: Challenges for paid and unpaid work

Part II of this book deals directly with work place issues. All of the chapters in the book relate to work or work-related policy issues, but this section is distinct in looking specifically at traditionally male sectors and occupations, gendered divisions in trade unions, or infrastructure support for workers. It begins with Linda Clarke, Colin Gleeson and Christine Wall's chapter "Women and low energy construction in Europe: A new opportunity?" The construction sector, and specifically the skilled trades, is one that has steadfastly resisted attempts to integrate females into the labour force. In North America women have accounted for less than 4 per cent of this labour force for the past forty years (Hegewisch & O'Farrell 2015; Piper 2007) and, as the authors show, in Europe women represent less than 1 per cent in the manual trades. Despite the dismal results from integration attempts in the past, the authors are more optimistic about the possibilities as Europe moves toward low carbon construction. She shows how employment relations are transforming, especially with the weakening of the "old boys' network", and the increased use of formal recruitment practices that favour women. Entry into the sector is shifting from an apprenticeship system to more recruitment from vocational colleges, where more women are found in construction training than as apprentices. These changes in conditions of employment are especially important in light of the policies pursued by the EU for greater low carbon construction.

The renewable energy sector is often portrayed as the way to reconcile job losses that appear inevitable with the switch away from fossil fuels toward clean energy (Dodge & Thompson 2015). The extent that this transformation is positive for labour is not uniformly acknowledged, and as the International Labour Organization (ILO) shows with regard to the jobs created "more often than not, they do not seem to compare well with more 'traditional' jobs" (2012, 134). Bipasha Baruah's chapter adds a very important gender analysis to the understanding of what is happening. The formal energy sector, like the construction sector analysed by the authors, is a male domain in both highly developed and emerging countries, although as Barauh shows, the variations are considerable when less developed nations are considered. In the chapter "Renewable inequity? Women's employment in clean energy in industrialized, emerging and developing economies" Baruah examines the current employment of women in the renewable energy sector to compare experiences in different parts of the world and shows that despite the widespread enthusiasm for renewables, there is a major blind spot in that women are marginalized globally in employment in the area. She also shows what conditions will be necessary for increasing women's participation and concludes that for any substantial progress to be made simply adjusting labour force practices in minor ways is not enough and transformative changes in gender relations are ultimately what are required.

Carl Mandy, in his chapter "UK environmental and trade union groups' struggles to integrate gender issues into climate change analysis and activism", analyses why progressive groups, particularly trade unions and environmental organizations, have little momentum on gender as part of their climate change initiatives. Trade unions in the UK explicitly address both gender issues in the workplace and climate change and in many cases both issues have high priority for action. The surprise is that these issues are usually treated separately, particularly within larger organizations. Environmental groups also are often progressive on gender issues, yet gender is seldom integral to the campaigns on climate change. This chapter examines the barriers faced by women and marginalized genders within UK activist communities across the intersection of climate change and labour movements. Consideration is given to both external barriers, such as gendered impacts of climate change policies, and internal barriers, such as sexism among activists and the gendered effects of organizational structures. Mandy also gives strategies for overcoming these barriers.

One of the crucial supports for work is transportation and it is also the area that is a major contributor to GHG emissions.[3] This means that public policy initiatives to reduce transportation contributions to climate change are at the forefront of change. Leonora Angeles in "Transporting difference at work: Taking gendered intersectionality seriously in climate change agendas" examines what would it mean in policy and practice if gendered intersectionality is a focus of consideration in transportation, work and climate change policy. She begins by explaining why gender matters in work and mobility issues and shows how it is possible to enrich the understanding of gendered work and transportation issues through an analysis of climate change. She concludes by identifying key policy and planning options and considerations that arise from bringing together gendered intersectional analysis, climate change and transportation issues.

The last chapter in this section, "The US example of integrating gender and climate change in training: Response to the 2008–9 recession" analyses the effectiveness of US programmes funded through the Department of Labor Women's Bureau. For the most part any training programmes for green jobs were primarily focused on males and training for "greening" the dirtiest industries, or those that have the highest GHG emissions. These are predominantly industries in the energy, manufacturing and building sectors where males predominate. When women were considered and initiatives to include women were designed, they were usually an afterthought to the focus on males' training and employment, and consisted of small, underfunded "pilot" programmes that were one-off in design and did not lead to wider participation for females. The programmes themselves were usually highly successful, but were constrained by time and funding limitation. This chapter shows what is effective in dealing with training for green jobs by gender.

Part III: Vulnerability, insecurity and work

The third major part of this book focuses on vulnerability and insecurity. This is a theme throughout the book, but this section looks specifically at those in increased peril through climate change. The effects of climate change disaster are more obvious to us than are other effects of climate change, as a result of the dramatic and sudden weather events that can occur. But less understood is how different populations in the same location experience these events. Margaret Alston's chapter "Gendered outcomes in post-disaster sites: Public policy and resource distribution" analyses the gendered differences as a result of the specific horrific fires that occurred on Black Saturday, 7 February 2009 in Australia where 178 towns were affected, 173 people lost their lives and over 2,000 homes were destroyed. These were the worst bushfire disasters in Australia's history. The research for Alston's chapter was conducted five years after the fires and shows that gender is a critical factor in shaping vulnerability and resilience both during and after a major event. In this case women were more likely to want to escape, while men were more likely to want to stay and defend homes and animals against the fire. In the post-disaster experience, men appear to have experienced significant post-traumatic stress that was demonstrated through a rise of health issues and in reports of increased violence against women. Policy responses focused on restoring communities quickly, a response many describe as "the secondary injury" for survivors, and which led to significant resources being distributed rapidly in ways that were gender-blind and that may have reduced adaptation and increased community divisions. There was little accommodation for what many describe as the need to adjust to "a new normal" nor time taken to assess how this new normal might address gender inequalities. Alston concludes that disaster policy needs to be undertaken with a gendered consciousness of the significant impacts on women and men in areas that have been irrevocably changed.

At the opposite side of the world very different types of crises arise that impinge on already vulnerable populations. One of the most obvious and visible effects of climate change is the shift in conditions in the Arctic, an area among those most threatened by rising temperatures. Sea ice has melted to such a dramatic extent that it was possible for a very large cruise ship to sail from Anchorage, Alaska through the Northwest Passage to New York City in mid-September 2016, something that was never possible previously (Migdal 2016). Michael Kim's chapter "Climate change, traditional roles and work: Interactions in the Inuit Nunangat" examines what this warming means for people living in the Canadian Arctic. These are Indigenous people whose existence is dependent on freezing temperatures, people who because of the remoteness of their communities are particularly vulnerable because there are few substitutes for species that decline through habitat loss. Kim shows the gendered nature of the Inuit Nanangat economy and especially the gendered food systems and how these are affected by climate change. While a division between men as hunters and women as involved in both food distribution and as primary wage earners was traditionally the division of labour, the decreasing viability of hunting food systems suggests that this division will be stressed. Kim shows that by taking the gendered economy into consideration, more effective adaptive strategies could be developed.

Kendra Coulter's chapter "Towards humane jobs: Recognizing gendered and multispecies intersections and possibilities" expands the idea of vulnerability to the intersectional consideration of human–animal relations particularly with regard to the labour politics of climate change. She begins by analysing the gendered and multispecies processes and effects of industrialized agriculture in the global north, a major contributor to climate change. She then proposes expanding ideas about green jobs to include labour that is humane: work that is good for both people and animals. The idea of humane labour extends beyond formal occupations and includes work-lives to include daily life when not working, lives over time, and well-being after formal employment has ended. These are aspects of life that matter for both people and animals, and connect the world of work to ideas of "the social", and the realm of public policy. Programmes and initiatives that are not workplace-based, but which improve work-lives bolster the prospects for humane jobs, and for a more humane, sustainable and just society.

Part IV: Rural and resource communities

The fourth major section of this book deals with rural and community work, specifically through the examination of the working lives and conditions of people whose work is close to nature either in agriculture, forest-based industries, other resource extracting industries or in rural communities. The challenges of these kinds of communities have had attention in resource-dependent developing countries, but some of the same kinds of problems arise in developed areas of the world.

Amber J. Fletcher analyses the gendered nature of farm work and the implications of the challenges of climate change in her chapter "'Maybe tomorrow will be better': Gender and farm work in a changing climate". Her research is based on the experiences of male and female farmers in the Canadian prairie region, which has one of the harshest and most variable climates in the world. Agriculture in Canada is a highly masculinized sector with rigid gendered divisions of labour and it is also a sector that is a focal point because of its contribution to GHG emissions. These political complexities affect attitudes toward climate change mitigation policy that tend to be differentiated by gender. Fletcher details the highly gender-differentiated experiences of climate change and provides policy recommendations to facilitate climate change preparedness and adaptation of farm communities. As she points out, these policies need to have a high level of gender awareness and should also be appropriate to the particular cultural context of rural Canada.

Maureen G. Reed's chapter "Understanding the gendered labours of adaptation to climate change in forest-based communities through different models of analysis" continues with the analysis of a rural resource-based economy and the gendered implications of climate change. She notes that surprisingly little is known about gender and climate change adaptation, primarily because forestry and climate change as a topic is framed primarily as scientific and technical issues and secondarily as economic and political ones. This, then, obscures the community-based impacts of climate change as well as the local capacities to deal with it. The heterogeneity of populations, which usually involve Aboriginal and settler inhabitants, means there is not a single gendered perspective to be understood and that feminist scholars and practitioners are well positioned to contribute to understanding the needs of populations in forest-based communities. The paper begins by considering research gaps and the challenges of applying feminist scholarship in the context of Canadian forestry. Her conclusions focus on the very real challenges and requirements for generating useful, applied research on gender and climate change adaptation founded on principles of feminist scholarship.

The final chapter in this section, entitled "The complex impacts of intensive resource extraction on women, children and Aboriginal peoples: Towards contextually-informed approaches to climate change and health", continues with some of the issues that were enumerated in the chapter on forest-based communities in that it deals with communities engaged in intensive resource extraction. This chapter is based on the work of a team of health-science researchers, Maya K. Gislason, Chris Buse, Shayna Dolan, Margot W. Parkes, Jemma Tosh and Bob Woollard, and examines the health implications of climate change in oil and gas resource extracting communities. The authors provide concrete examples of the ways that extractive industries and their related processes affect the health and lives of Aboriginal and non-Aboriginal women and children. The impacts on Aboriginal peoples are foregrounded because intensive resource development tends to occur on or close to the traditional territories of Aboriginal communities. This chapter presents both evidence and conceptual tools intended to strengthen research on the interrelated factors influencing vulnerability within communities by using three theoretical approaches: the ecosystem approaches to health (ecohealth), critical feminist theory and the Life Course Perspective (LCP).

Part V: Public policy and activism

The last section of the book includes chapters that are especially focused on government policy and political activism. It begins with an overview from Nathalie J. Chalifour on how governments at the national and sub-national levels deal with climate change from a gendered perspective. In "How a gendered understanding of climate change can help shape Canadian climate policy", Chalifour documents how gender has been ignored in discussions about policy responses to reduce GHG emissions and explains options for how gender could be brought into climate policy discussions. The main focus is on climate change mitigation policies and legal instruments to pursue meaningful results. The paper includes a discussion of the perspective of Indigenous women in Canada, and the extent to which their interests are accounted for in Canadian climate policy.

Annie Rochette's chapter on "The integration of gender in climate change mitigation and adaptation in Québec: Silos and possibilities" also uses a legal standpoint to look at the obligations of a sub-national government to integrate gender in climate change policy. She shows that governments in Canada have obligations to include gender considerations, although it is clear that this is still at a very elementary stage. In addition, Rochette analyses the integration of gender in climate change within Québec civil society and shows that civil society actors are still largely working in "silos" on these issues. However, information from local workshops shows that once participants saw some concrete examples of how climate change can be gendered, they are willing to learn more and to act. This is a positive picture that can lead to new strategies and alliances between the environmental and women's movements to build a more equal, ecological and carbon-neutral society.

Housing policies, particularly those related to cities, are identified as having a significant impact on climate change mitigation. Penny Gurstein and Sara Ortiz Escalante in "Urban form through the lens of gender relations and climate change: Cases from North America and Europe" analyse the significance of including gender in the planning for GHG reduction policies. Housing is as significant as transportation in supporting people in their working lives, either in the paid workforce or in unpaid labour within households. This chapter analyses North American and European housing policies with particular attention on the lived experience of women in their capacity as wage earners and caregivers. Gurstein and Escalante deal with "urban form", the structural and infrastructure elements – such as building heights and density, natural and open space features, transportation corridors, public facilities and activity centres – that physically define a city. They advocate dense, compact, mixed-use communities served by public transit to allow for greater opportunities for economic integration, lower transportation use and, as a result, lower GHG emissions. While this need for increased density is well understood by planners, less evident is the link between gender, urban form and climate change. If the gendered experiences of women in their productive and reproductive lives are addressed by lessening consumption and energy expenditure then climate change imperatives will also be addressed.

Patricia E. Perkins' chapter "Canadian Indigenous female leadership and political agency on climate change" details the ways that Indigenous women in Canada confront governments on issues of climate change. She shows that activist Indigenous women have built strong movements in opposition to the extraction of fossil fuels in Canada and examines the issues of distributive and procedural climate justice for women in the actions on climate change. The voices of Indigenous women are prominent because they are on the front lines of climate change, particularly those in communities related to fossil fuel extraction. Besides the gendered economic and social roles that all women face in a patriarchal society, cultural factors also lead Indigenous women to assert their voices and leadership on matters related to water, health, education and livelihoods.

My chapter "Using information about gender and climate change to inform green economic policies" is the last in the book. This chapter shows how the ideas and discussions about green jobs and the green economy use highly gendered assumptions about the ability to mitigate climate change and examines the ways that the inclusion of gender in the analysis of a green economy and concepts of green jobs could be transformational in thinking about the nature of the economy and climate change initiatives. The chapter shows how including the work of "social reproduction" could expand solutions to climate change to include a much wider range of options than currently seem possible. Ultimately it argues that a major shift in ideas about economic performance and economic success is essential in order to make the concepts of both a green economy and green jobs more inclusive and to take a truly meaningful direction in dealing with climate change.

Notes

1 For a bibliography of gender and climate change see: Whiteside & Cohen (2012).
2 In Canada Aboriginal people are also referred to as First Nations, Native, Inuit, Indigenous, Indians or Métis people, depending on the specific term each group prefers.
3 In the US transportation is the second largest sector contributing to GHG emissions, while in Canada it is the single largest contributor (EPA 2014).

References

Agarwal, B 2010, *Gender and green governance: The political economy of women's presence within and beyond community forestry*, Oxford University Press, London.

Ahmed, S & Fajber, E 2009, "Engendering adaptation to climate variability in Gujarat, India", *Gender & Development*, vol. 17, no. 1, pp. 33–50.

Arora-Jonsson, S 2011, "Virtue and vulnerability: Discourses on women, gender and climate change", *Global Environmental Change*, vol. 21, issue 2, pp. 744–775.

Bashevkin, S 1998, *Women on the defensive: Living through conservative times*, Chicago University Press, Chicago.

Clancy, J & Roehi, U 2003, "Gender and energy: Is there a northern perspective?" *Energy for Sustainable Development*, vol. 7, no. 3, September, pp. 44–49.
Cohen, MG 2015, "Gendered emissions: Counting greenhouse gas emissions by gender and why it matters", in C Lipsig-Mummé & S McBride (eds), *Working in a warming world*, pp. 59–81, McGill-Queen's Press, Montreal & Kingston.
Cohen, MG & Pulkingham, J (eds) 2009, *Public policy for women: The state, income security, and labour market issues*, University of Toronto Press, Toronto.
Crenshaw, KW 1991, "Mapping the margins: Intersectionality, identity politics, and violence against women of color", *Stanford Law Review*, vol. 43, pp. 1241–1299.
Daschuk, JS 2013, *Clearing the plains: Disease, politics of starvation, and the loss of aboriginal life*, University of Toronto Press, Toronto.
Daub, S & Klein, S 2016, "The new climate denialism: Time for an intervention", *Policynote*, 22 September, CCPA-BC, Vancouver, (http://www.policynote.ca/author/seth-klein-shannon-daub/#sthash.E48kjHer.dpuf).
Dodge, D & Thompson, D 2015, "Where are the renewable energy jobs? Mapping the geography of the green economy", Green Energy Futures, (http://www.greenenergyfutures.ca/episode/renewable-energy-jobs).
EPA 2014, "Sources of greenhouse gas emissions", US Environmental Protection Agency, (https://www.epa.gov/ghgemissions/sources-greenhouse-gas-emissions).
European Environmental Agency 2011, "Environmental tax reform in Europe: Implications for income distribution", no. 16, Copenhagen, (http://www.eea.europa.eu/publications/environmental-tax-reform-in-europe).
Halpern, DF 2012, *Sex differences in cognitive abilities: 4th edition*, Psychology Press, New York.
Hegewisch, A & O'Farrell, B 2015, "Women in the construction trades: Earnings, workplace discrimination, and the promise of green jobs", Institute for Women's Policy Research, (http://www.iwpr.org/publications/pubs/women-in-the-construction-trades-earnings-workplace-discrimination-and-the-promise-of-green-jobs).
ILO 2012, "Are 'green' jobs decent?" *International Journal of Labour Research*, vol. 4, issue 2, International Labour Organization, Geneva.
ITUC 2015, "Frontlines briefing, Climate justice: There are no jobs on a dead planet", International Trade Union Confederation, (http:/www.ituc-csi.org/ituc-frontlines-briefing-climate).
Johnasson-Latham, G 2007, "A study in gender equality as a prerequisite for sustainable development", Ministry of the Environment, Sweden.
Khamati-Njenga, B & Clancy, J 2006, "Concepts and issues in gender and energy", Energia, (http://www.africa-adapt.net/media/resources/80/Energia-gender-energy.pdf).
Lambrou, Y & Piana, G 2006, "Gender: The missing component of the response to climate change", Food and Agriculture Organization of the United Nations.
MacGregor, S 2010, "A stranger silence still: The need for feminist social research on climate change", *The Sociological Review*, vol. 57, pp. 124–140.
McNutt, K & Hawryluk, S 2009, "Women and climate-change policy: Integrating gender into the agenda", in A Dobrowolsky (ed), *Women and public policy in Canada: Neo-liberalism and after?* pp. 107–124, Oxford University Press, New York.
Mies, M & Shiva, V 1993, *Ecofeminism*, Zed Books, London.
Migdal, A 2016, "Cruise ship looks to make clean journey through Northwest Passage", *The Globe and Mail*, 19 August.
Milman, O 2015, "James Hansen, father of climate change awareness, calls Paris talks 'a fraud'", *The Guardian*, 12 December.

Mohai, P 1997, "Gender differences in perceptions of most important environmental problems", *Race, Gender & Class*, pp. 153–169.

Nagel, J 2016, *Gender and climate change: Impacts, science, policy*, Routledge, New York.

Nelson, JA 2009, "Between a rock and a soft place: Ecological and feminist economics in policy debates", *Ecological Economica,* vol. 69, issue 1, pp. 1–8.

Piper, W 2008, Skilled trades employment, *Perspectives,* October, Statistics Canada, Ottawa, (http:/www.statcan.gc.ca/pub/75-001-x/2008110/pdf/10710-eng.pdf).

Posner, E & Sunstein, CR 2008, "Climate change justice", *The Georgetown Law Journal,* vol. 96, p. 1565.

Sainsbury, D (ed.) 1999, *Gender and welfare state regimes,* Oxford University Press, New York.

Schubert, C 2016, Green nudges: Do they work? Are they ethical? Paper Series in Economics, no. 09, Universities of Aachen, Gießen, Göttingen Kassel, Marburg, Siegen, (https://www.uni-marburg.de/fb02/makro/forschung/magkspapers/paper_2016/09-2016_schubert.pdf).

SFU 2015, *Climate change, gender and work in rich countries,* International Research Network Workshop, 24–26 June, Vancouver, BC, (https://www.sfu.ca/climategender.html).

The Climate Examiner 2016, The Pacific Institute for Climate Solutions, September, Vancouver.

UN Paris Agreement 2015, *United Nations Framework Convention on Climate Change,* (http://unfccc.int/paris_agreement/items/9485.php).

UN Women n.d., "Climate change and the environment", (http://www.unwomen.org/en/how-we-work/intergovernmental-support/climate-change-and-the-environment).

UN Woman Watch n.d., Factsheet, (http://www.un.org/womenwatch).

Whiteside, H & Cohen, MG 2012, "Gender and climate-change bibliography", 1990–2011, (http://warming.apps01.yorku.ca/wp-content/uploads/Gender-and-Climate-Change-Bibliography.pdf).

Williams III, RC, Gordon, H, Burtraw, D, Carbone, JC & Morgenstern, RD 2014, *The initial incidence of a carbon tax across US states*, Resources for the Future, (http://www.rff.org/research/publications/impacts-us-carbon-tax-across-income-groups-and-states).

2

MASCULINITIES OF GLOBAL CLIMATE CHANGE

Exploring ecomodern, industrial and ecological masculinity

Martin Hultman and Jonas Anshelm

Introduction

The world is facing monumental challenges, including the spread of neo-fascism in Europe, world hunger and an unfair distribution of wealth in which a few per cent own the majority of wealth (Piketty 2014). At the same time, also faced is the very real possibility of the destruction of our common planet, as we know it. As gender scholars are aware: men in general are a big problem, especially white, middle-class, and rich men – those who travel too much, eat too much meat and live in energy consuming buildings (Nightingale 2006; Terry 2009). Following the solutions these men propose to handle climate change is akin to having an alcoholic draft drinking laws.

This chapter deals with historically shaped discourses and contemporarily enacted forms of masculinities in rich, extractive dependent countries with high per capita emissions. The empirics are drawn from Sweden. Since the late 1960s and the 1972 UN conference on the global environment in Stockholm, Sweden has claimed to be, and has also been widely recognized as one of the most environmentally progressive countries in the world, if not *the* most, by prominent scholars (Giddens 2009; Jänicke 2008; Urry 2011). This analysis will move beyond that stereotype as well as beyond binary categories of wo/men when searching for problems and proposing solutions, also making visible the difference among masculinities (Alaimo 2009). Of course, if we divide humans into well-confined categories and analyse per capita emissions and ecological footprints (e.g. Räty & Carlsson-Kanyama 2010) the truth is that white, middle-class, rich men are the problem. This is important research to keep in mind. But, in our analysis, masculinities are understood as always-in-the-making within and part of material-semiotic antagonistic discourses, which are the embodied nature of knowledge, materiality, meaning and power.

Masculinities are not fixed in biological terms, but shift according to various tensions in cultural and political material-semiotic discourses with bodies populating and changing such discourses (see Christensen & Jensen 2014).

Making masculinities visible in climate change politics

Almost all of the research on gender and climate change has been carried out in poor nations with low emissions per capita (Arora-Jonsson 2014). Especially prominent and enlightening has been gender analysis in relation to women affected by climate change (Neumayer & Plümper 2007) and female activism (Macgregor 2013). Of much less interest has been the male aspect, especially the question of how different masculinities enhance or influence environmental issues. While there has been research into gender roles and inequalities in relation to environmental and developmental goals, there has been little concern with constructions of hegemonic masculinity when examining how masculinity is embedded in and through environmental policy (Hultman 2013). A large field called masculinity studies from the 1990s has been evolving around the issue of different configurations of materiality, values and practices among men. Unexpectedly few scholars have thus far been interested in continuing the analysis of masculinities and environment that environmental historian Carolyn Merchant and particularly Raewyn Connell started in the 1980s. This lacuna of studies of male practices in rich fossil fuel dependent countries is surprising, not only because of the large role that men play in environmental politics, but also because one of the first studies in which the concept of "hegemonic masculinity" was used and which started the blooming field of masculinity studies dealt with men and transitional masculinity in environmental social movements (Connell 1990). Merchant, in the classic book *The Death of Nature: Women, Ecology, and the Scientific Revolution* (1980), initiated the analysis of nature-destructive masculinity. That perspective was somewhat lost, taken for granted or suppressed later on. After this book and others with the same theme perhaps men, as nature-destructive industrial masculinity, were firmly understood as doing bad (Shiva 1988). Thereby masculine practices shaping the environment were not closely analysed even though this is also an important part of gender and environmental studies (Alaimo 2012). The environmental political field thus has a paucity of masculinity studies (Dymén, Andersson & Langlais 2013). This chapter and currently ongoing research in this direction is thus an attempt to balance the situation.

Method and analytic categories

The analysis of masculinities in this chapter is based on a set of 3,500 articles found in the database Retriever using keywords such as climate change and greenhouse gas. The database contains articles published in all Swedish newspapers, all major regional newspapers and the vast majority of magazines. Once the material was compiled and arranged chronologically, as well as read through, it was sorted with discourse analytic tools (Anshelm & Hultman 2014a). Exploring different configurations of masculinity and climate change should shed further light on how the genders interact and are structured in the climate change debate. In so doing, this text both elaborates on a fairly new framework and broadens our understanding of the cultural formation of these configurations in the present form of global politics.

This chapter categorizes gender aspects of environmental positions as exhibiting one of three main tendencies: "industrial masculinity", "ecological masculinity" and "ecomodern masculinity". This chapter builds on work Hultman has done conceptualizing masculinities with his more than a decade long research into climate change, environmental history and energy politics to shape the forms of configurations that we re-use in this chapter (Hultman *forthcoming*). The ecomodern masculinity is the one that dominates today's climate change debate. This masculinity was part of the shift from the 1990s towards the recognition of environmental issues as an intrinsic part of politics for the future. Ecomodern masculinity can be defined as an asymmetric combination of determination and hardness from industrial modernity with appropriate moments of compassion and even vulnerability for the environment from the environmental movement in which the end result is merely 'green washing', as exemplified by Arnold Schwarzenegger (Hultman 2013). The ecomodern masculine character demonstrates caring and responsibility for the environment, while at the same time promoting economic growth and technological expansion. Ecomodern masculinity demonstrates an in-depth recognition of environmental problems, especially climate change, while at the same time supporting policies and technologies that conserve the structures of climate-destroying systems (Hultman *forthcoming*).

Industrial masculinity is a figuration that historically has been noted – by, for example, Merchant (1980) – as treating nature as both scary and a resource for extraction. Man has been presented as the chosen dominator, and engineering as the method of creating wealth for all humans. Talk in the climate change debate about a vulnerable earth transformed by anthropogenic emissions is handled with denial or strong scepticism by those enacting industrial masculinity, since in their idea the world is there for humans to conquer and extract resources from. This is a marginalized position today taken by climate denialists, although one perhaps simmering just below the surface, and is a much larger part of climate politics than is seen in the debate (Anshelm & Hultman 2014b).

The most marginalized masculinity in the climate change debate today is ecological masculinity. This evolved in the 1960s as antagonism to industrial masculinity. It incorporated practices such as the localization of economies, use of small-scale technologies, creation of renewable energy, decentralization of power structures and cohabitation with nature in everyday life (Connell 1990). Ecological masculinity today plays a small part in our present global climate change debate and is upheld by, for example, actors within MenEngage, 350.org or Indigenous social movements such as We Speak Earth and Idle No More.

We have started off this chapter by giving a background to the field and continued with presenting our method as well as the analytical concepts we are using.

We will now present the climate change debate and its discourses before we discuss the masculinities part of that debate.

Climate change debate

Masculinities of climate change need to be situated within a broader history. Energy and environmental politics in fossil fuel dependent rich countries over the last forty years are characterized by an intense conflict between an ecological discourse and a dominating industrial discourse that were both shoved to the periphery of the debate in the early 1990s as an ecomodern discourse began to dominate both national and international policies on energy, climate and environment (Hultman & Yaras 2012). Until 2006 the majority of politicians and other elite actors more or less treated climate change as just one of several environmental issues and as something to keep an eye on in the future (Zannakis 2015). This approach prevailed despite grave warnings from environmental organizations, individual researchers, research communities and the Intergovernmental Panel on Climate Change (IPCC), which stated as far back as 1990 that this was an issue not to be treated lightly (Knaggård 2014). This way of handling global climate change was about to change dramatically.

From the autumn of 2006 through 2009, the issue of global climate change was at the core of politics, and climate change was reinforced by both research and increased environmental activism: economist Nicolas Stern's report on climate change costs, Al Gore's film, *An Inconvenient Truth*, and the publication of the United Nations Intergovernmental Panel on Climate Change (IPCC)'s fourth report on global climate change made the issue urgent (Anshelm & Hultman 2014a). This was compounded by news reports in early 2007 in which the consequences of climate change – including melting ice, droughts and record summer heat in Southern Europe – on people and cultures were a recurring topic. Convincing and worrisome arguments emerged that climate change was an issue that needed to be taken very seriously and be considered in every political decision. The challenging framing at this time made it necessary to talk about the good society, the need for responsibility to future generations and the need to prevent climate catastrophe. It turned into an antagonistic dispute over what future society should look like that none could ignore.

The newspapers displayed a profound concern for the future, and it is significant that *Dagens Nyheter* portrayed this in a long series of articles entitled "Climate anxiety". *Aftonbladet*, in turn, called upon its readers to sign a petition to stop climate change and regularly announced how many people had followed the call. From the autumn of 2006 climate change took a central place in public political debate in Sweden. All of the parliamentary parties, as well as interest groups ranging from the Swedish Enterprise Organization to the Swedish Church, identified ambitious climate actions as a prerequisite for the survival of industrial civilization. Even though the urgency of climate change was understood in quite a similar way by almost all actors, the understanding of the causes and adequate solutions were far from the same and we will here discuss the four discourses that made up the debate, and which we have written extensively about elsewhere (Anshelm & Hultman 2014a).

The industrial fatalist discourse

The Swedish and global climate change debate is dominated by liberal-conservative ideas put forward together by actors from political parties, industry, liberal press and trade unions. During the years 2006 to 2009, one image of Sweden, as competitive, environmentally friendly and courageous, dominated. Sweden was said to be a country showing international leadership by example, and this drove a permeating fundamental belief that international agreements between states would make it possible to regulate emissions of greenhouse gases and, therefore, manage the risks of climate change.

For these actors only marginal changes to industrial capitalist society's fundamental economic and technical structures would be needed. Climate change was described as a temporary crisis phenomenon that required a complicated coordination of international efforts that had never been seen before. The climate crisis for industrial fatalists was described as a new market opportunity in which better-informed consumers create bigger markets. Climate change is thus incorporated smoothly into industrial modern growth-centred politics, where it is described as an "economic lever" for growth, enterprise and jobs. The solutions, however, were said to be large-scale nuclear power, a carbon market and flexible mechanisms. Since Sweden was already doing well according to this discourse other citizens in other countries were the real problem; even the poorest who wanted to include environmental justice were pointed out as a disturbance at the COP negotiations and described as rioting children.

The green Keynesianism discourse

The influential, but not dominating, green Keynesianism discourse was in some respects very different to the industrial fatalist discourse, even though they both shared a core belief in market mechanisms. Green Keynesianism made a rather profound reflection on the industrial capitalist mode of action. System modifications, behavioural changes and fundamental value changes were reported to be necessary to meet climate change. Such changes clarified that economic models must be reformed, growth concepts reclassified, ecological considerations internalized and a gentler approach to nature developed, and that demands for global justice be respected. Changes of this kind could not solely be left to the market, which is why policies responsible for promoting these changes were of great importance in the green Keynesianism discourse.

In Sweden, this required comprehensive changes in energy, transport and production systems towards sources of renewable energy, rail mass transit and energy efficient production. The green Keynesianism discourse emphasized the importance of binding international agreements. Sweden needed to implement its own policy of action before it could make demands on other countries. Actors used an ecomodern language that referred to the "untapped potential" of green jobs to save both the economy and the climate. The core idea was to stimulate consumption and investments towards a greener future with massive government stimulus, including a "green new deal" for Europe, without really questioning the overall consumption patterns or lifestyles in rich fossil fuel dependent countries.

The climate denial discourse

There was in Sweden a small group of climate denialists This group consisted, with only a few exceptions, of elderly men with elite positions in society either in academia or in large private companies. Even though they had a positivist standpoint, in relation to climate science a constructivist position was adopted and they described themselves as marginalized, banned and oppressed dissidents who felt compelled to speak up against what they saw to be a faith-based belief in climate science. They argued that the IPCC deliberately constructed its models in an alarmist direction and appealed to citizens' mistrust of the state and the establishment in a populist way. In this way, the climate denialist discourse quite paradoxically shaped a conflict between the people and the decision makers; between concrete, short-term and individual everyday problems on one hand and long-term, abstract and global issues on the other. Climate science for them was a thought-up construct with political goals. This position was paradoxical, because these men themselves were part of the elite, having been part of the ruling class their whole lives with successful careers in modern industrial society. They found many spaces to speak out and their arguments were disseminated through various types of media even though for them media climate reporting was brainwashing the whole world.

The eco-socialist discourse

The fourth discourse in Sweden in regards to climate change took the position that the research regarding climate change was real and that transformative action was required. Its starting point was that climate research is valid, but it hesitated to speak out regarding the consequences of a warming world, or discuss who the main emitters are. Climate justice is central in terms of both historical and contemporary differences in emissions and gains. There is a need for another relationship between nature and culture than the current extractivist form. Civil disobedience and direct action are recognized as important forms of politics in the name of the planet. This is a sidelined group, rarely even mentioned in mass media, government reports or so forth, that wants systemic change, not climate change. For eco-socialists climate change is the clearest sign of capitalism's inherent self-destructiveness. Its solutions are the localization of fossil dependent societies in which grassroots movements are crucial. For these eco-socialists it was somewhat absurd to discuss changes in individual consumption patterns while paying no attention at all to the capitalist system that demanded increased consumption. The possibility for change grew from a

TABLE 2.1 Climate discourses and masculinities

Discourse	*Masculinities*
Industrial fatalism	Ecomodern
Green Keynesianism	Ecomodern
Climate denialists	Industrial
Ecosocialist	Ecological

Source: Anshelm & Hultman (2014a).

global non-parliamentary grass-roots movement that, to some extent, affirmed the use of civil disobedience and direct action. The eco-socialist discourse described the outline of an alternative society, not just an alternative technology.

Eco-socialists said that the dominant climate politics of industrial fatalism and green Keynesianism were fake politics designed to assure citizens that everything was under control. They criticized the dominant environmental movement organizations for their gradual adaptation to the political and economic growth agenda. Management of resources should be done within planetary boundaries, they argued. Organic food needed to become a part of everyday life, and carbon rationing, as a morally good action, should become popular (Jonstad 2009).

Ecomodern masculinity

The first configuration of masculinity to play a role in shaping global climate change is ecomodern masculinity. The ecomodern discourse has enabled economic growth to be placed back squarely at the centre of the environmental debate, and it claims that there is no conflict between economic growth and environmental problems. In fact, it declares that environmental problems actually foster growth, innovation and competitiveness (Hajer 1996; Hultman & Nordlund 2013). The ecomodern discourse – or ecological modernization as it has been called elsewhere – emphasizes a continuation of industrial modernization instead of a hegemonic shift, and it brought major changes to energy and environmental policies in the early 1990s that still figure in environmental politics today. Ecomodern masculinity is part of this dominant ecomodern discourse.

Greening of modernity

Ecomodern masculinity is part of both the industrial fatalism and green Keynesianism discourses of global climate change (Anshelm & Hultman 2014b). Both discourses today say that there is no immediate need to change industrial capitalist society's fundamental economic and technical structures in order to combat climate change. Climate change politics was presented as a competition and Sweden portrayed as competitive and a courageous frontrunner country in a global environmental race. This is an image not restricted to Sweden. Politicians and actors in the climate change debate all over the world try to pose as the most environmentally friendly (Boykoff 2011; Carvalho 2007).

In the industrial fatalist approach significant changes to industrial society's way of life, economic growth, the use of natural resources, the production of energy and goods, transportation, the flow of materials or any other aspect of industrial society's metabolism, are unnecessary or even counter-productive to dealing with climate change. The green Keynesianism discourse has a slightly different approach. It contains the classic social democratic confidence in the market as the engine of wealth creation but only if the market is properly regulated by strong government enforcement to reduce inherent dysfunctions (Jackson 2011). Proponents argue that economic models must be reformed, growth concepts reclassified, ecological considerations internalized, a gentler approach to nature developed and demands for global justice respected. These seemingly different ways of performing eco-modern discourse were understood early on by Hajer (1996), who wrote that ecological modernization could be executed and understood both as a technocratic project (industrial fatalism) and institutional learning (green Keynesianism).

One of man*kind's most important decisions*

Men in Sweden from the Liberal-Conservative government as well as the Social Democratic Party, Green Party and most of the Left Party enacted ecomodern masculinities. Prime Minister Fredrik Reinfeldt from the conservative political party Moderaterna stressed that the upcoming 2009 Copenhagen negotiations on global climate change meant that we all faced "one of mankind's most important decisions" and that it was Sweden's task to "show leadership and take the initiative". He stressed in the speech that he had learned to "respect" that human behaviour in large regions of the world was unsustainable, that he understood the global warming threat and that it required political decisions to manage it (Reinfeldt 2008).

With ecomodern masculinity the primary solution to climate change was new technology – not very different from the technology in systems that caused the problems. Emissions from the transport system were not a problem of having a private car, for example – it was the emissions of that particular car that were the problem. A fundamental, unrestricted and under-problematized confidence in rationality and progress characterized the reaction to the dangers that industrial civilization had brought upon itself (Anshelm & Hultman 2014a).

The Prime Minister and his allies declared several times that Sweden should be a "prototype", "leading country" and "good example", and had the obligation to "show leadership and take the lead" (Anshelm & Hultman 2014a). The government declared that Sweden's ability to combine low emissions with economic growth ought to convince other countries that emission decreases do not jeopardize national economies but are actually a prerequisite for continued industrial and economic development. In this fashion the government, with heavy support from the daily press, created a narrative about Sweden as a frontrunner in an international competition of climate politics.

Asymmetric combination

According to enactors of ecomodern masculinity only minor corrections are needed to deal with global climate change. The answer to the crisis created by industrial modernization was more industrial modernization with ecological conditions taken into consideration. There was consequently no crisis of the system, only marginal dysfunctions that could be managed through innovation, new knowledge, improved information, enlightenment and so on.

Sweden has invested heavily in hydro power and nuclear power over the years, and more recently in bio energy. As a result, the country today, if some well curated statistics are used, has low carbon emissions in relation to its GDP. This is only the case, however, provided that the emissions related to the vast consumption of goods produced abroad are not taken into consideration, and that ecological footprints as a measurement are generally disregarded. Sweden's position as a front-runner in GHG reductions is based on dammed rivers, racialized politics against the Indigenous Sámi population, large farms of industrialized forests and heavy investment in nuclear power not reflected upon by ecomodern masculinity. Moreover, Swedes export low-carbon technology and import high-carbon consumer products in order to maintain the national standard of living, enhance economic growth and decrease emissions (Lidskog & Elander 2012). Some call this a progressive national politics that leads the way in the global combat against climate change and climate mitigation. Others would simply call it an obvious asymmetry that outsources its environmental problems while claiming moral superiority on the issue of climate change.

Industrial masculinity

The second masculinity to be discussed here is industrial masculinity. Industrial masculinity has a strong foothold in the world as shown by, for example, Carolyn Merchant (1980, 1996). In the climate change debate this figuration is seen most clearly when analysing climate denialists. The mere talk in climate debate about a vulnerable earth transformed by anthropogenic emissions is handled with denial or strong scepticism by those enacting industrial masculinity.

Making modernity

Industrial masculinity contains values from engineering and economics and favours large-scale and centralized energy technologies and the practice of patriarchy. In relation to nature the most important idea is to separate it from humans and value it as a resource for human extraction. Carolyn Merchant identified a kind of masculinity that accuses others of religious fervour at the same time that it uses faith as a basis for its embrace of a modern industrial society. Since the Enlightenment, a separation between man/woman and culture/nature has been created that leads to the dichotomy of men/culture as rulers over women/nature.

Merchant detects an important change from organic metaphors of nature, which were dominant up until the sixteenth century in Europe, to mechanical metaphors; eventually nature was regarded as building blocks useful for the purpose of creating a human-made Eden on Earth. This shift coincided with the rise of industrial-scale operators who viewed nature as a resource – the mining, water and timber industries, for example (1980).

Climate denialism

In the twenty-first century, industrial masculinity forms the basis for climate denialism. When industrial modernization once again was truly challenged in the wake of the climate change debate, industrial masculinity in the form of climate denialism appeared on the environmental political scene again (Anshelm & Hultman 2014b). McCright and Dunlap (2003) identify the conservative political movement in the US as a central actor that is influenced by a small group of "dissident" or "contrarian" scientists who lend credentials and authority to conservative think tanks. It is well recognized that in order to maintain an illusion of intense controversy, industries, special interest groups and public relations firms have manipulated climate science and exploited the US media. This is not a social movement; it is a project of a few influential men (Lahsen 2013). In research based on Gallup surveys in the US McCright and Dunlap (2011), who take gender into consideration in their analysis, have found a correlation between self-reported understanding of global warming and climate change denial among conservative white men. This suggests that climate change denial is a form of identity-protective cognition. This denialism is articulated in Sweden by a small, homogeneous group of, almost exclusively, men and conservative think tanks. These men have successful careers in academia or private industry, strong beliefs in a market society and a great mistrust of government regulation (Anshelm & Hultman 2014b).

The denialists' arguments are strengthened by references to the authority of titles found in a variety of academic disciplines and thereby demonstrate a general belief in the positivistic industrial modern science underpinning these academic disciplines. In relation to climate science, however, these denialists adopt a constructivist position. They dismiss climate-science as a mix of science and politics so entwined that they can no longer be distinguished (Anshelm & Hultman 2014b).

In Sweden denialists have connections to associations where representatives of business, scientific and technology research meet. One clear example is of Per-Olof Eriksson, a former board member of Volvo and former president of the multinational steel company Sandvik, who wrote an article in the leading Swedish business paper *Dagens Industri* declaring his doubts that carbon emissions affect the climate. He said that the Earth's average temperature has risen due to natural variations (Eriksson 2008). Ingemar Nordin, professor of philosophy of science, continued by saying that the IPCC's selection and review of scientific evidence was consistent with what politicians wanted. Nordin claimed that politics shaped basic scientific research, and that only scientists who submitted politically acceptable truths were awarded funding (2008). Economy Professors Marian Radetzki and Nils Lundgren claimed that the IPCC deliberately constructed their models "in an alarmist direction" using feedback mechanisms that gave the impression that significant climate change was taking place (2009). Later on fifteen Swedish professors, all men, proclaimed themselves as climate denialists (Einarsson et al. 2008).

Their rhetoric is a typical patriarchal line in which men, particularly, with engineering and/or science backgrounds claim to have the knowledge to care for an ill-educated working class and developing nations (Anshelm 2010). Instead of understanding climate denialists as anti-science or anti-political, we argue that it is important to understand how their masculine identity has been shaped and how this figuration is co-constructed with the challenges they make towards climate science.

They make use of faith-based conservative rhetoric which has a long tradition among industrial engineers, economists and scientists (overwhelmingly male fields), as witnessed in the debate over nuclear power in which the coming nuclear age was described as a Garden of Eden. In Sweden the first reactor was named Adam and the second Eve; engineers even told their building as a religious story in which Eve was created from the rib of Adam (Anshelm 2010). The connection between Christian faith and the masculine control of nature goes back even further, over several hundred years. Their rationality of domination over nature, instrumentality, economic growth and linearity has been hegemonic throughout the industrial modern era (Merchant 1996).

Ecological masculinity

We have looked at the ecomodern and industrial masculinities. The third figuration we will analyse in connection to global climate change is that of ecological masculinity, which today plays a sidelined role in the climate change debate. The localization of economies, use of small-scale technologies, creation of renewable energy, decentralization of power structures and cohabitation with nature were proposed by an ecological masculinity as activities that should be part of everyday practices.

Entangled nature

From the mid-1970s onwards, criticism has – with various degrees of success – been raised against modern industrial society's flaws and shortcomings, including a challenge to industrial masculinity (Melosi 1987; Rome 2003). A vision of another society was then formulated and practised; these visions, challenging the dominant modern industrial energy and environmental politics, were seriously discussed throughout the 1980s. The rise of the Green Party in the parliamentary assembly, new regulations and small-scale renewable energy projects are examples of this change. This vision was against large-scale industrial socio-technical solutions and in favour of small-scale renewables and decentralization. In opposition to large segments of the political and scientific elite, initiatives such as eco-villages, labelling requirements, and cooperative wind and solar projects were begun (Hultman 2014a). Existing knowledge about society's impact could now be translated into practical projects and arranged in new kinds of communities. These change agents did not shut themselves off from society, but created alternative projects amidst the dominant model. During this period, a masculinity of a more caring, humble and sharing sort was presented as being more adequate to an ecologically sound society. This masculinity was created, among other places, within the environmental movement that challenged the hegemonic masculinity (Connell 1990).

In the recent years of the climate change debate ecological masculinity has merged with ideas from Indigenous people, eco-socialism and Transition Towns to make up an important association (Anshelm & Hultman 2014a). The central idea was that the problem of climate change could not be resolved without creating a different global socio-ecological system. Ecological masculinity was intertwined with a discourse that rested on the assumption that climate change is a productive force that fosters change in the pathological growth ideology of the industrial capitalist society and unjust global exchange relationships. This possibility grew out of a global non-parliamentary movement proposing an alternative society, not just climate-friendly technology (Anshelm & Hultman 2014a).

Within the eco-socialist discourse is where we find elements of ecological masculinity. Author David Jonstad noted that it was the "hunt for economic GDP growth" that drove the consumption of fossil fuels and that only when this was abandoned would opportunities arise to build global agreement on emissions cuts. The big problem was that virtually the entire human race had become dependent on economic growth. Jonstad (2009) saw it as inevitable that the ecological consequences of GDP-thinking would sooner or later force a re-examination of "the social logic of the consumption society", of the global distribution of resources, and of ideas about what the "good life" actually meant. Here Jonstad enacted an ecological masculine position in which the asymmetry of the ecomodern masculinity was criticized. He said we cannot both have the consumption patterns of today and at the same time be sustainable.

Common to the calls for a different politics and an ecological masculinity was the implication that extensive social structural changes were needed because of the climate crisis, and that this was not something that could be achieved through voluntary, individual consumer choices and market solutions. They would require extensive democratic participation, equality politics and politicians who assumed long-term responsibility for the biosphere, even if this meant interference in citizens' consumption habits and behaviour. A leading author and intellectual in Sweden, Göran Greider, said for example, "Transition programs, even utopian, need to be formulated to downshift industrial civilisation to ecologically sustainable levels" (2008, 4). Local experiments of eco-villages, organic food and zero-energy housing were included in a network of transitions inspired by the movement in the UK (Bradley & Hédren 2014).

As outlined by, for example, the Swedish author and scholar Björn Forsberg (2007), the contours of a sustainable social system must rest on the principle that all economic activity that impoverishes ecosystems must end and that the economy needs to be adapted to minimize environmental burdens. The consequence of this was that a number of carbon and energy-intensive phenomena such as air traffic, mining and the long-distance importing of vegetables must end. The economy must return to a locally defined context where the power over production and consumption would be held by members of the local society and not by global market forces. This required circular flows and local and small-scale solutions. This did not mean that national or transnational trade would be banned, but that the needs of the local economy would take preference.

Ecological masculinity and transitional agency

Among others Forsberg stressed that a localized economy was realistic and reasonable for handling the challenge of climate change and that it also had a much longer history; it was also already practised by the majority of the world's people. The big problem was how the downsizing of a fossil fuel based economy would proceed. Two complementary strategies were needed: reform from the inside of the growth economy and the development of "pockets of alternative economic thinking" that could serve as good examples (Forsberg 2007). Ecological masculinity within the climate change debate involved the message "system change, not climate change" and advocated civil disobedience and direct action in order to speed up the system change. The goal was to move the issue of climate change from the closed and paralysed UN negotiations to the alternative forum created by the global climate movement, thereby creating new conditions for achieving equitable and effective climate action (Anshelm & Hultman 2014a). Johan Ehrenberg (2007) touched upon similar ideas when he emphasized the changes made by individual citizens with regard to their lifestyles and consumption in the light of a growing awareness of the need. He also emphasized citizens' increasingly radical and democratic demands on politicians to use public investment to make changes to energy, transport and production systems that threatened basic living conditions.

Discussion: Masculinities in a fossil fuel burning world

Men, and a few women, are travelling around meeting each other on various occasions, in various settings trying to find solutions to global climate change. We argue that continuing this practice is made possible by the hegemony of ecomodern masculinity which recognizes global climate change but at the same time is engaged in different enactments of large scale solutions such as nuclear power, carbon capture and sequestration, carbon markets and geoengineering. What if that is not taking us closer to the solutions, but is the actual problem? It might also be that if the ecomodern masculinity dominates this much we might tend not to recognize other forms of masculinities and the enactments of, for example, urban gardening, permaculture, r-economy, collaborative economy, etc.?

In this chapter we use an analytical framework proposed by Hultman (*forthcoming*) that suggests that there is more than one masculinity within environmental politics but less than many. We have employed this to analyse the climate change debate. Ecomodern masculinities dominate global climate change debates. This figuration proposes that environmental problems such as climate change should be handled with only a slightly revised industrial modernity, rather than a complete overhaul. Ecomodern masculinities – where toughness, determination and hardness go hand in hand with well-chosen moments of compassion, vulnerability and eco-friendly technology – appear to be the ultimate win-win figuration. But looking more closely reveals a cover-up to continue down the same modern industrial path that created the problems in the first place.

Even though the ecomodern discourse maintains a "business as usual in the form of industrial modernization" attitude, such a discourse is still troublesome for industrial masculinity because ecomodern discourse opens up the debate of climate change as a societal issue that needs to be addressed by industry, politicians and the public. The industrial masculinities figuration has dominated industrial modernization, but this climate change position is not possible to take up without denying all, or most of, the research findings regarding climate change. While industrial masculinity portrays nature as a resource that needs to be tamed and worked with accordingly, ecomodern masculinity is able to depict nature as alive and in need of the care of the market. In both cases nature thus becomes something possible to dominate with masculine practices.

Ecological masculinity does present itself as a possibility in our time of great need for paths to a liveable earth that are in contrast to the industrial and ecomodern masculinities. An ecological masculinity would be part of remaking the economy and facilitate the transition towards a more environmentally benign way of being part of the world.

We suggest there is a need for more research into understanding the values and practices of men, not least because of the large importance men play in shaping, formulating and deciding environmental issues globally. Finding ways of moving beyond the binary of men and women towards masculinities and femininities connected to discourses may create new ways of getting closer to the actual problems and finding durable solutions.

References

Alaimo, S 2009, "Insurgent vulnerability and the carbon footprint of gender", *Kvinder, Køn & Forskning*, vol. 3, no. 4, pp. 22–35.

Anshelm, J 2010, "Among demons and wizards: The nuclear energy discourse in Sweden and the re-enchantment of the world", *Bulletin of Science, Technology & Society*, vol. 30, no. 1, pp. 43–53.

Anshelm, J & Hultman, M 2014a, *Discourses of global climate change*, Routledge, London.

Anshelm, J & Hultman 2014b, "A Green fatwā? Climate change as a threat to the masculinity of industrial modernity", *Norma*, vol. 9, no. 2, 84–96.

Arora-Jonsson, S 2014, "Forty years of gender research and environmental policy: Where do we stand?" *Women's Studies International Forum*, vol. 47, December, pp. 295–308.

Boykoff, MT 2011, *Who speaks for the climate? Making sense of media reporting on climate change*, Cambridge University Press, Cambridge.

Bradley, K & Hedrén J 2014, "Utopian thought in the making of green futures", in K Bradley & J Hedrén (eds), *Green utopianism: Perspectives, politics and micro-practices*, pp. 1–22, Routledge, NY.

Carvalho, A 2007, "Ideological cultures and media discourses on scientific knowledge: Re-reading news on climate change", *Public Understanding of Science*, vol. 16, no. 2, pp. 223–243.

Christensen, AD & Jensen, SO 2014, "Combining hegemonic masculinity and intersectionality", *NORMA: International Journal for Masculinity Studies*, vol. 9, no. 1, pp. 60–75.

Connell, RW 1990, "A whole new world: Remaking masculinity in the context of the environmental movement", *Gender & Society*, vol. 4, no. 4, 452–478.

Dymén, C, Andersson, M & Langlais, R 2013, "Gendered dimensions of climate change response in Swedish municipalities", *Local Environment* 18, no. 9, pp. 1066–1078.

Ehrenberg, J 2007, "Medan klimatet skenar gör politikerna – ingenting", *Dagens ETC*, p. 49.

Einarsson, G, Franzén, LG, Gee, D, Holmberg, K, Jönsson, B, Kaijser, S, Karlén, W, Liljenzin, JO, Norin, T, Nydén, M, Petersson, G, Ribbing, CG, Stigebrandt, A, Stilbs, P, Ulfvarson, A, Walin, G, Andersson, T, Gustafsson, SG, Einarsson, O & Hellström, T 2008, "20 toppforskare i unikt upprop: koldioxiden påverkar inte klimatet", *Newsmill*, 17 December.

Eriksson, P 2008, "Jorden går inte under av utsläpp och uppvärmning", *Dagens Industri*.

Forsberg, B 2007, *Tillväxtens sista dagar*, Karneval förlag, Stockholm.

Giddens, A 2009, *The politics of climate change*, Cambridge University Press, Cambridge.

Greider, G 2008, "Ökad tillväxt och ökad lycka följs inte längre åt", *Friluftsliv*, p. 4.

Hajer, MA 1996, *The politics of environmental discourse: Ecological modernization and the policy process*, Clarendon Press, Oxford.

Hultman, M 2013, "The making of an environmental hero: A history of ecomodern masculinity, fuel cells and Arnold Schwarzenegger", *Environmental Humanities*, vol. 2, pp. 83–103.

Hultman, M 2014a, "Transition delayed: The 1980s ecotopia of a decentralized renewable energy systems", in K. Bradley & J Hedrén (eds), *Green utopianism: Perspectives, politics and micro practices*, pp. 243–257, Routledge, London.

Hultman, M 2014b, "How to meet? Research on Ecopreneurship with Sámi and Māori", International Workshop: Ethics in Indigenous Research – Past Experiences, Future Challenges, 3–5 March, Umeå University, Sweden.

Hultman, M 2016, "Gröna män? Konceptualisering av industrimodern, ekomodern och ekologisk maskulinitet", Special Issue Environmental Humanities, *Kulturella Perspektiv*, vol. 25, no. 1, pp. 28–39.

Hultman, M (*forthcoming*), "Industrial, ecological and ecomodern masculinity: Conceptualising forms of masculinities in the environmental field", in S Buckingham (ed), *Understanding climate change through gender relations*, Routledge, London.

Hultman, M & Yaras, A 2012, "The socio-technological history of hydrogen and fuel cells in Sweden 1978–2005: Mapping the innovation trajectory", *International Journal of Hydrogen Energy*, vol. 37, no. 17, September, pp. 12043–12053.

Jackson, T 2011, *Prosperity without growth: Economics for a finite planet*, Routledge, London.
Jänicke, M 2008, "Ecological modernisation: New perspectives", *Journal of Cleaner Production*, vol. 16, no. 5, pp. 557–565.
Jonstad, D 2009, "Sista chansen att tvärnita", *Göteborgs-Posten*, 23 November.
Knaggård, Å 2014, "What do policy-makers do with scientific uncertainty? The incremental character of Swedish climate change policy-making", *Policy Studies*, vol. 35, no. 1, pp. 22–39.
Lahsen, M 2013, "Anatomy of dissent: A cultural analysis of climate skepticism", *American Behavioral Scientist*, vol. 57, no. 6, pp. 732–753.
Lidskog, R & Elander, I 2012, "Ecological modernization in practice? The case of sustainable development in Sweden", *Journal of Environmental Policy & Planning*, vol. 14, no. 4, pp. 411–427.
McCright, A & Dunlap, R 2003, "Defeating Kyoto: The Conservative movement's impact on U.S. climate change policy", *Social Problems*, vol. 50, no. 3, pp. 348–73.
McCright, A & Dunlap, R 2011, "Cool dudes: The denial of climate change among conservative white males in the United States", *Global Environmental Change*, vol. 21, no. 4, pp. 1163–1172.
Macgregor, S 2013, "Only resist: Feminist ecological citizenship and the post-politics of climate change", *Hypatia*, vol. 29, no. 3, pp. 617–633.
Melosi, MV 1987, "Lyndon Johnson and environmental policy", in RA Divine (ed.), *The Johnson years: Vietnam, the environment, and science*, pp. 119–120, University Press of Kansas, Lawrence.
Merchant, C 1980, *The death of nature: Women, ecology, and the scientific revolution*, Harper & Row, San Francisco.
Merchant, C 1996, *Earthcare: Women and the environment*, Routledge, NY.
Neumayer, E & Plümper, T 1997, "The gendered nature of natural disasters: The impact of catastrophic events on the gender gap in life expectancy, 1981–2002", *Annals of the American Association of Geographers*, vol. 97, no. 3, pp. 551–566.
Nightingale, AJ 2006, "The nature of gender: Work, gender and environment", *Environment and Planning D: Society and Space*, pp. 165–185.
Nordin, I 2008, "Klimatdebatten – en röra", *Östgöta Correspondenten*, vol. 9, no. 6, 6 June.
Piketty, T 2014, *Capital in the twenty-first century*, Belknap Press, Cambridge, MA.
Radetzki, M & Lundgren, N 2009, "En grön fatwa har utfärdats", *Ekonomisk debatt*, vol. 37, no. 5, pp. 57–65.
Räty, R & Carlsson-Kanyama, A 2010, "Energy consumption by gender in some European countries", *Energy Policy*, vol. 38, no.1, pp. 646–649.
Reinfeldt, F 2008, "Speech in Almedalen", Visby, Sweden.
Rome, A 2003, "'Give earth a chance': The environmental movement and the sixties", *Journal of American History*, vol. 90, September, pp. 534–541.
Shiva, V 1988, *Staying alive: Women, ecology and development*, Zed Books, London.
Terry, G 2009, "No climate justice without gender justice: An overview of the issues", *Gender & Development*, vol. 17, no. 1, pp. 5–18.
Urry, J 2011, *Climate change and society*, Polity Press, Cambridge.
Zannakis, M 2015, "The blending of discourses in Sweden's 'urge to go ahead' in climate politics", *International Environmental Agreements: Politics, Law and Economics*, vol. 15, no. 2, pp. 217–236.

3

IT'S NOT JUST THE NUMBERS

Challenging masculinist working practices in climate change decision-making in UK government and environmental non-governmental organizations

Susan Buckingham and Rakibe Külcür

Introduction

Tenacious campaigning by women's groups worldwide since the 1980s has raised international awareness that gendered and women's concerns are thoroughly imbricated with environmental issues and increasingly established in international agreements since the 1990s. In 1992, the United Nations Conference on Environment and Development agreed a sustainability agenda for the twenty-first century that required the inclusion of women as a major stakeholder group. Chapter 24 of Agenda 21[1] clearly made the case for this on the basis that women are most highly represented amongst those who, worldwide, have the worst environmental experiences. Chapter 24 also acknowledged that women's experiences of subsistence food growing and other provisioning enhances their environmental knowledge and concern, yet they have very little control over environmental decision-making. Likewise, the United Nations Conference on Women held in Beijing in 1995 stipulated the importance of linking women's rights to their participation in environmental decision-making. Nevertheless, it wasn't until 2012 that the United Nations Framework Convention on Climate Change (UNFCCC) eventually agreed to start a formal process to move towards gender equality in climate change decision-making positions (UNFCCC 2012).[2] Such a lacuna is only explicable if we consider that the UN's own work on gender mainstreaming has not been taken sufficiently seriously, nor embedded, at the highest scale. It is also tempting to speculate that masculinist ways of doing business prevail over a desire to address the problems of climate change which may have more success with a broader, more diverse decision-making process and participants (Buckingham 2010; Ergas & York 2012). Research into women's involvement on company boards suggests that this is the case (Davies 2011, 2015). That it is now under consideration for climate change decision-making is the result of further vigorous campaigning by women (WEDO 2010).

This chapter explores why, despite international legislation, gender mainstreaming has failed to be integrated into climate change related decision-making. It first outlines and summarizes critiques of gender mainstreaming. While having been established in the region of twenty years, it is argued to have been hijacked as a pro-business argument, rather than being used as a mechanism for achieving gender equality in decision-making. It is clear from a brief review of professions most influential in climate change decision-making that women are in the minority by a considerable margin and that a gender pay gap persists. The chapter then discusses the UK evidence from three research projects undertaken by the authors over the course of twelve years. While the first two of these discrete research projects have particular findings, which have generated particular analyses (see Buckingham, Reeves & Batchelor 2005 on gender mainstreaming waste management, and Külcür 2012 on gendering environmental non-governmental organizations), it was through the third – smaller – research project, which interviewed staff in UK government departments and quangos involved in some way in climate change related decision-making, that we understood how ubiquitous is the institutional masculinization of environmental decision-making. It is for this reason that we have taken the decision to consider all three projects from the position of their decision-making structures, in order to better understand the gendered structures of decision-making, and how these might be overcome. The earliest research project investigated how municipal waste management was gendered in three EU member states. It was funded in 2002 by the European Commission as a pilot project in gender mainstreaming.[3] The second research project, started in 2006, was undertaken in Turkey and the UK for the award of a PhD and investigated how environmental non-governmental organizations (ENGOs) are gendered.[4] For this chapter, data have been drawn from the UK case studies of these two projects. The third piece of research was undertaken in connection with a European Union COST Action on Gender and Science, Technology and Environment, and comprised interviews with managers and women's network coordinators in three UK government departments/quangos with a climate change remit.[5] In parallel with this, a selection of equality impact assessments of energy policies was scrutinized.

Gender mainstreaming

The Beijing conference established the principle of "gender mainstreaming" in which all government policies and practices should, as a matter of course, be evaluated for their impacts on men and women, so that neither group is disproportionately discriminated against. The EU legally incorporated this in 1996, binding all member states to its enactment. According to Walby (2005), although there are "weaknesses in implementation", the EU has "become a transnational actor that is very important for the contemporary development of gender mainstreaming with strengths in promoting the policy in abstract" (461). The Amsterdam Treaty, signed in 1997, formalized fundamental rights and provided mechanisms by which these must be upheld by member states. Where previously it was illegal to discriminate other than on grounds of nationality, the new Article 13 "enabled the Council to take appropriate action to combat discrimination based on sex, racial or ethnic origin, religion or belief, disability, age or sexual orientation" (European Union 1997). Whereas previous legislation referred to equality in rates of pay between men and women, the new Treaty introduced two additional articles. These included the promotion of equality between men and women (Article 2 amendment), and a new paragraph which was added to Article 3 stating that: "In all the other activities referred to in this Article, the Community shall aim to eliminate inequalities, and to promote equality, between men and women." Within the same Treaty, the principle of sustainable development was enshrined for the first time with "the clause calling for environmental protection requirements to be integrated into the definition and implementation of other policies", as a way of achieving this (European Union 1997).

The UK adopted gender mainstreaming into policy in 1998, but this has had limited impact. In pilot research commissioned by the European Commission to investigate the uptake of gender mainstreaming in municipal waste management, very little evidence was found of how this was taking place at the local scale in a sample of three European countries. Neither was it clear how national government offices were disseminating good practice, or even guidelines (Buckingham, Reeves & Batchelor 2005).

Given that gender equality is so embedded in European policy, and that the mechanism by which this can be achieved – gender mainstreaming – has been identified and legislated for, it seems instructive to explore how this has been used to address climate change issues. Before turning to evidence from two projects involving, respectively, municipal waste management authorities and national level government departments, it is useful to consider the potential and problems of gender mainstreaming in Europe. Angela McRobbie (2009) has questioned how adequate gender mainstreaming is as a justice objective, suggesting that it replaces rather than supports feminism, and bolsters neoliberal managerialism by focusing on how decision-making can benefit from women's skills and knowledge, rather than on the gender justice of equal opportunities (and outcomes) for women (Buckingham 2015). Further, Grosser and Moon (2005) argue that gender mainstreaming benefits the corporate social responsibility agenda, "simultaneously good for both business and wider society" (in Walby 2005, 457). This could be argued to be the equivalent of the "ecological modernization" argument made by government and business to try to contain changes needed for more sustainable ways of managing the environment in the prevailing business model. Calls within Europe (Davies Report 2011; European Parliament 2012) for better women's representation on the boards of FTSE listed companies are also justified on the basis of more prudent, responsible and productive business performance, although progress on European legislation has been disappointingly slow (Duncan 2015). Indeed, a recent briefing from the UK Government Department for Business Innovation and Skills puts "The Business Case for Equality and Diversity" (BIS 2013).

Walby (2005), more optimistically, places gender mainstreaming variously as "a leading-edge example of the potential implications of globalisation for gender politics" (464) although she appears to be conflicted as to the use of gender mainstreaming as she also places is it "in the tension between the mainstream and interventions to secure gender equality" (463). The ambivalence of gender mainstreaming is again illustrated by Walby who articulates it not only as a "specialized policy tool" but also as a "feminist strategy that draws on and can inform feminist theory" (466). Within the context of climate change, Röhr et al. (2008) argue that gender mainstreaming has been institutionalized very tenuously, and attributes this to the androcentricism of institutions. They go on to state that while it is necessary to have an equal number of male and female decisions makers, alone it will not guarantee "[gender] justice in institutional orientation". Moreover, they argue for the need to go beyond a consideration of gender differences to understand and question gender relationships which are underpinned by masculinist power structures. Magnusdottir and Kronsell (2015) have found that numerical equality between men and women in Swedish climate change planning organizations does not necessarily lead to greater gender sensitivity in decision-making and they propose that one of the reasons for this is that the institutional structures in which women make their careers encourage conformation with prevailing masculinist working practices. Without dismantling these structural dimensions of gender inequality it is impossible to make any fundamental changes, which may explain the limited success of gender mainstreaming.

A review of UK strategic planning authorities' capacity to embed gender mainstreaming found it to be limited (Reeves 2002), which, given the work on women's committees and on women in planning in urban authorities in the 1980s, was disappointing. Reeves' work with the Royal Town Planning Institute (RTPI) in the early 2000s was "committed to ensuring that planning practitioners understand the important contribution spatial planning can make in achieving gender equality" (Reeves 2003). In early work on gender mainstreaming Reeves viewed it as a redefinition of "the approach to gender equality beyond legislative and positive action" to consider "the relationship between men and women" (Reeves 2000, 9). Supported by Reeves' work, the RTPI called for greater consideration of gendered use of and needs in planning through the publication of a gender mainstreaming toolkit for planners (RTPI 2003).

The lack of women working in climate change related sectors in the UK

By 1990, Clara Greed had exposed the scarcity of women in the built environment professions of surveying and planning, and subsequent reports show scant improvement. By 2004, Howatt, Olwafemi and Reeves had noted for the RTPI that the proportion of its corporate members who were women was 23 per cent. Nor is there necessarily linear growth in such numbers. Merlin Fulcer (2010) has reported that the proportion of practising women architects in the UK dropped from 25 per cent in 2008 to 19 per cent in 2010 as architectural practices were impacted by the economic recession, and women employees appeared to be the first to be laid off work. Data for women working in other climate change related professions, such as waste management, energy, transport and water planning are equally stark. Here, women rarely exceed 25 per cent of the workforce, with the exception of renewable energy, in which 28 per cent of staff are women compared to the energy sector as a whole, in which 12.8 per cent of jobs are held by women. In 2014, Ernst and Young launched their "Women in Power and Utilities" review, which identified women as comprising a low 5 per cent of board members of the top earning power and utilities companies worldwide (Ernst and Young 2016). Seven per cent of executive directors are women, and 23 per cent are non-executive directors in both Europe and North America. In Europe women comprise 12 per cent of the senior management team, compared to 21 per cent in North America. A review of gender differences in its membership by the Institution of Environmental Sciences (IES)[6] identified that women working full time were more likely than their male counterparts to be on fixed term or temporary contracts; that members who were men were more likely to be employed in industry than women members; and that women were more likely to earn salaries in the lower band, and less likely to be earning bonuses. Women members were more likely than men to have taken parental leave, but less likely to have the support of their employers to do so, and one third returned to work at lower pro rata salaries and less responsible positions than held previously. This led to one of the report's conclusions that a "leaky pipeline" resulted in women being lost in the environmental science sector (Heaton 2010).

A pay gap in which women in full time work earn less than men in full time work exists across all industry sectors, from 3.8 per cent for conservation and environment professionals to 14.1 per cent for architects, town planners and surveyors. The gender pay gap for engineering professionals, in which energy, waste, transport and water related professions sit, was 12.5 per cent (Perfect 2012).

Meanwhile, work on gender mainstreaming in the municipal waste management sector undertaken in Ireland, Portugal and the United Kingdom in the early 2000s found isolated pockets of good practice which were overshadowed by a general mystification of what a gender sensitive waste management practice might look like (Buckingham, Reeves and Batchelor 2005). Interviewing waste managers, non-governmental organizations and local residents established that where waste professionals had come from an engineering background (and were primarily male), waste strategies were heavily weighted towards management and disposal methods, whereas professionals who came from other backgrounds (in the case studies, these comprised education, economics and business) were more likely to try to understand and engage with residents' attitudes and behaviour towards waste (for example, what are the barriers to recycling), and to use this to develop effective waste management strategies. What was also notable was a failure of municipalities to transfer their own good practice in, for example, ethnic minority residents' participation, or gender oriented human resources practice, into what could become gender sensitive waste management practice. For example, one municipality studied had put a lot of effort into recruiting local residents with South Asian heritage to meetings which would discuss how recycling, for example, could most effectively be developed. However, the same council officers had not thought how they might ensure the gender balance at these meetings – even though women were found, in the research, to be the main waste managers in the household. There did not appear to be mechanisms within or between municipalities through which examples of gender sensitivity could be transmitted. One of the conclusions from this research was that gender mainstreaming is hampered by a lack of necessary gender expertise in fields such as urban planning and waste management to dismantle structures of gender inequality. Writing in the context of the Netherlands in 1999, Mieke Verloo was positive about the aims of gender mainstreaming to prioritize the lives and experiences of individuals, its potential to lead to better government, its aim to involve men as well as women, its acknowledgement of the diversity amongst men and women, and its aim to make gender equality issues visible "in the mainstream of society" (1999, 8). However, she was sceptical that sufficient expertise amongst planning professionals was available to challenge prevailing discourses, and to align the necessary interests from those "at the top" and those "down under" in the planning system. She refers to this as "frame alignment" (8).

Exploring how gender is embedded in climate change decision-making in the UK

Under the UNFCCC decision 18/CP.20, also known as the *Lima Work Programme on Gender*, parties are required to move towards gender equality in decision-making and to periodically report to the COP on progress (UNFCCC, 2015). It can therefore be reasonably surmised that this goal should be set for all levels of climate change decision-making, including at the city/municipal level.

In the UK, no urban planning can now be undertaken without consideration of how it will impact on climate change, or will contribute to mitigating for or adapting to climate change. The UK Government Planning Policy Statement (PPS) on Climate Change (Department for Communities and Local Government 2007) has been issued as a supplement to PPS1 (1997), which "sets out the overarching planning policies on the delivery of sustainable development through the planning system". Where there are differences between these two PPSs, the one on climate change takes precedence. The Department of Energy and Climate Change (DECC) and the Local Government Association have signed a Memorandum of Understanding to provide leadership to meet UK energy and climate change objectives. Meanwhile, gender mainstreaming requires "the (re)organisation, improvement, development and evaluation of policy processes, so that a gender equality perspective is incorporated in all policies at all levels and at all stages, by the actors normally involved in policy-making" (Council of Europe 1998). In the UK this is currently translated through the *Equality Act* (2010), which requires local authorities to, amongst other things, know and involve their local communities in ways that eradicate inequalities between those in "protected groups" and the rest of the community. By this legislation, there are nine protected groups, identified by age, disability, gender reassignment, marriage/civil partnership, pregnancy/maternity, race, religion/belief, sex and sexual orientation (Equality and Human Rights Commission 2015). The *Act* also requires these inequalities to be eliminated in the workforce. It would therefore seem appropriate that climate change related planning policies are screened for their impact on gender fairness and equity.

As part of the third research project introduced at the start of this chapter, a number of climate change related policies introduced since the government passed the 2008 Climate Change Bill and the 2010 *Equality Act* have been reviewed, to ascertain the extent of gender sensitivity in the mandatory consideration of "main affected groups" required by the legislation. Interviews with officers in three government departments/quangos further explored the extent to which a consideration of gender is a factor in both policy making and staffing, and how these two elements might be inter-related. While the first strategy enables a consideration of how far policy is evaluated in relation to gender, the second was most effective in providing an idea of the gendered context in which decision-making takes place. The extent to which gender is taken into account in both strategies is explored below.

Women working in government

In their review of the Swedish experience of gender mainstreaming, Sainsbury and Bergqvist identified gender training as a "prerequisite for implementing mainstreaming since decision makers and administrators responsible for all policies are to integrate a gender equality perspective in their work" (2009, 221). This did not appear to be the case from the interviews undertaken in the UK and reported on here. Evidence for this is drawn from four in-depth semi-structured interviews[7] conducted with four individuals/pairs of individuals working in government and government related departments on policies associated with climate change, and from an analysis of five publicly available impact assessments on proposed energy related interventions. Equality analyses across the nine protected characteristics, as identified above, are a required element of the impact assessments.

In an interview with staff members in one government department, it was explained that the policy architects undertake equality analyses. While staff in human resources departments receive equality and diversity training, this does not extend to staff who are required to undertake equality analyses of policy, who, instead, are provided with templates designed to assist them. It did not therefore appear that the training received by human resources staff was transferred across to policy departments. Elsewhere in the department, it was suggested that equality analyses were most likely to be undertaken by junior staff on the policy architect's team. These analyses were not thought to be given a high priority, or likely to be built in from an early stage of policy design. To the best of the interviewees' knowledge, no explicit training was provided for equality analysis. A review of impact assessments undertaken on proposed energy related policies revealed missed opportunities to address gender inequalities, and a tendency for claiming that a policy "does not discriminate against any of the above protected characteristics", and "equality impacts of the scheme should be neutral". In many cases, the "protected characteristics" which are deemed not to be impacted are simply not listed at all, so it is not possible to see what kind of consideration may or may not have taken place. For example, the quality assessment on the Green Deal (a policy designed to improve energy efficiency through the supply chain) apparently found no evidence that gender would affect people's experience of the intervention regarding its impact, the impact of the affordable warmth obligation to support households (other than referring to pregnant women and recent mothers) and the carbon saving obligation. However, references to age mask gender differences, and recent government information provided in an Office for National Statistics report (ONS 2013) states that excess winter mortality claims a higher proportion of the female population than of the male, and that this gender gap is increasing. Moreover, the greatest increase was found amongst females under 65. Women are particularly vulnerable, as they are most likely to be heading the poorest households (The Poverty Site 2016).

Another indication of the superficiality of equality analyses is revealed through a discussion with a senior decision-maker and a human resources manager about the link between mental ill health and fuel poverty, and yet this was not considered under the protected characteristic of disability. It could also have been cross-referenced to gender as, according to the Mental Health Foundation (2016), substantially more women are diagnosed with mental health problems. It is hard not to conclude that effective training on gender equality would have enabled the assessors to incorporate these points.

One gesture made to address the energy concerns of women appears to have been made in the setting up of a "Green Deal Women's Panel" in 2010, comprising representatives from Parliament, media and industry.[8] Its brief includes: "To consider the best channels and opportunities for promoting the Green Deal to women", and:

> to provide government with recommendations for effective communication and for influencing public opinion on the benefits of installing Green Deal measures in their homes.

It is unclear how this eminent group of women engage with the real and everyday energy concerns of less privileged women, and, linked to the analysis of Magnusdottir and Kronsell referred to earlier, it is by no means assured that the concerns of women more broadly will be served by an elite, homogenous group, members of which represent well established, masculinized institutions.

The wider context in which this apparent lack of well-informed assessment took place was explored with a senior staff member who regretted the lack of any women directors in his government department. He was unable to account for this through specific professional expertise as promotion was made largely through the civil service route, where women were well represented in the workforce overall (45 per cent of all staff, and best represented at lower levels). From a frank and "off the record" discussion, it appeared that strategic collaborative work was most difficult with those departments with clear climate change remits, all of which appeared to perform worst across Whitehall with regard to equal opportunities in senior appointments. Indeed, one interview with a senior manager indicated some despair with a related Government Department where gender was "off the radar". The Cabinet Office had targets to achieve a 39 per cent representation of women in the Senior Civil Service, and 34 per cent of grades 3 (Director General) and 2 (Director) to be women by April 2013. According to data presented in the interview, this target had been missed by one government department studied by 11.3 and 18 percentage points, respectively.

Barriers to women progressing within organizations were thought to include a lack of opportunities for part time and job share working, a lack of formal accountability for equal opportunities and gender equality, and an organizational culture in which the young and child free are apparently the preferred choice over women who have child care responsibilities. The head of the women's network in another department, in which women had made greater progress, cited flexible working as a frequently stated advantage given by staff working there, although she noted that most senior posts were still held by men. Conditions of work, which do not support the caring responsibilities of staff, have been openly stated in the Institution of Environmental Sciences survey, referred to above, as contributing to the loss of women to the profession.

While staff involved in a women's network established at one government department felt well supported by the department's Head of Operations and a Minister, this was felt to be contingent on particular personalities and vulnerable to staff changes at senior levels. At the time of this research (2014), the department had twice failed in attempts to set up a "Diversity Strategy" and one recommendation from the local women's network was to appoint a senior manager responsible for equality and diversity.

Two convenors of a women's network in a government department observed that the current image of climate change-related work, reflected in the wording of recruitment advertising that is "dynamic, thrusting, hardworking, demanding, full-on" appealing to the "young and passionate" may not appeal to the widest and most diverse community of applicants. This echoes the machismo of climate change negotiating, evident from late night sessions and last minute brokered deals common in COPs (Grubb & Yarmani 2001; Röhr et al. 2008). It also echoes the prevailing ethos in environmental NGOs, as demonstrated below.

Women working in environmental non-governmental organizations

One might assume that while male-dominated senior staffing and masculinist organizational structures are endemic in business and government, non-governmental organizations may challenge this. NGOs may be expected to be more women-friendly workplaces than some other organizations (Vázquez 2011), because they "work upon sets of principled ideas or values" (Gillespie 2006, 327) where profit-making is not the aim. This was highlighted by a male CEO interviewed:

> I think the ENGO world is one of the easiest sectors where women succeed in. I think that's because NGOs are based on values, so that they are driven by an ideological belief such as fairness, decency, quality and therefore it would be inconsistent for them to discriminate against someone, I don't know, on gender or physical capacity . . . so that is why I think easier. (Külcür 2012, 161)

However, in research undertaken in ENGOs in the UK and Turkey by Rakibe Külcür, the second author of this chapter, as a part of a PhD thesis, a gendered division of labour was observed (2012). The UK data was collected through 38 interviews and one focus group interview across nine ENGOs. The interviews were conducted mainly with senior managers, but also with employees and volunteers, since more women were represented in these job segments and interviewing employees enabled data to be gathered not only from dominant groups (mainly senior men).

Furthermore, the researcher placed herself as a volunteer and researcher in one ENGO in the UK in order to observe the organizational practices and to enable triangulation of data. In addition, secondary data on gender compositions and the campaigns of the ENGOs were collected from annual reports, staff charts, publications and the websites of the organizations. This research found that women

TABLE 3.1 The total number of interviewees in the British ENGOs

Seniority	*Numbers*
CEOs	6
Senior Managers	14 (+1)*
Middle Managers	11
Non-Managers	13 (+5)*
Total	38 and 1 Focus group

Source: Külcür (2012).

* The numbers in brackets indicate the participants in the focus group interview.

TABLE 3.2 The total number of interviewees in the British ENGOs by gender

Gender	*Numbers*
Female	23 (+4)
Male	15 (+2)

Source: Külcür (2012).

were represented in stereotypically female occupations including as secretarial and administrative staff, CEO assistants as well as in job segments such as human resources, marketing, communication and fundraising (Külcür 2012). On the other hand, men dominated the most senior management positions (CEOs and board members) and science related roles such as in conservation departments, and in job segments such as IT. Surprisingly, many of the senior managers thought that the management of the environmental sector especially at the top is masculinized, as was expressed by one male CEO of an ENGO:

> I think, environmental NGOs are very male, I think . . . if you go through them . . . we have dinner every two months at the [. . .][9] some of the main ENGOs, I would say there are (X-CEO-F) from the (ENGO), and (Y-CEO-F) from the (ENGO), if X sends the deputy sometimes it's a woman sometimes it's a man, and if Y sends her deputy, then it is a man, if men send deputies then is a man and uhff . . . It is very very male. (Külcür 2012, 168)

The environmental NGO sector was, however, numerically dominated by women. When exploring the reasons for this, certain stereotypical assumptions concerning gendered roles and the patriarchal divisions of labour emerged. Research participants assigned women to traditional care-giving roles and men to career development. Just as women's contribution and women's work in the unpaid private sphere have been invisible in society, so these gendered assumptions were repeated in the ENGOs.

Through the research, the reason for the domination of the sector by women employees and gender-imbalanced management was concluded to be most likely due to the ENGOs' gendered organizational practices. Working practices were incompatible with work–life balance and family life, especially where, commensurate with gendered roles in society, women were the main homemakers and carers. Such gendered organizational practices included recruitment and promotion; the expectation that staff work long and irregular hours, evenings and weekends; frequent travelling; and a lack of flexible working options (including a lack of management support for working from home). One of the surprising findings was that although large ENGOs had more possibilities to provide a flexible working environment than smaller ENGOs, they were not always willing to do so.

Such working practices tended to favour the young and single, and it was notable how women who opted to work part time after having had children did not progress in their careers. A prevalence of staff on short-term contracts as a result of limited term project funding with low wages was also found to be common. As a consequence, many women working in ENGOs reduced their working hours or left their jobs once they became mothers. These work practices, and women ENGO employees' responses, reflect the situation in other organizational structures in which limited career prospects and low salaries do not seem to provide an impetus for women to work continuously, particularly when they have a family (Verbakel & de Graaf 2008). These short term employment practices have a significant importance as they seem to result in poor job prospects and high labour turnover, which in turn have negative implications for women's careers, since continuous work experience is important for career advancement (Heidenreich 2010).

The research in ENGOs also demonstrates how women in management positions have felt excluded and passed over for promotion. Generally, those interviewed reported how having children was generally negative for career development in environmental NGOs, and those women who had achieved the highest positions tended to have no children and/or male partners who played a supporting role (for example, had retired, or worked part time). In the participant observation ENGO, although women outnumbered men, there was only one female middle manager (working part time) based in the office who had a child, while four out of five male managers that we interviewed had children. One male CEO confirmed this trend:

> (How many female managers have children?) Only one [out of five] has children, the others don't. So, may be that proves the point that the women who don't have children go higher up in the career ladder . . . whereas if you look at the men . . . well . . . two out of three have children. (Külcür 2012, 215)

Examining organizational culture also revealed some of the factors behind the under-representation of women in senior roles. It appeared that men as the dominant group determined the group culture in the ENGO sector, as elsewhere. Although mentors have been identified as important by organizational scholars for career progression, as pointed out by one of the female research participants in a senior position, role models for women in top positions are very rare and as a result some women holding senior positions may feel more unsupported compared to their male counterparts. One woman who had left a management position at a large international ENGO complained that:

> In the environmental movement generally at the senior level, you don't really see many female managers at all. I've been quite alone in that regard. Very male . . . but even when they are women, the culture itself felt very male . . . there is a way of working in the environmental movement that is very male, very macho, and . . . that's not very appealing to women actually . . . There is, the maleness of a clubby culture, you know, it is not very family friendly in terms of having a work–life balance, so there is certainly that issue that comes into it . . . and . . . I have had so many sexist remarks. (Külcür 2012, 206)

This provided evidence that women in senior roles, where there is male domination, are required to put in additional effort to survive (Lyness & Thompson 1997) while they often feel like "outsiders on the inside" (Moore 1988, 568).

Networking is a significant factor for career progression, as Rastetter and Cornils (2012) have noted: "[w]omen in executive positions are disadvantaged due to their minority position concerning their access to networks and their ability to build a mentoring relationship" (43). Thus, the under-representation of women, particularly for executive and board positions, may be due to the limited networking possibilities for women and the existence of an old boys' network. It was pointed out by one interviewee that the:

> number of men [in ENGOs] who've been to private schools, you know the kind of boarding schools background in these organizations and I do feel sometimes that that leaves a big legacy in terms of how men deal with women and the recognition of a sort of male tribe . . . there are lots of public school educated men in this area. (Külcür 2012, 224)

It was also reported that career chances in ENGOs seemed to benefit from having a science degree:

> I think that women are passionate about the environment, but the more you go to senior levels it is about numbers and statistics and science and I don't think you have as many women who are interested in that side or have that background . . . (Külcür 2012, 220)

If this perception is true, given the domination of most science subjects in UK universities by male students, this limits the extent to which women are likely to progress in ENGOs. The eco-feminist writer Carolyn Merchant observed how, since the Enlightenment, knowledge has increasingly become the preserve of experts, and experts (predominantly men) have become increasingly specialized. In the UK we specialize early: those who stay on beyond compulsory education are limited to study three or four subjects at advanced (A) level, and if they apply to university, these will determine what courses they will be able to follow. This is particularly so in subjects which have an explicitly environmental bias, such as environmental management, engineering (within which transport, energy, waste management and water expertise mostly lies), ecology and so on. These undergraduate degrees mostly demand A levels in maths and sciences, and these subjects are heavily dominated by male students, biology being the exception (59 per cent of students taking biology in 2014 were female). Other data reported by the Joint Qualifications Council in 2014 (and reported in WISE 2015, cited here) include the fact that 39 per cent of students taking maths A level exams were female; and more than four times as many males as females sat physics A level (79 per cent compared to 21 per cent). The gap is narrower in chemistry, and closing, with 48 per cent of all students taking the A level exam being female (WISE 2015).

While ENGOs mimic institutional structures in many respects, the low pay and insecure contracts make it difficult for women to pay for professional help with childcare, a strategy which is available to higher paid women in the private and government funded sectors. So, in some ways regarding gender equity, it appears that ENGOs emulate the gendered organizational structures of government and business, while at the same time compounding this lack of opportunities for women with much lower pay and less secure contracts.

Conclusions

Despite the range of organizations considered above, common features can be identified: the failure to transfer good practice in gender equality, where it exists, from one department to another (particularly from human resources, where it is more commonly found); the failure to recognize the need for gender equality and/or gender equality training; and a failure to train those involved in assessing the gendered impacts of policies in how to recognize these impacts. These findings support the more sceptical critics of gender mainstreaming, both in the climate change sector (such as Röhr et al. 2008), and more generally (such as McRobbie 2009). Well-embedded discrimination against women appears to exist both through prevailing masculinist attitudes, and through gendered working practices.

However, gender mainstreaming may have a fundamental flaw in the way it is being used to justify the business case. If the case for including women in equal numbers to men rests on the results that this is likely to achieve, what happens if these results don't materialize? Are women "disinvited" from the decision-making process? (And it is curious, isn't it, that if women are necessary for more effective, risk sensitive decision-making, that decision-making structures are still so heavily weighted towards men.) The gendered rights case has also to be made, regardless of outcome. And if the case for opening out decision-making to include equal representation of women is accepted, then, by extension, women of colour, disability, women who are carers, who are from different religions, cultures, ethnic groups, social classes, generations, and who are lesbian or transgender, must be well represented as well. For only if such diversity of women – and men – is achieved can the stranglehold of masculinist ways of thinking, and working, be broken. And that, surely, is necessary in order to have a chance of diverting catastrophic climate change.

Notes

1 Agenda 21 is a plan of action to be taken globally, nationally and locally by organizations of the United Nations System, governments, and major groups in every area in which there are human impacts on the environment (UN 1992).
2 Discussion on the role of gender in climate change decision-making had been initiated in 2001, and further raised by a women's lobby at the 2007 Conference of the Parties (COP), and in 2010, but it was not until Doha in 2012 that a process to achieve change was put into place.
3 The lead author, Susan Buckingham, was the Principal Investigator on the project; the Co-Investigator was Dory Reeves.
4 The PhD was undertaken by the second author, Rakibe Külcür, and supervised by the first author, Susan Buckingham. Rakibe Külcür was awarded her PhD in 2012.

5 This was undertaken by Susan Buckingham in 2014 and 2015.
6 The IES (Institute of Environmental Sciences) is a professional organization open to those working in the environment sector.
7 By their request, it was agreed that the interviewees would remain anonymous.
8 The panel included two MPs, one Baroness, RIBA, BBC Presenter, Ashden Awards for Sustainable Energy, Think 2050 (environmental social media), U-Switch (comparison site energy etc suppliers), Mark Group (energy saving advice), QED-UK (electricals) and ICE Org (loyalty scheme).
9 By their request, it was agreed that the interviewees and their organizations would remain anonymous.

References

BIS (Business Innovation & Skills) 2013, "The business case for equality and diversity: A survey of the academic literature", (https://www.gov.uk/government/uploads/system/uploads/attachment_data/file/49638/the_business_case_for_equality_and_diversity.pdf).

Buckingham, S 2010, "Call in the women", *Nature*, vol. 468, p. 502.

Buckingham, S 2015, *Gender and Environment*, Routledge, London.

Buckingham, S, Reeves, D & Batchelor A 2005, "Wasting women: The environmental justice of including women in municipal waste management", *Local Environment*, vol. 10, no. 4, pp. 427–444.

Davies, Lord of Abersoch 2011, "Women on boards", UK Government, (https://www.gov.uk/government/uploads/system/uploads/attachment_data/file/31480/11-745-women-on-boards.pdf).

Davies, Lord of Abersoch 2015, "Women on boards", UK Government, Davies Review Annual Report, (https://www.gov.uk/government/uploads/system/uploads/attachment_data/file/415454/bis-15-134-women-on-boards-2015-report.pdf).

Department for Communities and Local Government 2007, "Planning policy statement: Planning and climate change, supplement to planning policy statement", vol. 1, TSO, London.

Duncan, J 2015, "Women on boards: MEPs frustrated at lack of progress", *The Parliament Magazines*, (https://www.theparliamentmagazine.eu/articles/news/women-boards-meps-frustrated-lack-progress).

Ergas, C & York, R 2012, "Women's status and carbon dioxide emissions: A quantitative cross-national analysis", *Social Science Research*, vol. 41, pp. 965–976.

European Parliament 2012, "Gender balance among non-executive directors of companies listed on stock exchanges", (http://www.europarl.europa.eu/sides/getDoc.do?type=TA&language=EN&reference=P7-TA-2013-0488).

European Union 1997, Treaty of Amsterdam, (http://europa.eu/eu-law/decision-making/treaties/pdf/treaty_of_amsterdam/treaty_of_amsterdam_en.pdf).

Equality Act 2010, (http://www.local.gov.uk/equality-frameworks/ /journal_content/56/10180/3476575/ARTICLE).

Equality and Human Rights Commission 2015, "Protected characteristics", (http://www.equalityhumanrights.com/private-and-public-sector-guidance/guidance-all/protected-characteristics).

Ernst and Young 2016, "Talent at the table: Women in power and utilities index", (http://www.ey.com/GL/en/Industries/Power—Utilities/Women-power-and-utilities).

Gillespie, A 2006, "Facilitating and controlling civil society in international environmental law", *RECIEL*, vol. 15, no. 3, pp. 327–338.

Greed, C 1990, *Surveying sisters: Women in a traditional male profession*, Routledge, London.
Grosser, K & Moon J 2005, "The role of corporate social responsibility in gender mainstreaming", *International Feminist Journal of Politics*, vol. 7, no. 4, pp. 532–554.
Grubb, M & Yamin F 2001, "Climatic collapse at the Hague; What happened, why, and where do we go from here?", *International Affairs*, vol. 77, no. 2, pp. 261–276.
Heaton, J 2010, "Membership employment survey, Part 2: Gender and the environmental sciences", (https://www.the-ies.org/sites/default/files/reports/IES%20Employment%20Report%202010%20ii.pdf).
Heidenreich, V 2010, "Recruitment processes and effects on gender representation in the boardroom", Gender, Work and Organisation 6th Biennial International Interdisciplinary Conference, 21–23 June, Keele University, UK.
Howatt, H, Olafemi, S & Reeves, D 2004, "RTPI News: Tackling global gender inequality: Advancing the status of women planning resource", (http://www.planningresource.co.uk/article/424913/rtpi-news-tackling-global-gender-inequality-advancing-status-women).
Külcür, R 2012, "Environmental injustice? An analysis of gender in environmental non-governmental organisations (ENGOS) in the United Kingdom and Turkey", PhD thesis School of Health Sciences and Social Care, Brunel University, (http://bura.brunel.ac.uk/handle/2438/7680).
Lyness, KS & Thompson, DE 1997, "Above the glass ceiling? A comparison of a matched sample of female and male executives", *Journal of Applied Psychology*, vol. 82, no. 3, pp. 359–375.
Magnusdottir, G & Kronsell, A 2015, "The (in)visibility of gender in Scandinavian climate policy-making", *International Feminist Journal of Politics*, vol. 17, no. 2, pp. 308–326.
McRobbie, A 2009, *The aftermath of feminism: Gender, culture and social change*, Sage, London.
Mental Health Foundation, "2016 Statistics", (http://www.mentalhealth.org.uk/help-information/mental-health-statistics/men-women/).
Merchant, C 1996, "Earthcare", *Women and the Environment*, Routledge, London.
Moore, G 1988, "Women in elite positions: Insiders or outsiders?", *Sociological Forum*, vol. 3, no. 4, pp. 566–585.
Office for National Statistics 2013, "Excess winter mortality in England and Wales", London.
Perfect, D 2012, "Gender pay gaps, Equality and Human Rights Commission", (https://www.equalityhumanrights.com/sites/default/files/bp_6_final.pdf).
Poverty Site, The 2016, United Kingdom (http://www.poverty.org.uk/80/index.shtml?2).
Rastetter, D & Cornils, D 2012, "Networking: Aufstiegsförderliche Strategien für Frauen in Führungspositionen", *Gruppendynamik und Organisationsberatung*, vol. 43, pp. 43–60.
Reeves, D 2000, "Mainstreaming equality to achieve socially sustainable development: An examination of the gender sensitivity of strategic plans in the UK with implications for practice and theory", Glasgow, UK.
Reeves, D 2002, "Mainstreaming gender equality", *Town Planning Review*, vol. 73, no. 2, pp. 197–214.
Reeves, D 2003, "Gender equality and plan making: The gender mainstreaming tool kit", (http://www.rtpi.org.uk/media/6338/GenderEquality-PlanMaking.pdf).
Röhr, U, Spitzner, M, Stiefel, E & Winterfeld, U 2008, "Gender justice as the basis for sustainable climate policies: A feminist background paper", German NGO Forum on Environment and Development, Bonn.

RTPI 2003, "Gender equality and plan making: The gender mainstreaming toolkit", (http://www.gstcouncil.org/images/library/45_Gender-Equality-and-Plan-Making-Mainstreaming-Toolkit.pdf).

Sainsbury, D & Bergqvist C 2009, "The promise and pitfalls of gender mainstreaming", *International Feminist Journal of Politics*, vol. 11, no. 2, pp. 216–234.

UNFCCC 2012, "Promoting gender balance and improving the participation of women in UNFCCC negotiations and in the representation of parties in bodies established pursuant to the Convention or the Kyoto Protocol", (https://unfccc.int/files/bodies/election_and_membership/application/pdf/cop18_gender_balance.pdf).

UNFCCC 2015, "Report on gender composition", Conference of the Parties Twenty-first Session Paris, (http://unfccc.int/resource/docs/2015/cop21/eng/06.pdf).

United Nations 1992, "Sustainable development knowledge platform agenda 21", (https://sustainabledevelopment.un.org/outcomedocuments/agenda21).

Vázquez, JJ 2011, "Attitudes toward non-governmental organisations in Central America", *Nonprofit and Voluntary Sector Quarterly*, vol. 40, no. 1, pp. 166–184.

Verbakel, E & de Graaf P M 2008, "Resources of the partner: Support or restriction in the occupational career? Developments in the Netherlands between 1940 and 2003", *European Sociological Review*, vol. 24, pp. 81–95.

Verloo, M 1999, "On the conceptual and theoretical roots of gender mainstreaming", ESRC Seminar Series, *The interface between public policy and gender equality: Mainstreaming gender in public policy-making: Theoretical and conceptual issues*, pp. 1–10, Sheffield Hallam University, 12 March.

Walby, S 2005, "Introduction: Comparative gender mainstreaming in a global era", *International Feminist Journal of Politics*, vol. 7, no. 4, pp. 453–470.

WEDO 2010, "Open Letter to UN Secretary General Ban Ki Moon", (http://www.wedo.org/wp-content/uploads/Open-Letter-to-UN-SG5.pdf).

WISE 2015, "Women in science, technology, engineering and mathematics: The talent pipeline from classroom to boardroom", Bradford, UK, (https://www.wisecampaign.org.uk/uploads/wise/files/WISE_UK_Statistics_2014.pdf).

PART II

Challenges for paid and unpaid work

4

WOMEN AND LOW ENERGY CONSTRUCTION IN EUROPE

A new opportunity?

Linda Clarke, Colin Gleeson and Christine Wall

Introduction: Meeting energy targets

The European Union (EU) 20/20/20 targets to reduce energy use, increase renewable energy, and reduce carbon dioxide (CO_2) emissions by 20 per cent by 2020 particularly impact on the construction sector, given that it is responsible for 40 per cent of EU CO_2 end-use emissions. From the United Nations to European Directives and national building regulations, it is recognized that low energy buildings will have a significant impact on a low carbon future. Buildings have also been identified by the Intergovernmental Panel on Climate Change (IPCC) as the most economic sector for CO_2 savings, leading to a carbon reduction roadmap to low carbon construction and retrofit based on an assessment of labour requirements for new standards of construction and on meeting EU energy directives, including the 2010 Energy Performance of Buildings Directive (EPBD), the 2009 Renewable Energy Sources (RES) Directive and the 2012 Energy Efficiency Directive (EED 2012) (EC 2011a). This roadmap is intended to result in "nearly-zero energy buildings", new and retrofitted, achieved through energy efficient envelopes and on-site renewables supported by new qualifications, quality assurance schemes and "Green Deals". To facilitate this, the EU has supported an audit of labour availability through the Build-up Skills initiative with National Roadmaps, each of which should integrate into the broader EU employment strategy. This latter includes raising "employment rates substantially, particularly for women and young and older workers", focusing on "a more skilled workforce, capable of contributing and adjusting to technological change with new patterns of work organization" and implementing "lifelong-learning strategies and competence development" (EC 2010, 2).

The difficulties in meeting such targets for construction in Europe, an industry employing 14.5 million workers and contributing to 10.4 per cent of GDP, are more and more evident (EuroAce 2013). The imposition of stringent control measures through building regulations should mean that new buildings achieve higher energy efficiency than previously but, when actual energy demand is measured through air permeability or coheating tests of the envelope or through thermal imaging, a difference is revealed between what was modelled and anticipated at the design stage and what occurs in practice (BSI 2012; Gleeson 2016; Johnston, Wingfield & Miles-Shenton 2010). This performance "gap" affects all buildings, which means that its reduction is critical to meeting the UK *Climate Change Act* of 2008 and European Union low carbon targets of 80 per cent reduction by 2050 (EC 2011a). Social relations of production and the organization of the building process are central to understanding this difference between the energy loss envisaged and actual building performance. Different occupations are so contractually divided from each other that there are real problems with respect to bridging interfaces between them and to achieving the integrated team-working required to achieve air tightness. This implies changes not only in how labour is organized and employed but also in the quality of the labour involved, including the qualifications of building workers and the vocational education and training (VET) system in place. The knowledge, skill and competences required for low energy construction (LEC) include knowledge of building physics, mathematics, engineering or material behaviour, as well as more abstract competences such as reading off drawings, setting out, bridging interfaces and constructing to high precision. Given the changes required in the labour process and in VET systems and given EU policy to comply with a more inclusive employment strategy if energy targets are to be met and the performance gap to be overcome, perhaps now is really the time for the construction industry to become less exclusive. This chapter focuses particularly on the UK and argues that the social obstacles to fulfilling these targets are much the same as those confronting the greater participation and integration of women in the construction industry.

The need for a new strategy

The *Energy Efficiency Plan 2011* (EC 2011b) specifically addresses the need for qualified workers, the lack of appropriate training for architects, engineers, auditors, craftsmen, technicians and installers, notably for those involved in refurbishment, and the requirements for new skills and environment-conscious VET in construction and for adapting curricula to reflect new qualification needs in order to transition to energy-efficient technologies. The overview report of the *Build Up Skills* national reports, which are focussed on these social obstacles and especially on upskilling the existing workforce through continuing training, notes weaknesses in national VET systems and a "shortage of cross-trade knowledge and skills including insufficient coordination between occupations and their 'borderline' skills and unsatisfactory interdisciplinary training opportunities within upper secondary and continuing education and training systems" (EC 2014, 65). LEC is qualitatively different from the traditional construction process in requiring a singular approach to the building envelope by all site occupations, going beyond their immediate scope of responsibilities and calling for an understanding of the building fabric as a unified system with the emphasis on insulation continuity, treatment of thermal bridges and targeted air tightness. This is apparent in the German *Build Up Skills* report which locates the main problem in reducing emissions in: "interfaces between trades and lack of any understanding for a house/building as one integrated system" (Build Up Skills 2012, 7).

The suggestion is that requirements can only be met if obstacles are overcome that lie a) in the VET system (achieving broad and comprehensive know-how) and b) in the building production process (bridging trade interfaces). Yet studies across Europe indicate a lack of thermal and energy literacy and a growing need for transversal abilities within those areas critical to achieving energy efficiency (Zero Carbon Hub 2014). The imperatives for low energy construction to meet EU 20/20/20 targets introduce new VET requirements, including: greater educational input to achieve thermal literacy for all workers concerned; broader qualification profiles to overcome interfaces between the activities of different occupations, which are precisely those areas where the main heat losses occur; and, given the complex work processes involved, integrated team working and communication, learning from feedback and achieving greater cross-occupational understanding of what low energy construction entails. Sealing the building envelope, particularly through insulation, is critical to achieving energy efficiency in buildings, just as is overcoming those occupational interfaces that contribute to energy loss, for instance between the bricklayer and groundworker or concretor. Such occupational interfaces are not just socially but contractually divided, as bricklaying and groundwork come under separate subcontracts. This suggests a major transformation of the current fragmented labour process, characterized by extensive subcontracting, the use of agency labour and the self-employed, widespread non-formal on-the-job learning, high labour mobility, exclusively white male social networks and a sharp separation between operatives and professionals. Not only is it important for each occupation to know what the other is doing, but the integrated teamwork needed to prevent energy loss implies less subcontracting and more direct employment if the different occupations are to work more closely together. Surely the need to transform the construction VET system and production process opens up the possibility to resolve one of the most intransigent problems of the sector across the EU: its white male character, which has changed little over the past thirty years?

Women in construction

In Britain, as in most countries in Europe, although women represent about 11 per cent of the workforce for the construction sector for the manual trades the figure is less than 1 per cent (Smith Institute 2014). Yet the consistently higher numbers of women undertaking construction-related education and training in universities and colleges in Britain and in other European countries than are found in construction employment indicate that many women do want to work in the industry but fail to obtain entry (CITB 2015). In 2014, for instance, 71 per cent of male graduates with engineering and technology degrees entered industry employment, whereas only 56 per cent of women in the same cohort did so (Engineering UK 2014).

Current studies on climate change and occupational and skill requirements in the construction and energy sectors make little reference to gender. The International Labour Office (ILO) in 2011, for instance, pointed to the gender imbalance in the renewable energy sector, citing the case of Germany where female participation is 24 per cent, much lower than the percentage for the whole economy of 45 per cent and equal to the percentage of women in the energy and water supply sector as a whole. The conclusion is drawn that:

> This reflects a pattern that women are typically underrepresented in the energy sector (including in renewable energy) because many of the jobs are in engineering and durable goods manufacturing. (ILO 2011, 28–29)

When it comes to the construction sector, however, the situation is very much worse, with a very low participation of women in construction, architectural and engineering activities. Only in administrative, technical and clerical functions are women represented in any significant number, whilst their employment participation in some building professions, such as architecture, is rather higher than others, such as electrical and civil engineering.

The UK compares particularly unfavourably with other European countries in terms of women's participation in STEMM (science, technology, engineering, mathematics and medical) professions. Engineering UK (2011) reported that, while a few East European countries have around 20 per cent women in STEMM, the West European percentage is lower (17 per cent) and the UK (9 per cent) is close to the bottom of the league table. In terms of membership of professional institutions in the UK, in 2014 women comprised 16 per cent of the Royal Institute of British Architects' (RIBA) members, 15 per cent of the Royal Institute of Chartered Surveyors' (RICS) members, 8.6 per cent of the Institute of Civil Engineers' members, and 3 per cent of the Chartered Institute of Building (CIOB) members (TCI 2012). This low membership is mirrored in terms of employment at professional level, as revealed in the 2011 Census for construction-related profession managers and professionals in the UK. Here women represent only 10 per cent of managers and professionals, with the lowest category being electric engineers where they represent only 2.7 per cent and mechanical engineers (4.7 per cent), followed by design and development engineers (6.7 per cent), construction project managers and related professionals (6.9 per cent), and civil engineers (7.2 per cent). Significantly, however, in terms of the occupations important for LEC, the proportion of women is much higher in the technical occupations, at 24 per cent overall, especially quality assurance technicians (38.5 per cent) and architectural and town planning technicians (28.8 per cent), whilst only 8.9 per cent of engineering technicians are female (Clarke et al. 2015).

The Census confirms the considerable divide between operative and professional levels, with women's presence at operative level below 3 per cent, compared with 10 per cent for the professions – still way below average for the workforce as a whole (51 per cent). Comparing the different operative occupations, higher proportions of women are evident in painting and decorating (4.8 per cent), and electrical trades (4.3 per cent), whilst for bricklayers and masons (1.5 per cent), carpenters and joiners (1.4 per cent), and plasterers (1.4 per cent) the proportions are very much lower, though not nearly as low as for pipefitters (1 per cent). Relatively high proportions of women are found at supervisory level, especially skilled metal, electrical and electronic trades supervisors (11 per cent), many of whom may not be in the construction sector.

Though a number of initiatives have been taken to improve the participation of women, these have been largely confined to individual European member states and usually to particular occupations and appear to have had little impact. It is over ten years since a survey of European construction social partners (trade unions and employers) was conducted to investigate the presence of women in skilled trades and the policies, collective agreements and practices that play a role in women's integration (Clarke et al. 2005). The most notable successes reported were by the painters' unions in Finland and in Denmark, where painters long ago reached critical mass (Pedersen 2004). Overall, the survey showed that, though the social partners pandered to a discourse of gender equality, this was not a priority issue and did not lead to equal opportunity policies or programmes. The conclusion drawn was that "the social partners have the platform to start to make inroads and to change the industry from within, but still need to be encouraged to put women in construction on their agenda" (Clarke et al. 2005, 174). This remains the situation today, as revealed in two European Federation of Building and Woodworkers (EFBWW) women's network questionnaires, on working conditions and trade union participation, with women retaining a significant presence in the construction sector only in East European countries (CLR News 2015).

Historical precedents: Women in construction

Despite this gloomy contemporary picture, women have always maintained a presence in the building trades, particularly during the two World Wars, when they were encouraged to take up jobs opened up by male conscription, resulting in over 25,000 working in the construction industry with a participation rate, in 1943, of 3.8 per cent (Clarke & Wall 2011). However, it was not until the late twentieth century in Britain, in the 1970s and 1980s, that a combination of political change and grassroots campaigning created a set of circumstances that supported considerable numbers of women to train and work in construction (Wall 2004). After the *Sex Discrimination Act* (1975) became law, making it illegal to discriminate on the grounds of sex in employment or education, small numbers of women accessed training in government training centres (Payne 1991). Inner London local authorities recruited women into their building departments, known as Direct Labour

Organizations (DLOs), and, in the early 1980s, women-only training workshops were also set up, funded by London local authorities and providing introductory courses in the trades before women entered the male-dominated construction environment; many trainees then consolidated their training by joining DLOs as adult trainees (Michielsens, Wall & Clarke 1997; Wall & Clarke 1996, WAMT 2001). For example, in the mid-1980s, Hackney DLO was running one of the largest training schemes for building workers in Britain, backed by the construction union UCATT (Union of Construction Allied Trades and Technicians). Over 50 per cent of the adult trainees were women, many going on to permanent jobs in construction. Local authorities committed to changing their male dominated construction workforce created a framework of support for women in the trades through the provision of designated women's officers, regular meetings, the placing of more than one woman on any site, flexible hours of work, and a clear and transparent set of equal opportunity guidelines backed up by internal procedures to address grievances. The success of these measures is evident from the presence of 266 women in construction manual occupations in seven Inner London DLOs alone in 1989.

The legacy of the systems set up in the 1980s in Britain survived, and DLOs continued to address the very low numbers of young women applying for apprenticeships (Fuller, Beck & Unwin 2005). Recruitment prioritized the importance of positive female images and a diverse range of advertising outlets, as well as having women form part of the selection team. For example, of the 283 apprentices appointed by Leicester DLO between 1985 and 2002, 84 (30 per cent) were women. Leicester continues this legacy, employing 123 women in 2012 as part of its 431 strong workforce, with 18 out of 75 craft apprenticeships held by women (Clarke, Michielsens & Wall 2006; Craig & Oates 2014).

Another place where women are to be found is on mega projects, which open up the possibility of developing initiatives to tackle the gender balance of the workforce. Such projects have the advantage of highlighting practical as well as policy steps that can be taken to increase gender inclusivity. Some examples of good practice include: the Vancouver Island Highway Project in Canada; US projects such as the Century Freeway Project Los Angeles, Portland Main Bridge Project and the New York Times Building; and UK projects such as Terminal 5, the Olympic Park and Crossrail (Clarke & Gribling 2008; Clarke et al. 2015). In most of these, the emphasis has been on the operative rather than the professional workforce, with the exception of the £14.8bn, 26 mile, Crossrail scheme, which has largely focussed on employing woman engineers, driven in part by engineering skill shortages and by the so-called "business case" that the more gender balanced a team, the better it performs. Crossrail's initiative responds to an Equality Impact Assessment, which required the development of a procurement policy to encourage local sourcing of goods services and labour (Crossrail 2006). The steps taken included training those involved in recruitment and promotion not to have any "unconscious bias"; encouraging contractors to recruit a more diverse workforce; carrying out "blind" recruitment by removing names and gender from applications; setting up a women's forum, diversity working group and mentoring programme, and creating opportunities for senior women to act as role models; working with one hundred schools to encourage more women into engineering; establishing more inclusive maternity leave and flexible working policies; assisting in organizing pre-employment training; and organizing competitions and meetings to promote and support women engineers. As a result of these various measures, Crossrail's statistics for the percentage of women in the organization are 29 per cent of project managers, 12 per cent of apprentices and 19 per cent of graduates (Kitching 2014).

Taken together, these various projects all set out to use their size, capacity and profile to make a difference and set ambitious targets and aspirations as a prerequisite for action. They indicate particular factors critical to the achievement of securing greater inclusivity, and hence a higher proportion of women, including: the roles of public procurement (Wright 2013); the significance of Major Project Agreements (or Project Labour or Community Workforce Agreements) (Moir 2014), collective agreements (Calvert & Redlin 2003) and local labour agreements (GLA 2007); involving the workforce; controlled recruitment (National Women's Law Centre 2014); childcare provision; systematic monitoring (Moir, Thomson & Kelleher 2011); and training, including as a means of siphoning recruitment (Cohen & Braid 2000a, 2000b). The projects thus provide important lessons concerning measures that can be effective in including women. First and foremost is to secure an overriding agreement, particularly with a guarantee of direct employment, with all stakeholders – including contractors, subcontractors, trade unions, clients and local authorities – on working conditions, direct employment and a preference for local residents. Coupled with this is contract compliance, including an insistence on equality measures and employment goals in all tender documents, carried through by pre-job compliance meetings and continued monitoring. Once these are secured, clear targets need to be set for the recruitment process, to ensure that the workforce employed is reflective of the local population, and equal opportunity policies have to be proactively applied by all contractors and subcontractors. Training needs to be put in place, including pre-employment training and special training to facilitate broader occupational profiles, with formal links established with colleges and universities in the vicinity, and work placements and work experience provided, including specific aspects of the project set aside for training purposes. Perhaps most important of all, working conditions need to be carefully regulated, including structured working hours, childcare provision, inclusive maternity leave, flexible working arrangements and mentoring. And, finally, securing support at senior level and from women's groups and trade unions, through organizing meetings, conferences and so on, as well as careful monitoring, are essential.

VET transformation

The possibility of a more energy efficient and at the same time inclusive construction sector is that much greater given changes occurring in the industry. Entry is changing, so that in some countries, including the Netherlands, recruitment is made directly from vocational colleges, where generally a higher proportion of women are to be found than in the labour market. With the decline in apprenticeships in many countries and increasing reliance on placements and internships to obtain work experience, this form of recruitment is likely to increase. So too are occupational labour markets, whereby workers' status depends as much on their qualifications as on experience (Marsden 2007). This is conducive to greater female participation, as women are reliant on formal qualifications for entry into construction. In addition, the employment relation is undergoing transformation, including through the use of agencies, so that the "old boys' network" on which much recruitment has up to now depended is weakening and the use of formal recruitment practices, which are more favourable to women, is increasing. Policy at European level pursued by the EU and the social partners, including the gender dimension of the 2020 Strategy and the European Trade Union Congress (ETUC) gender equality policy, gives an added impetus to widening participation. And, finally, in Britain as in other countries, the acute need for new affordable energy-efficient social housing and for retrofitting existing properties opens up opportunities for women to train and enter the industry.

Detrimental to low energy performance, especially in a country such as Britain, are the often low level of skills and qualifications of many employed in the industry and the lack of initial and further training. The decline in VET has been steady since the 1980s and has gone together with a decline in direct employment. At the same time as the numbers of those self-employed and employed by agencies has increased, the already low number of first year trainees and apprentices has plummeted since 2007, with those in the wood trades halving from 13,743 to 4,536, bricklaying trainees falling from about 9,000 to 2,364 and plant operatives from 4,747 to 485 (CITB 2015). In 2015, 11,586 construction trainees were recorded in Britain, a historical low, only 35 per cent of whom were undertaking some kind of work-based training, and only about 3,000 were following an apprenticeship programme. Most VET provision continues to be concentrated in the traditional trade areas and in the north of the country, with the majority of apprenticeships in the four main building trades – wood, bricklaying, painting and decorating, and plastering and dry lining – though these constituted only about half of the forecast requirement for skilled manual trades in 2015. In addition the vast majority of construction trainees only achieve a National Vocational Qualification (NVQ) Level 2, a qualification far lower than that in other leading European countries and too low to then progress to supervisory or managerial levels.

In this situation, it is not surprising that few builders take responsibility for training; 73 per cent of construction companies have no training plan, 81 per cent have no training budget and only 19 per cent invest in training (BIS 2013). Declining employer responsibility for training is understandable in the light of the high levels of self-employment and the fragmentation of firms and degree of subcontracting in the industry. It is, however, that much more serious given the employer-led nature of the VET system, where trainees depend on employer goodwill to acquire work experience, qualifications and VET (including for green construction), where lobbying by employer trade associations is critical to new qualifications being developed, and where government policy is focussed on work-based apprenticeship.

Despite the fact that the building envelope is so important to low energy performance, with labour requirements relating to solid ground floor slabs, cavity brick walls and timber roofs, there are significant envelope production roles in Britain with no formal VET. For those areas with formal training, whilst insulation and energy efficiency is included in bricklaying, site carpentry and plastering qualification profiles, there is no knowledge requirement to understand the envelope as a single system, no reference to air barriers or thermal bridging and no requirement to understand the interplay between the separate envelope occupations or the final energy performance needed. And what is entirely missing is any celebration of the "thermal literacy" of the construction worker, central to achieving a low carbon future. Low and zero carbon technologies – such as heat pumps, solar thermal and photovoltaics – have proven to be especially sensitive to poor design, installation, commissioning and operation and, along with envelope construction, require enhanced technical knowledge and soft skills associated with communication, team working and self-management, as well as thermal literacy. Whilst the lack of thermal literacy has been recognized by researchers and industry, "specialist" low energy or "sustainable construction" training schemes developed to cover air tightness, insulation continuity and so on are "add-on" courses and not an integral part of initial VET schemes for construction (West 2010). The same issues apply in higher education, where some university programmes are described as "specialising" in low energy design and construction.

Labour process transformation

The debate concerning the possibilities for a low energy and inclusive construction process, in particular forming the integrated teams of mixed occupations needed in order to bridge the sharp interfaces that give rise to heat loss, becomes yet more complex when employment conditions in the British construction sector are considered. Not only is much of the workforce self-employed and/or employed through agencies, but the construction process is driven by versions of "compulsory competitive tendering" with sub-contracting and sub-sub-contracting dominant, producing up to ten tiers in the supply chain. Labour-only subcontracting, self-employment and piecework are especially evident in a number of occupations, from bricklaying through to plumbing, all heavily male-dominated occupations. These employment conditions impose additional constraints on achieving energy efficient performance and on increasing female participation, for instance when the workforce knows that the more work produced, the more money will be paid so that reasonable working time and quality control come second to achieving completion targets.

What is apparent from UK statistics on the number and size of private contractor companies is the large number of micro-firms, with 39 per cent of the total 194,000 firms being sole operators and 48 per cent employing between two and seven people. Altogether 94 per cent of companies employ fewer than 14 people and only 0.06 per cent over 600 (ONS 2013). A large proportion of sole proprietors are self-employed, many small firms are simply acting as subcontractors to the larger firms, and many self-employed are with labour-only subcontractors, usually divided on a trade basis, whether for groundwork, bricklaying, carpentry, painting, plumbing or electrical work. This has dramatic repercussions for LEC, giving rise to sharp contractual divisions between the different areas of work, whether between the bricklayer erecting the walls, the carpenter the doors, the groundworker the foundations, or the roofer the roof structure; these are the very interfaces through which the greatest heat losses occur, as revealed in the various measures of energy performance. Alongside these obstacles associated with a fragmented construction labour process go others, including difficulties to transit from college to work, employer disengagement with training and the decline in apprenticeships. These difficulties imply the need for a more open, inclusive and permeable labour market and construction process, facilitating the transfer and development of skills and knowledge and the deployment of transversal abilities.

In this respect it is revealing that an exemplary social housing scheme in the North of England, composed of 91 units and achieving energy efficiency above Passivhaus standards, was carried out by a direct labour force, with all trades working together as a team and organized into trade unions. Green technologies and traditional building techniques were combined and part of the scheme was set aside for apprentice training, in collaboration with a local FE College, with subsequent maintenance the responsibility of the housing association's own (90-strong) repair and maintenance team. The good employment, working and training conditions evident were ones fitted to the fulfilment of low energy targets as well as to the employment of large numbers of women. In this case the construction process is potentially more inclusive given that the structural and cultural obstacles to greater female participation were either weak or did not exist. These obstacles include: inappropriate selection criteria; male dominated training courses; lack of formalized recruitment practices and procedures; fragmentation of employment; traditional stereotypes and sexist attitudes; a male dominated culture, networking and environment; and lack of work-life balance possibilities (Fielden et al. 2000; Sang & Powell 2012).

There is extensive literature outlining the problems of self-employment in the construction sector (e.g. Harvey & Behling 2008; House of Commons 2016) and it is apparent that much of the success in improving diversity on projects, such as the Olympics and T5, was attributable to the insistence on direct employment. As emphasized by the Greater London Authority (2007):

> The prevalence of self-employment and temporary agency working (in particular of migrant workers) on short projects on sites, often under different

> terms and conditions even on the same site and in the same trade, hampers the development of a stable workforce with clear paths of recruitment, retention and progression that a wider, more diverse, pool of workers can enter. (93–94)

One of the key problems identified with working practices in construction is also the long-hours culture and expectation of total availability and "presenteeism". This is one of the main issues concerning gender equality because, as Ness (2012, 668) argues: "the exclusion of women both enables and condemns men to work long hours". Long hours working is therefore not only an obstacle to women's participation, but the product of their historical exclusion from the world of work. Research on major projects (including Wembley and T5) in London confirms the long hours of work. As stressed by the Greater London Authority (2007):

> The long, irregular working hours and travel times often required in construction act to exclude many people from working in the industry due to the difficulty of combining work with domestic and other responsibilities. These work patterns underpin the preference for engaging mobile workers . . . and hampers the development of a sustainable London labour market. (94)

The GLA recommended that regional public authorities, industry lead bodies and unions should discuss:

> How it may be appropriate on major projects to promote stable working hours and shorter travel times in conformity with Working Time Directive requirements, clean environment and transport policies, effective health and safety procedures and measures, and minimising disturbance to the general public. (94)

People in the industry work many hours more than their contracts state and "face time" is still a core part of the working culture, with long hours seen as an indicator of commitment and therefore used as a pre-requisite for promotion (Wright et al. 2014). Experienced women often leave the industry in order to "escape" the long working hours, especially after having children, and the UK science and engineering sectors "lose" (i.e. leavers from the workforce) their female workforce at a much higher rate than other sectors (Hart & Roberts 2011). The lack of flexible working, including part-time work, as an opportunity to balance work and caring roles is a key element in explaining female retention problems.

The "masculine" nature of the work environment, especially on-site work, characterized by the dominant "'football and families' culture", with humour and sexualized banter in the workplace common, can also prevent female inclusion (Faulkner 2009). In some cases successful integration is perceived as dependent on assuming "male" behavioural norms and intensified work patterns because "'belonging' in construction workplace cultures is highly gendered" (Watts 2012, 1). Language used in everyday talk and in reports etc. can also reinforce stereotypes around gender and engineering (Faulkner 2006).

With respect to the reliance on established informal networks and the lack of transparency and accountability in relation to diversity, subcontracting and retention processes, the Greater London Authority (GLA 2007, 91) made the interesting recommendation that regional public authorities work with the CITB, employers and unions to promote the employment of dedicated managers/coordinators to work on large sites with a remit for promoting different methods of sourcing applicants and ensuring equal opportunities in recruitment and subcontracting. They should also consider ways of incorporating suppliers' track records in equal opportunities and diversity performance as part of tendering processes for contracts, as well as promoting the appointment of equality representatives to be kept informed of recruitment and retention processes and to liaise with workers from target groups on issues of concern such as discrimination.

In terms of policy, the UK Workplace Employment Relations Survey (WERS) provides evidence that fewer STEMM workplaces have formal diversity policies than in the some other parts of the economy, although it is important to acknowledge that policy statements do not necessarily reflect practice (they could be "empty shells", as shown by Hoque & Noon 2004). Ensuring the enforcement of policies is critical and to achieve this consideration should be given to setting robust equality targets against which progress can be monitored, as well as increased monitoring of information on employment practices and career progression such as the use of fixed-term contracts, flexible working arrangements and requests, and the progression of different groups through the organization. The gender balance of teams should be routinely evaluated, a "gender audit template" introduced, and equity champions nominated. Specific equity requirements should be included in contracts and an evaluation of suppliers' diversity track records incorporated as part of the tender evaluation process. Finally, where appropriate, wider institutional pressures on the equality agenda should be mobilized, particularly in relation to discriminatory behaviour or breaches in codes of professional conduct for which externally enforced sanctions already exist.

Conclusions

In summary, the barriers to gender diversity in construction are linked to the training and education context; employment policies and practices, including the fragmented nature of sector and procurement, long working hours and the lack of flexible working; weak human resource policies and practices, with respect to recruitment and selection, retention, networking, mentoring and role models; and the "macho" environment. Yet many of these are also barriers to achieving effective low energy construction, including the need for a more holistic and high standard VET system, less fragmented employment and occupational structure, stable and direct employment to allow for integrated team working, and good working conditions. The suggestion is that meeting the challenge of low carbon construction also opens up the possibility to include more women, especially considering their generally higher educational achievements, their greater presence in environmentally-oriented subject courses and the persistent reports of skill shortages in construction in many European countries.

References

BSI 2012, PAS 2030, "Improving the energy efficiency of existing buildings: Specification for installation process, process management and service provision, Edition 2", British Standards Institute (BSI), London.

BSI 2013, "UK construction, an economic analysis for the sector", Department for Business and Skills, July, (https://www.gov.uk/government/uploads/system/uploads/attachment_data/file/210060/bis-13-958-uk-construction-an-economic-analysis-of-sector.pdf).

Build Up Skills 2012, "Build Up Skills Germany: Analysis of the national status quo", Intelligent Energy Europe, European Commission, (https://ec.europa.eu/energy/intelligent/projects/en/projects/build-skills-de).

Calvert, J & Redlin, B 2003, "Achieving public policy objectives through collective agreements: The project agreement model for public construction in British Columbia's transportation sector", *Just Labour*, vol. 2, spring, pp. 1–13.

CITB 2015, "Training and the built environment", Construction Industry Training Board (CITB), Norfolk.

Clarke, L, Pederson, EF, Michielsens, E, Susman, B & Wall, C 2005, "The European social partners for construction: Force for exclusion or inclusion?" *European Journal of Industrial Relations*, vol. 11, no. 2, July, pp. 151–178.

Clarke, L & Gribling, M 2008, "Obstacles to diversity in construction: The example of Heathrow Terminal 5", *Construction Management and Economics*, 26 October, pp. 1055–1065.

Clarke, L, Michielsens, E, Snijders, S & Wall, C 2015, "No more softly, softly: Review of women in the construction workforce", Probe Publications, University of Westminster, UK.

Clarke, L, Michielsens, E & Wall, C 2006, "Women in manual trades", in A Gale & M Davidson (eds), *Managing diversity and equality in construction: Initiatives and practice*, pp. 151–168, Routledge, London.

Clarke, L & Wall, C 2011, "Skilled versus qualified labour: The exclusion of women from the construction industry", in M Davis (ed.), *Class and gender in British labour history, Renewing the debate (Or starting it?)*, pp. 96–116, Merlin Press, Pontypool.

CLR News 2015, "Women in construction", no. 3, European Institute for Construction Labour Research, (http://www.clr-news.org/).

Cohen, MG & Braid, K 2000a, "The road to equity: Training women and First Nations on the Vancouver Island Highway – A model for large-scale construction projects", in MG Cohen (ed.), *Training the excluded for work: Access and equity for women, immigrants, First Nations, youth and people with low income*, pp. 53–74, UBC Press, Vancouver.

Cohen, MG & Braid, K 2000b, "Training and equity initiatives on the British Columbia Vancouver Island Highway Project: A model for large-scale construction projects", *Labour Studies*, vol. 25, no. 3, fall, pp. 70–103.

Craig, S & Oates, A 2014, "Empowering women in construction", in M Munn (ed.), *Building the future: Women in construction*, pp. 77–86, The Smith Institute, London.

Crossrail 2006, "Equality impact assessment, project and policy assessment report", January, Department for Transport, UK.

Engineering UK 2011, "An investigation into why the UK has the lowest proportion of female engineers in the EU", London.
Engineering UK 2014, "Engineering UK 2014: Synopsis, recommendations and calls for collaborative action", London.
EuroAce 2013, "Implementing the cost-optimal methodology in EU countries: Lessons learned from three case studies", The Buildings Performance Institute Europe (BPIE), Brussels.
EC 2010, "An agenda for new skills and jobs: A European contribution towards full employment", Final communication 682, European Commission (EC), Strasbourg.
EC 2011a, "A roadmap for moving to a competitive low carbon economy in 2050", Final Communication 112, European Commission (EC), Brussels.
EC 2011b, "Energy efficiency plan 2011", Final Communication 109, European Commission (EC), Brussels.
EC 2014, "Build up skills: EU overview report", Staff working document, written by N Cliquot & S Gausas, European Commission (EC), Brussels.
Faulkner, W 2006, "Genders in/of engineering: A research report", University of Edinburgh, UK.
Faulkner, W 2009, "Doing gender in engineering workplace cultures: Observations from the field", *Engineering Studies*, vol. 1, no. 1, pp. 3–18.
Fielden, SL, Davidson, MJ, Gale, AW & Davey, CL 2000, "Women in construction: The untapped resource", *Construction Management and Economics*, vol. 18, no. 1, pp. 113–121.
Fuller, A, Beck, V & Unwin, L 2005, "The gendered nature of apprenticeship", *Education and Training*, vol. 47, issue 4/5, pp. 298–311.
Gleeson, C 2016, "Residential heat pump installations: The role of vocational education and training", *Building Research and Information*, vol. 44, issue 4, pp. 394–406.
GLA 2007, "The construction industry in London and diversity performance", February, Greater London Authority (GLA), London.
Hart, R & Roberts, E 2011, "British women in Science and Engineering: The problem of employment loss rates", Working paper, Division of Economics, University of Stirling, Scotland.
Harvey, M & Behling, F 2008, *The evasion economy: False self-employment in the UK construction industry*, UCATT, London.
Hoque, K & Noon, M 2004, "Equal opportunities policy and practice in Britain: Evaluating the 'empty shell' hypothesis", *Work, Employment & Society*, vol. 18, no. 3, pp. 481–506.
House of Commons 2016, "Self-employment in the construction industry", Business and Transport Section, UK, (http://researchbriefings.files.parliament.uk/documents/SN00196/SN00196.pdf).
ILO 2011, "Skills and occupational needs in renewable energy", International Labour Office (ILO), Geneva.
Johnston, D, Wingfield, J & Miles-Shenton, D 2010, "Measuring the fabric performance of UK dwellings", Proceedings of the Association of Researchers in Construction Management (ARCOM) Twenty-Sixth Annual Conference, 6–8 September, Rose Bowl, Leeds, UK.
Kitching, R 2014, "Women in engineering: Leading the charge", *New Civil Engineer*, 1 July.
Marsden, D 2007, "Labour market segmentation in Britain: The decline of occupational labour markets and the spread of 'entry tournaments'", *Economies et Sociétés*, vol. 41, no. 6, pp. 965–998.
Michielsens, E, Wall, C & Clarke, L 1997, *A fair day's work: Women in the direct labour organisations*, London Women and the Manual Trades, London.

Moir, S 2014, "Gender segregation in the construction trades: Lessons from thirty-five years of US policy failures", International Labour Process Conference, Kings College, London.

Moir, S, Thomson, M, & Kelleher, C 2011, "Unfinished business: Building equality for women in the construction trades", Labor Resource Centre Publications, (http://scholarworks.umb.edu/cgi/viewcontent.cgi?article=1004&context=lrc_pubs).

National Women's Law Center 2014, "Women in construction: Still breaking ground", NWLC, Washington.

Ness, K 2012, "Constructing masculinity in the building trades: 'Most jobs in the construction industry can be done by women'", *Gender, Work & Organization*, vol. 19, no. 6, pp. 654–676.

ONS 2013, "Construction statistics: No. 14, 2013 Edition", Office for National Statistics (ONS), UK.

Payne, J 1991, *Women, training and the skills shortage: The case for public investment*, Policy Studies Institute, London.

Pedersen, EF 2004, "Painters in Denmark: A woman's trade?" in L Clarke, E Frydendal Pedersen, E Michielsens, B Susman & C Wall (eds), *Women in construction*, pp. 135–146, CLR Studies 2, Reed Business Information, The Hague.

Sang, K & Powell, A 2013, "Equality, diversity, inclusion and work–life balance in construction", in A Dainty & M Loosemore (eds), *Human resource management in construction projects: Critical perspectives* (2nd edn), pp. 163–196, Routledge, London.

Smith Institute 2014, "Building the future: Women in construction", London.

TCI 2012, "Women still under-represented in construction", The Construction Index (TCI), Construction Institute, UK.

Wall, C 2004, "'Any women can': 20 years of campaigning for access to training and employment in construction'", in L Clarke, E Frydendal Pedersen, E Michielsens, B Susman & C Wall (eds), *Women in construction*, pp. 158–172, CLR Studies 2, Reed Business Information, The Hague.

Wall, C & Clarke, L 1996, *Staying power: Women in direct labour building teams,* London Women and the Manual Trades, London.

WAMT 2001, "Building the future: 25 years of women and manual trades", Women and Manual Trades (WAMT) Newletter, London.

Watts, JH 2012, "Women working in construction management roles: Is it worth it?" *Global Journal of Management Science and Technology*, vol. 1, no. 3, pp. 38–44.

West, S 2010, "Airtight and super-insulation construction techniques", International Specialised Skills Institute Inc, Department of Education, Employment and Workplace Relations, Australian Government.

Wright, T 2013, "Promoting employment equality through public procurement: Report of a workshop held by the Centre for Equality and Diversity (CRED)", Queen Mary, University of London.

Wright, A, Clarke, L, Michielsens, E, Snijders, S, Urwin, P & Williamson, M, 2014, "Diversity in STEMM: Establishing a business case", Report of research by the University of Westminster for the Royal Society's Diversity Programme, June.

Zero Carbon Hub 2014, "Closing the gap between design and as-built performance", Evidence Review Report.

5

RENEWABLE INEQUITY?

Women's employment in clean energy in industrialized, emerging and developing economies

Bipasha Baruah

Introduction

Concerns about environmental sustainability and fossil fuel insecurity have convinced many countries to transition to low-carbon energy supplies derived from solar, hydro, bioenergy, wind and other renewables. Since producing and distributing renewables is more labor-intensive than producing and distributing fossil fuels, this shift is creating new employment opportunities and addressing energy poverty in remote or underserved communities.

Renewable energy (RE) employed 7.7 million people globally in 2014, an 18 per cent increase from the previous year, and employment in the sector is expected to continue growing in the future (IRENA 2015). Applying a gender lens to the enthusiasm for renewables reveals a major blind spot since women are marginalized globally in employment in RE. Worldwide, women constitute fewer than 6 per cent of technical staff and below one percent of top managers (UN Women 2012) in RE. In the absence of appropriately targeted training, education, apprenticeships, employment placement, financial tools and supportive social policies, transitioning to renewables may exacerbate existing gender inequities and hinder human development goals.

This chapter synthesizes and analyzes existing research on women's employment in RE in industrialized, emerging and developing economies. It highlights similarities and differences in occupational patterns in women's employment in renewables in different parts of the world and makes recommendations for optimizing women's participation in the sector.

Because jobs in RE tend to be dispersed across different sectors of employment (such as construction, manufacturing, installations, fuel processing, operations and maintenance), specific information related to the sector is seldom captured in national statistics. Sex-disaggregated data on employment in renewables is even harder to find. The peer-reviewed literature on this topic is currently very limited but there is a significant amount of "working knowledge" available from practitioner sources. I drew upon and amalgamated both types of knowledge in order to document the nature, magnitude, nuance and complexity of the issues involved. Because sex-disaggregated statistics on employment in renewables is so scarce, I also relied on data from the broader categories of "green jobs" and "clean jobs", as well as related science, technology, engineering and math (STEM) fields in order to write this chapter. The inclusion of the latter is justified by the fact that employment in the most well-paid sectors of RE, namely, installations, construction, engineering and architecture (Antoni, Janser & Lehmer 2015), tend to require STEM training.

A global summary of women's employment in RE

Although sex-disaggregated data on RE employment in industrialized countries is scarce, the numbers we are able to put together points to the severe underrepresentation of women. In OECD countries, women hold a minority of jobs in the energy industry in general. The share of female employees in RE has been estimated at about 20 to 25 per cent, with most women working in administrative and public relations positions. The available data from Canada, the US, Spain, Germany and Italy indicate a general trend of women in the RE sector being employed mostly in non-technical occupations. The greatest representation of women in RE in OECD countries is in sales, followed by administrative positions and then engineers and technicians (IRENA 2013). In absolute numbers, the largest sources of RE employment for women in industrialized countries are solar photovoltaics, solar heating/cooling, wind power, biomass and biofuels (IRENA). The underrepresentation of women in RE employment in many OECD countries is part of a bigger problem of the underrepresentation of women in STEM fields. There is an obvious economic benefit for women who choose these careers. While gender-based wage inequality also exists in STEM jobs, it is smaller. Women in STEM jobs earn 33 per cent more than those in non-STEM occupations. The gender wage gap in STEM jobs is roughly 14 per cent, while the gender wage gap for non-STEM jobs is 21 per cent (TNC 2014).

RE deployment continues to grow globally as a sustainable and increasingly economically viable alternative to conventional sources of energy. There is also growing recognition of the positive social and economic impacts of RE deployment. The employment effects of RE investment, in particular, are increasingly gaining prominence in the global RE debate, but specific analytical work and empirical evidence is relatively limited. Even general employment data on RE employment is unavailable or unreliable in many settings. Sex-disaggregated data on employment in renewables is particularly spotty everywhere in the world. This makes analyzing trends and making comparisons challenging. We do have a better sense of the numbers of women who hold formal jobs or earn commissions in developing countries and emerging economies from activities such as manufacturing, constructing, operating, maintaining and selling solar lights and improved cookstoves because these initiatives tend to be driven by governments, NGOs, private sector organizations and social enterprises (see Baruah 2015a, 2015b).

Estimates on women's employment in renewables can also differ dramatically depending on whether the analysis includes or excludes large hydropower and informal employment – particularly in traditional biomass and fuel-crop production (IRENA 2013). There are no national data sets that can tell us the exact or even approximate numbers of women who are employed in informal jobs in activities such as fuelwood collection and charcoal production. Extrapolations from regional data sets suggest that the numbers are significant. For example, up to 13 million people may be employed in commercial biomass in sub-Saharan Africa (Openshaw 2010). Fuelwood and charcoal represent between 50–90 per cent of all energy needs in developing countries and 60–80 per cent of total wood consumption (ESMAP 2012). Since women are largely responsible for procuring fuelwood for household needs (Ilahi 2000) it is reasonable to expect that millions of women are engaged formally or informally in the biomass sector. More broadly speaking, it is well established that women represent 50 per cent of the agricultural workforce in sub-Saharan Africa (FAO 2011). Given the similarities in the nature of activities, it is reasonable to infer that women's participation in biomass production is also significant.

Production of fuel crops is also an important source of formal and informal RE employment in developing countries (Kammen 2011). Farming communities in the developing world have always known how to use oilseeds for lighting purposes. Jatropha, for example, is increasingly integrated into farm production systems in India, Cambodia and Mexico as a complement rather than as a substitute for food crops. A review of community biofuel projects carried out by the International Network on Gender and Sustainable Energy (ENERGIA) in 2009 discovered that village-level production of biofuels can be economically sustainable, create employment and improve energy access in underserved rural and remote communities. Such systems can reduce the drudgery of fuel collection for women and improve their participation in community affairs and decision-making. However, the ENERGIA review does warn about the danger of shifting from integrated fuel and food crop production to mono-cropping for commercial production as it may result in loss of land, income and food security.

Opportunities and constraints for women's employment in RE

A review of the available literature reveals the following as factors that may either impede or facilitate women's meaningful participation in the RE sector in industrialized, emerging and developing economies: both societal and self-misperceptions about women's technical abilities; opportunities and constraints associated with self-employment and entrepreneurship; the pros and cons presented by part-time work and arrangements like job sharing; the limitations and opportunities women face in managing work-related travel; skill shortages in the RE sector; and public sector involvement in framing policy to enable employment equity in RE. It is important to emphasize that the line between opportunity and constraint is quite fuzzy since some constraints may potentially become opportunities with appropriate policy interventions, shifts in societal attitudes, and economic and political changes within specific countries. The implications of part-time work and job sharing, which I include in the next section, do not apply solely to women in the RE sector. These are growing employment trends around the world and have deep implications for gender equity in all fields, not just in renewables. I include it for discussion in this chapter because it is important to understand women's access to employment in renewables within the context of broader global trends in employment and social policy.

Misperceptions

A combination of women's self-perception as well as societal perceptions of women's incompetence in technical occupations has been identified quite frequently in the literature as an impediment to women's optimal participation. Although there are references to issues of perception and its consequences in studies of RE projects in different contexts (see, for example, Fernández-Baldor et al. 2014; Lehr 2008) there does not appear to be any research devoted specifically towards documenting effects of such misperceptions upon women in RE. Findings drawn from similar fields such as construction and engineering occupations in various locations in the Global North and South does provide an understanding of how women are perceived, and often perceived themselves, in non-traditional male-dominated occupations (see, for example, research in India (Baruah 2010), Nigeria (Adeyemi 2006), Togo (Adubra 2005), the US (Paap 2006), the UK (Greed 2000) and Canada (Little 2005)). These studies reveal a common finding that women in these fields are deemed less competent than men even with the same or superior qualifications and work experience.

Women may also be discouraged from entering occupations in RE and more broadly in engineering and technology fields because of misperceptions of the work involved in these fields. Because the technological aspects of these occupations get so much attention, women are often led to believe they are not socially useful. The message that these occupations can improve lives is often overshadowed by the technical aspects of building things (Tyseer Aboulnasr, Dean of the Faculty of Applied Science at the University of British Columbia, quoted in Myers 2010). This may explain why women are a significant presence, and often even the majority, in medical and biological sciences as well as in certain engineering disciplines (biosystems, environmental, chemical) in which they can clearly see how their work makes a difference, but less well-represented in fields like civil, electronic and computer engineering that are perceived as more technically focused and socially isolating (ibid.). Of course, not all women aspire to be socially useful; many may just as easily be convinced by money, prestige or other factors. It is important not to ascribe essentialist feminized attributes to all women.

A 2009 Engineers Canada survey of female high school students found that many had negative perceptions of engineering and technology occupations, often seeing them as "dirty work that tradespeople would do" (Valerie Davidson, engineering professor at the University of Guelph, quoted in Myers 2010). The same survey reports that women were more likely to equate engineering and technology (but especially engineering) with construction work, outdoor work, working in a cubicle, and relating primarily to computers and machines, rather than people. The result is that women attribute lower status to engineering and technology occupations compared with, for example, health and social sciences. The professional community of engineers, particularly in OECD countries, does not appear to have done well at leveraging the message that engineering is prestigious and socially useful work. By contrast, much larger numbers of middle-class women study engineering and other technical fields in some developing countries and emerging economies, at least partially because they are perceived as well-paid high-status occupations. I discuss this in more detail later in this chapter.

Self-employment and entrepreneurship

The issues of self- and societal perceptions discussed in the previous section may be less of a constraint when women own enterprises or are otherwise self-employed. Provided that there is adequate business training, financial support and social safety nets in place, women seem to do well with self-employment. Women are establishing new RE enterprises both in industrialized countries and emerging economies. The founder of Women in Renewable Energy (WiRE), a Canadian non-profit organization dedicated to advancing the role and recognition of women in the RE sector, revealed that of its membership base of over a thousand women in the province of Ontario, at least 20 per cent were either entrepreneurs or aspiring entrepreneurs (Black 2015). She could not, however, provide additional information about whether the enterprises were in areas like communications and advertising, the actual generation of RE or the sale of RE technologies. Women are also very well represented in the manufacturing, stocking and selling of improved cookstoves in Asia and Africa (GVEP 2012). Of course, when gender inequality is viewed as a structural issue, as it is and should be, it is difficult not to be intellectually uncomfortable with the instrumental and stereotypical deployment of women in marketing and dissemination initiatives for improved cookstoves. At the same time, it is important to recognize the creation of better-paid and less menial livelihoods for poor women (Baruah 2015b).

It is important to provide the right type of support for women's entrepreneurial abilities but it is also important to be cognizant that entrepreneurship is often not a realistic livelihood strategy for some – particularly low-income – women, and even well-intentioned and progressive interventions by governments and social enterprises fail to level the playing field for them (see Baruah 2015a for details). Low-income women in both industrialized countries and emerging economies do not become entrepreneurs because the burden of entrepreneurship and the risk associated with loans is simply too high for them. Poorer women everywhere in the world tend generally to be more interested in stable wage employment rather than entrepreneurship (Baruah 2015a).

Research conducted in various locations in developing countries, emerging economies and countries that were part of the former Soviet Union, has demonstrated that microcredit is not an appropriate tool to support the development of small and medium enterprises (Bateman & Chang 2012). More specifically, research in India demonstrates that especially in the absence of subsidies, the higher cost of the technologies makes microcredit loans an inadequate tool for enterprise development in renewables (Baruah 2015a). Most poor women in developing countries and emerging economies are interested in the energy sector because of the potential for income generation but they are also extremely averse to financial risk. They are much more likely to pursue opportunities in the energy sector if they can earn incomes without becoming indebted. Acquiring new skills – such as learning to build and repair RE technologies – may be better suited for their economic realities and limitations. Organizations like the Self-Employed Women's Association (SEWA) in India and Grameen Shakti in Bangladesh are actively trying to meet these needs through various solar and biomass initiatives but the creation of permanent and stable sources of income remains a challenge. Women who have been trained to build, install and repair solar technology, for example, continue to face the challenge of finding permanent employment with their newly acquired skills as they are often only able to earn incomes on an intermittent basis through contracts and orders placed by social enterprises, non-profits and government agencies. This shortcoming is common to most livelihood initiatives in the RE sector in developing countries and emerging economies. It highlights the need for governments to provide adequate social security to protect against vagaries in the market, natural disasters, illness, maternity, old age, job losses and other risks to people's well-being. Providing social protection within a human rights framework and delinking social security from employment status is a strategy worth pursuing worldwide. Women can gain optimal traction from RE initiatives only if there are wider socially progressive policies in place. Since women's ability to take advantage of new energy-related employment options is, to begin with, often constrained by legal or social barriers that limit their education, property rights, land tenure and access to credit, it is crucial that government policies go beyond energy sector planning to optimize economic opportunities for women.

Part-time work

A study conducted by a labor union organization in Spain revealed that in 2008, women accounted for just over 26 per cent of the country's RE workforce (Arregui et al. 2010). This is slightly higher than the 24 per cent average for Spain's broader industrial sector. Only 2 per cent of RE jobs in Spain are part-time but women hold 67 per cent of them (Arregui et al.). It is important to remember, though, that part-time jobs are not necessarily always "bad" jobs. Some recent research suggests that creating more part-time jobs and arrangements like work-sharing, provided they have job security as well as health and pension benefits, may be a feasible way to restructure work in the future while creating both economic security and ecological sustainability in all sectors of the economy (Malleson 2015). Since overproduction and overconsumption, particularly by the wealthy in all global settings, remains the biggest impediment to environmental sustainability, transitioning to clean energy sources, or to a green economy more broadly, is not going to be enough in and of itself to prevent climate change and address other environmental problems. However, some recent research suggests that restructuring work in innovative ways while expanding social security nets may present some solutions for balancing economic needs and environmental concerns (see, for example, Malleson 2014; Nedelsky 2014).

Work-time reduction has not been very prevalent in North America, but it has been pursued with limited success in some European countries. Unions in France fought for and successfully won a 35-hour working week. Unions in the Netherlands have played a pivotal role in creating quality part-time jobs. And unlike in North America, those part-time jobs are actually good jobs. They have roughly the same hourly pay as full-time work and similar benefits and security. The average American works about 1,900 hours per year, while the average Dutch person works about 1,350 hours per year – about 30 per cent less (Malleson 2014).

While I agree with the potential of part-time work and job-sharing to promote economic security and environmental sustainability, I am ambivalent about the assertion (see Nedelsky 2014) that creating more part-time jobs for all will lead to a more equitable division of household labor between women and men. This assumption is based on the fact that women do a disproportionate amount of household and caregiving work everywhere in the world while also working outside the home almost as much as men who have not reciprocated in a commensurate way in sharing caregiving work. The idea that larger numbers of men will spend more time on caregiving if they have to work less and/or have access to flexible working schedules is hopeful but has not been supported with much empirical evidence. Countries with more equitable gender norms do tend to have a better-established tradition of flextime policies so perhaps there is reason for optimism. In the US, for example, only 27 per cent of firms offer more than 50 per cent of their employees flextime. By contrast, 68 per cent of Swedish workplaces offer flextime to 80 per cent of employees (Malleson 2015). Even if such policies were in place in more countries, we would be left with the more significant challenge of changing the perception of caregiving from being a burden to being considered as a deeply satisfying and important aspect of human existence. Governments can certainly play a role in enabling such a shift by instituting guaranteed annual income and "living wage" regulations; by changing labor laws, perhaps including maximum hours and minimum wage regulations; and by ensuring that part-time work is good work, with prohibitions against lower pay and fewer benefits. However, the deeper political and social consciousness required for a transformation of the intra-household gender division of labor would have to be enabled informally and socially, perhaps through collective action, but not through legal sanctions or other government actions. Policy by itself cannot make men want to spend more time caregiving if care work continues to be perceived as low-status feminized work. Neither can policy require women to give up their control over care, particularly over the raising of children, if they have been socialized to believe that children are their primary responsibility. Until more transformative social change takes place in gender relations, flexible working schedules may reinforce rather than subvert existing gender imbalances in employment and care (Nedelsky (2014) acknowledges this even as she promotes the possibility of part-time work for all).

Will workers' unions remain relevant if part-time jobs become more of a norm, or will new modes for organizing, mobilizing and collective bargaining emerge in the future? These are also important questions to ask. Unions are generally much stronger in European (especially Scandinavian) countries and even in Canada and the UK than they are in the US (OECD Stats Database 2013). Regardless of what form representative organizations may take in the future, promoting gender equity must feature as a core principle. Countries that have the highest union densities (Denmark, Netherlands, Sweden, Norway, France) also have strong feminist movements and feminist contingents within the big unions. These movements have managed to rearticulate what contemporary unions should be, and brought back to prominence some of the union movement's original causes, as well as broader societal questions about the importance of delinking social entitlements from employment status (Malleson 2014). Other OECD countries (the US and Canada, for example) and emerging economies that do not have strong feminist contingents within unions might benefit from such organizing and strategizing. The level of unionization in new green jobs tends to be low to begin with in most countries. Whether new or reconfigured modes for organizing, mobilizing and collective bargaining will emerge in the future remains a matter of conjecture.

Travel and mobility

Much like jobs in the conventional fossil fuel industry, employment in RE can require significant travel and time away from home. This can be challenging for men too, but women with caregiving responsibilities may be put at a particular disadvantage. The locations of large RE construction projects are determined in part by the geography of natural resources and are often in isolated areas, with no provisions for the families of workers (IRENA 2013). Although such factors may explain women's underrepresentation in energy sector employment to some extent, many women may already work in less-than-optimal environments for much less pay than they would make in the energy industry and, given the option, would probably prefer work in the energy sector for higher wages. Because of entrenched male-biased hiring norms in the energy sector, women are also frequently not given the option to choose between undesirable working conditions with low pay and similar conditions with higher pay (McKee 2014).

Women in developing countries and emerging economies also face mobility constraints owing to social responsibilities and the traditional division of labor but some do find creative solutions. For example, a group of low-income women trained by a social enterprise called Technology Informatics Design Endeavour (TIDE) in India to construct biogas cookstoves set themselves up as a cooperative. They travel in groups of two or more to build stoves in distant rural areas (Baruah 2015a). Although women continued to shoulder much of the responsibility for childcare and household maintenance, family members often became more willing to share responsibility for domestic chores once the women started earning higher incomes. Researchers working in India (see, for example, Patel 2014 and other contributions to Nielsen & Waldrop 2014) have corroborated that gender roles tend to be malleable, that the intra-household division of labor is dynamic and negotiable, and that economics often trumps "culture and tradition". Extended, joint and intergenerational family set-ups, where grandparents live with married children and grandchildren, can be particularly helpful for women in occupations that require extensive travel and time away from children. I have collaborated frequently with middle-class professional women in India on research projects. Many have small children but are able to travel for work within India and internationally on a far more frequent basis than most professional women in North America and Europe would be able to precisely because they live in "traditional" joint family settings where grandparents are always available to take care of children. The availability of affordable live-in caregivers also often gives middle-class professional women in developing countries and emerging economies a mobility advantage that most of their counterparts in industrialized countries cannot afford. Although the gender division of intra-household labor may remain intact in both industrialized countries and emerging economies, insomuch as it is still mainly women doing the caregiving, middle-class women in some settings in the Global South often have a comparative career advantage due to these factors. Since there is often a stronger sense of collective and social parenting in many non-Western settings, women are also judged less harshly, if at all, for "leaving" their children with other family members or caregivers for work-related travel. Several researchers and journalists have written about the role such factors can play in enabling women's careers in some developing countries. As part of the New York Times' *Female Factor* series, Timmons (2010), for example, explores why women remain scarce among top bankers in New York and London despite decades of struggle to climb the corporate ladder but hold some of the most prestigious portfolios in India's relatively young financial industry.

Skill shortages

Skill shortages have been reported in RE in all OECD countries (IRENA 2013). Women's underrepresentation in RE is often an outcome of the low numbers of women who graduate as engineers in industrialized countries. For example, in Germany it was estimated in 2011 that out of one million engineers, only 13 per cent were women (Blau 2011). Out of 384,000 engineering students, only 79,000 (21 per cent) were women. Low as they are, these numbers for Germany are actually 10 per cent higher than they were in 2001 (ibid.). These numbers are consistent with trends in STEM fields in other EU countries (VDI 2009). The National Science Foundation (NSF) in the US reports that between 2003 and 2008, the total number of four-year engineering degrees awarded annually increased by about 10,000 to 69,895, with almost all the increases going to male graduates. This effectively reduced the percentage of women receiving undergraduate engineering degrees in the US from 20.5 per cent to 18.5 per cent (Mahmud 2012).

The number of women getting into engineering in Canada has also been on the decline, despite decades of efforts to encourage more girls to think of technical careers. Even though women currently make up more than half of the undergraduate populations across Canada, the number of women enrolled in engineering programs dropped from a high of 21 per cent in 2001 to 17 per cent in 2009 (Myers 2010). The numbers of licensed engineers in Canada who are women has grown from 7 per cent in 2000, but the figure still sits at only 10 per cent, according to Engineers Canada (ibid.). Women comprised 47 per cent of the Canadian workforce in the 2006 census. The participation rate of women in the engineering field averaged 13 per cent (Calnan & Valiquette 2010). Meanwhile, employment growth in engineering and technology occupations overall surged by 45 per cent between 1997 and 2008, according to Census data, compared to a growth rate of 24 per cent for all other occupations. Despite the fact that there has been a dramatic increase in the number of new jobs in engineering and technology, the vast majority are still taken by men (ibid.). The numbers for women in engineering in Australia are equally unimpressive (Engineers Australia 2012).

In the past few years, universities, technical schools and community colleges in Europe and North America have tried to integrate RE topics into their course offerings, and many have developed specialized RE courses and programs. An increasing number of companies have also begun partnering with higher education institutions and vocational colleges to develop tailor-made education and practical training for junior and specialized professionals (see, for example, Katz 2012). The US Department of Labor's Think Women in Green Jobs initiative is a promising example of a program offering training on renewables specifically to women (Women's Bureau 2010). However, very few other initiatives have any stated commitment to, or goals for, gender equity.

Even if equity policies were in place for training programs, women may find themselves unable to access apprenticeship opportunities at par with male graduates. Since completing an apprenticeship is often considered an essential part of qualifying for most trades, women are left at a disadvantage even if they have completed the formal training for some occupations. The challenges women face in seeking apprenticeships has been documented most extensively for construction workers (see, for example, National Women's Law Center 2014). Responsive strategies for leveling the playing field for women have also been made within the context of the construction industry (Cohen & Baird 2003). They may be worth adapting for the RE sector.

Skill shortages in the RE sector are also being reported in developing countries and emerging economies (IRENA 2013). Especially among women, there is often a lack of technical and business skills required for employment and enterprise development in renewables. Low levels of literacy and limited access to basic education make such skills particularly challenging for poor and rural women to acquire (Baruah 2010). Organizations like the Indira Gandhi National Open University, Solar Sister, Grameen Shakti, Barefoot College, Selco Solar and SEWA that provide customized solutions for training for individual women and opportunities for cross-mentoring among local entrepreneurs are attempting to close some of these gaps.

Unlike North America and Europe, where women remain a minority in engineering programs, comparatively large numbers of middle-class women in some emerging economies – India and China, for example – study engineering (Paris Tech Review 2010). Although women may continue to experience glass ceilings and employment discrimination in various forms in such countries, recruitment, especially for entry-level positions, is not a challenge because of the large numbers of women earning engineering degrees. In China, 40 per cent of engineers are women (ibid.). From less than 1 per cent in the 1970s, enrolment of women in engineering degrees in India had grown to 15 per cent in the early 2000s (Parikh & Sukhatme 2004). The most popular specializations for Indian women also bode well for RE employment. Thirty-seven per cent of electronics engineers in India are women. The figures for civil, computer, electrical and mechanical engineering are 19.7 per cent, 17.8 per cent, 16.1 per cent and 9.3 per cent, respectively (Parikh & Sukhatme 2004).

In the 1980s, 58 per cent of engineers in the USSR were women, but a well-established tradition of state-enforced gender diversity disintegrated in the 1990s and 2000s with the collapse of the USSR and its industrial model. In 1998, women accounted for 43.3 per cent of engineers in Russia; in 2002, only 40.9 per cent (Paris Tech Review 2010). And the numbers have continued to decline further. The Baltic nations (Estonia, Latvia, Lithuania) that were formerly part of the USSR, but joined the EU in the 1990s, revealed similar patterns of comparably high but declining rates of participation by women in engineering and technology fields. The World Economic Forum (WEF) reports that in Estonia, for example, female professional and technical workers still outnumber men two to one – 68 per cent compared to 32 per cent (Anderssen 2013). Estonia offers significant tuition incentives to draw high-school graduates into fields such as engineering and continues to be identified by the WEF as the country with the highest per capita number of female engineers, even as the numbers of women joining the field have declined over the decades (Anderssen). One does not have to advocate for a return to Soviet-style central planning to emphasize that state initiatives aimed at improving representation and removing barriers for career advancement for women in engineering and policy-making do work, and they can benefit the RE sector in both OECD countries and emerging economies. The potential and need for state intervention is discussed in more detail in the next section.

Public sector involvement

Generally speaking, the energy workforce globally represents a vertically and horizontally gender-stratified labor market, with women concentrated in the lowest-paid positions, closest to the most menial and tedious components and furthest from the creative design of technology and the authority of management or policymaking (Baruah 2015a). However, there are quantitative and qualitative differences in women's employment in renewables in different contexts. Much of the expansion of renewables in developing countries and emerging economies has occurred because large numbers of rural, urban poor and remote communities either have no access to the grid, or they have unreliable or inadequate access to electricity. A large volume of employment has been generated for both men and women in these contexts because organizations serving such communities (see, for example, the initiatives of Solar Sister in various African countries and Char Montaz in Bangladesh) have actively sought to use RE technologies to also secure and improve livelihoods. Such off-grid, mini-grid and standalone RE initiatives have offered women a larger volume of employment (albeit often poorly compensated and insecure) as well as opportunities to participate in decision-making. These initiatives are deployed at the local level where women are more likely to be involved in the procurement, design, installation, operation, maintenance and consumption of energy (Smith 2000). Decision-making within bigger energy utility systems in both the Global North and South are, on the other hand, made by higher-level professional staff within the spheres of generation, transmission and distribution where women are almost always severely underrepresented.

The importance of public sector involvement in creating a policy framework to enable the sustainable development and dissemination of renewables as well to as ensure employment equity has been made in OECD countries as well as in developing countries and emerging economies (Calvert & Cohen 2011; ENERGIA 2006). The ten countries with the largest RE employment in 2014 (in order of number of jobs) were China, Brazil, the US, India, Germany, Indonesia, Japan, France, Bangladesh and Colombia (IRENA 2015). These countries have become major manufacturers of RE equipment, producers of bioenergy feedstock and installers of production capacity. An array of industrial and trade policies continues to shape employment, with stable and predictable government interventions favoring job creation. Although governments in these countries may not be directly involved in developing and disseminating renewables, they have put incentives and subsidy structures in place that direct private investment to areas that would otherwise not be prioritized. However, many of these countries have not introduced meaningful policies to promote employment equity in the RE sector. Policy intervention aimed at gender equity in employment is the most important pre-requisite for optimizing women's employment in renewables and in STEM more broadly.

Brazil is a good example of an emerging economy in which women's participation in STEM has risen dramatically in recent years due to substantial investment and progressive social policies that include state-funded tuition (Huyer & Hafkin 2013). Brazil is the largest national investor in STEM in Latin America and the Caribbean – at about 1.4 per cent of its GDP (ibid.). Brazil is particularly notable for the prominent role played by women in education and in research. The availability and transparency of scholarship awards, particularly at graduate level and in science and technology, have aided women's substantial participation.

Equity and access policies adopted to promote gender equality are often linear and positivist in both industrialized countries and emerging economies (Baruah 2015a). They do not seek any special privileges for women and simply demand that everyone receive consideration without discrimination on the basis of sex. They are inadequate because they fail to address the wide range of social and institutional factors that prevent women from succeeding, and also because they do not demand preferential pro-women hiring practices to correct historical and current injustices and inequalities. However, I would argue that even such simplistic liberal policies can improve women's access to opportunities in sectors that are almost completely male-dominated. Other researchers (for example, Clancy & Roehr 2003) agree that even straightforward liberal employment equity policies would serve as a good starting point to improve women's access to employment in the energy sector in North America and Europe.

More comprehensive and finely tuned policies that take structural constraints into consideration will optimize women's performance and advancement in the RE sector. Government spending through stimulus packages and public procurement can also address gender inequality (Stevens 2009). Contractors for public agencies should be required to adopt affirmative action goals to correct the underrepresentation of women in their workforce. Green stimulus spending should come with conditional requirements for the recruitment and retention of women. Although countries like Canada, the US, Australia, France and the UK earmarked significant stimulus funding in the aftermath of the 2008 financial crisis for green initiatives, very little, if any, funds were allocated for the integration of women into green occupations (Cohen 2015). The US did allocate minor funds – out of the $27 billion in total allocated for energy efficiency and RE research and investments – for training women for green occupations in its *American Recovery and Investment Act* of 2009. Even this minor injection of funds resulted in several short-term pilot initiatives to demonstrate the potential for women in high-growth green occupations (see Cohen 2015 for examples). Despite the constant lip service paid to the importance of green jobs in industrialized economies, even boutique initiatives of the kind enabled by stimulus funding in the US are hard to find in many OECD countries. In Canada, a few promising initiatives aimed at training and employing First

Nations and inner city workers in RE and building retrofits have emerged recently from collaborations between provincial governments, publicly-owned utilities and social enterprises (Fernandez 2015). Such initiatives remain rare and the possibilities for replication in other settings are unclear.

Conclusion

There are similarities and differences between industrialized countries and emerging economies in the patterns of women's employment in the RE sector. A much larger volume of employment has been generated for women in developing countries and emerging economies through RE initiatives that also address energy poverty in remote or underserved communities. There is tremendous additional potential to create livelihoods for women in the RE sector. However, women can gain optimal traction from RE initiatives only within the context of wider socially progressive pro-women policies and more transformative shifts in societal attitudes about gender roles. This is as true for developing countries and emerging economies as it is for industrialized nations. The growth of the RE sector should benefit both women and men but we must be proactive about enabling women to establish a stronger equity stake to compensate for historical and contemporary economic injustices and unequal outcomes. This will require more concrete and proactive actions and policies. Simply creating opportunities for training and employment in new fields and suggesting that women are not unwelcome in them is obviously not enough.

References

Adeyemi, A 2006, "Empirical evidence of women's under-representation in the construction industry in Nigeria", *Women in Management Review*, vol. 21, no. 7, pp. 567–577.

Adubra, A 2005, *Non-traditional occupations, empowerment and women: A case of Togolese women*, Routledge, New York.

Anderssen, E 2013, "Gender geography: Where's the best place in the world to be a woman", *The Globe and Mail*, 8 March, (http://www.theglobeandmail.com/news/national/gender-geography-wheres-the-best-place-in-the-world-to-be-a-woman/article9488293/).

Antoni, M, Janser, M & Lehmer, F 2015, "The hidden winners of re-promotion: Insights into sector-specific wage differentials", *Energy Policy*, vol. 86, pp. 595–613.

Arregui, G, Candela, J, Estrada, B & Sara Pérez, B 2010, "Study on employment associated to the promotion of RE in Spain", ISTAS, (www.istas.net/web/abreenlace.asp?idenlace=8769).

Baruah, B 2010, "Gender and globalization: Opportunities and constraints faced by women in the construction industry in India", *Labor Studies Journal*, vol. 35, no. 2, pp. 198–221.

Baruah, B 2015a. "Creating opportunities for women in the RE sector: Findings from India", *Feminist Economics*, (http://dx.doi.org/10.1080/13545701.2014.990912).

Baruah, B 2015b. "Opportunities and constraints for women in the RE sector in India", *Women and Environments International*, vol. 94/95, pp. 7–10.

Bateman, M & Chang, H 2012, "Microfinance and the illusion of development: From hubris to nemesis in thirty years", *World Economic Review*, vol. 1, pp. 13–36.

Black, R 2015, "Personal interview", 7 May, Toronto, Canada.

Blau, J 2011, "Germany faces a shortage of engineers", *IEEE Spectrum*, (http://spectrum.ieee.org/at-work/tech-careers/germanyfaces-a-shortage-of-engineers).

Calnan, J & Valiquette, L 2010, "Paying heed to the canaries in the coal mine: Strategies to attract and retain more women in the engineering profession through Green Light Leadership", Engineers Canada, Ottawa.

Calvert, J & Cohen M 2011, "Climate change and the Canadian energy sector: Implications for labour and trade unions", CCPA, Ottawa.

Clancy, J & Roehr, R 2003, "Gender and energy: Is there a northern perspective?" *Energy for Sustainable Development*, vol. 7, no. 3, pp. 44–49.

Cohen, M 2015, "Gender in government actions on climate change and work", *Women & Environments International*, vol. 94/95, pp. 11–16.

Cohen, M & Baird, K 2003, "The road to equity: Training women and First Nations on the Vancouver Island Highway", in M Cohen (ed.), *Training the excluded for work*, pp. 53–74, University of British Columbia Press, Vancouver.

ENERGIA 2006, "Incorporating women's concerns into energy policies", Leusden, The Netherlands, (http://www.energia.org/fileadmin/files/media/factsheets/factsheet_policies.pdf).

ENERGIA 2009, "Biofuels for sustainable rural development and empowerment of women: Cases studies from Africa and Asia", Leusden, The Netherlands, (www.theworkingcentre.org/sites/default/files/ENERGIA_Biofuels_book_text_pages.pdf).

Engineers Australia 2012, "The engineering profession: A statistical overview", (https://www.engineersaustralia.org.au/sites/default/files/shado/Resources/statistical_overview_2015.pdf).

ESMAP (Energy Sector Management Assistance Program) 2012, "Commercial Woodfuel Production", (www.esmap.org/sites/esmap.org/files/FINAL-Commercial Woodfuel-KS12-12_Optimized.pdf).

FAO 2011, "The state of food and agriculture in 2010–2011", Food and Agriculture Organization of the United Nations, Rome.

Fernandez, L 2015, "How government support for social enterprise can reduce poverty and green house gases", CCPA, Manitoba.

Fernández-Baldor, A, Boni, A, Lillo, P & Hueso, A 2014, "Are technological projects reducing social inequalities and improving people's well-being? A capability approach analysis of RE-based electrification projects in Cajamarca, Peru," *Journal of Human Development and Capabilities*, vol. 15, no. 1, pp. 13–27.

Greed, C 2000, "Women in the construction professions: Achieving critical mass", *Gender, Work and Organization*, vol. 7, no. 3, pp. 181–96.

GVEP (Global Village Energy Partnership) 2012, "The improved cookstove sector in East Africa: Experience from the Developing Energy Enterprise Programme (DEEP)", (www.gvepinternational.org/sites/default/files/deep_cookstoves_report_lq_for_web.pdf).

Huyer, S & Hafkin, N 2013, "Brazilian women lead in science, technology and innovation, study shows", *Elsevier Connect*, (http://www.elsevier.com/connect/brazilian-women-lead-in-science-technology-and-innovation-study-shows).

Ilahi, N 2000, "The intra-household allocation of time and tasks: What have we learnt from the empirical literature?" World Bank, Washington, DC, (http://siteresources.worldbank.org/INTGENDER/Resources/wp13.pdf).

IRENA 2013, *Renewable energy and jobs: Annual review 2013*, International Renewable Energy Agency, Abu Dhabi.

IRENA 2015, *Renewable energy and jobs: Annual review 2015*, International Renewable Energy Agency, Abu Dhabi.

Kammen, D 2011, "Biofuels: Threat or opportunity for women?" (http://blogs.worldbank.org/climatechange/biofuelsthreat-or-opportunity-women).

Katz, J 2012, "Emerging green jobs in Canada: Insights for employment counsellors into the changing labour market and its potential for entry-level employment", Green Skills Network, Toronto.

Lehr, U 2008, "RE and employment in Germany", *Energy Policy*, vol. 36, no. 1, pp. 108–17.

Little, M 2005, *If I had a hammer: Retraining that really works*, University of British Columbia Press, Vancouver.

McKee, L 2014, "Women in American energy: De-feminizing poverty in the oil and gas industries", *Journal of International Women's Studies*, vol. 15, no. 1, pp. 167–178.

Mahmud, A 2012, "Graduate studies spur success in engineering", *ASME*, (https://www.asme.org/career-education/articles/graduate-students/graduate-studies-spur-success-in-engineering).

Malleson, T 2014, *After occupy: Economic democracy for the 21st century*, Oxford University Press, New York.

Malleson, T 2015, "Interview: How shorter work hours can help the climate and women's equality", *LaborNotes*, 8 March, (http://labornotes.org/2015/03/interview-how-shorter-work-hours-can-help-climate-womens-equality).

Myers, J 2010, "Why more women aren't becoming engineers", *The Globe and Mail*, 9 November, (http://www.theglobeandmail.com/report-on-business/careers/career-advice/why-more-women-arent-becoming-engineers/article1216432/).

National Women's Law Centre 2014, "Women in construction: Still breaking ground", NWLC, Washington, DC.

Nedelsky, J 2014, "Part-time for all: Creating new norms of work and care", Natural Law Colloquium Fall 2014 Lecture, Fordham University, NY.

Nielsen, KB & Waldrop, A 2014, *Women, gender and everyday social transformation in India*, Anthem Press, London.

OECD Stats Database 2013, "Trade union density", (https://stats.oecd.org/Index.aspx?DataSetCode=UN_DEN).

Openshaw, K 2010, "Biomass energy: Employment generation and its contribution to poverty alleviation", *Biomass and Bioenergy*, vol. 34, pp. 365–378.

Paap, K 2006, *Why white working-class men put themselves – and the labor movement – in harm's way*, Cornell University Press, New York.

Parikh, P & Sukhatme, S 2004, "Women engineers in India", *Economic and Political Weekly*, vol. 39, no. 2, pp. 193–201.

Paris Tech Review 2010, "Why aren't there more women engineers?" (http://www.paristechreview.com/2010/09/29/why-more-women-engineers/).

Patel, R 2014, "Today's 'Good Girl': The women behind India's BPO Industry", in K Nielsen & A Waldrop (eds), *Women, gender and everyday social transformation in India*, pp. 21–32, Anthem Press, London.

Smith, J 2000, "Solar-based rural electrification and microenterprise development in Latin America: A gender analysis", National Renewable Energy Laboratory, Boston, (www.nrel.gov/docs/fy01osti/28995.pdf).

Stevens, C 2009, "Green jobs and women workers: Employment, equity and equality", *Sustainlabour*, (www.sustainlabour.org/IMG/pdf/women.en.pdf).

Timmons, H 2010, "Female bankers in India earn chances to rule", *New York Times*, 27 January, (http://www.nytimes.com/2010/01/28/world/asia/28iht-windia.html?_r=0).

TNC (The Nature Conservancy) 2014, "The untapped potential of young women in natural sciences", *Treehugger*, 16 June, (http://www.treehugger.com/green-jobs/untapped-potential-young-women-natural-sciences.html).

UN Women 2012, "Fast-forwarding women's leadership in the green economy", (http://www.unwomen.org/2012/06/fast-forwarding-womens-leadership-in-the-green-economy/).

VDI (Association of German Engineers) 2009, "European Engineering Report", (www.vdi.de/uploads/media/2010-04_IW_European_Engineering_Report_02.pdf).

Women's Bureau 2010, "Think women in green jobs", US Department of Labor, (https://www.dol.gov/wb/media/Greenprojects.htm).

6

UK ENVIRONMENTAL AND TRADE UNION GROUPS' STRUGGLES TO INTEGRATE GENDER ISSUES INTO CLIMATE CHANGE ANALYSIS AND ACTIVISM

Carl Mandy

Introduction and methodology

Climate change and gender issues are explicitly addressed by many UK trade unions and environmental groups, with high priority often given to one or both. Nonetheless, they are often treated separately, particularly within larger organizations. This essay suggests possible reasons why a more integrated approach is not more widespread, which may need to be addressed if combined development in these two areas is to occur more extensively.

Research for this paper relies on information and critiques from academic publications, and information from the organizations themselves. The organizational material used includes published material and documents, and interviews with key informants from a sample of trade union and environmental groups. Grey literature (the name often given to materials not published widely) research includes website material, declarations of policy and activist objectives, press releases, online and internally produced journal articles, and blog postings. The typical purposes of the literature relates to informing current and potential members about group intentions and actions, providing material for media coverage, and detailing organizational structures and operations. This was invaluable in determining the extent to which campaigns supported policies, officers and committees focused on gender and/or climate change.

The academic literature used for analysis was frequently more analytical and critical of both trade unions and environmental organizations' ability to deal with gender and climate change. This suggests a more common awareness of the struggle to integrate gender and climate change issues. Similarly, some grey literature, such as Manchester Climate Monthly (2014), also provided critiques of groups' handling of these issues.

Given that trade unions and environmental groups do not publicize all of the details of their activities, and also have many projects in development at any time, interviews were also conducted with feminist trade union and environmental activists. The purpose was to discover the perspectives of people directly involved about their experiences of how decisions are made on these issues, and whether they perceived any gendered barriers to successful climate change activism arising either internally and/or externally to their organizations. Feminist activists were particularly well placed to comment on the complexities and possible tensions in integrating climate change and gender analysis and activism.

UK trade unions' grey literature on gender and climate change

All of the trade unions in this study have programmes related to improving working conditions for women, and also indicate commitments to integrating labour issues with measures to deal with climate change. While a number of major UK trade unions recognizably address one or the other of these two issues, those outlined in this section have measures and initiatives focused on both. Nonetheless, a notable feature here is a minimal integration between gender and climate change, as these issues are typically attended to in isolation.

The Trades Union Congress (TUC) encompasses 52 unions and 5.5 million workers (TUC 2015b). They have conducted studies with gendered analyses, covering topics such as pay gaps (Thiranagama 2008), career stereotyping (TUC 2013) and austerity cuts on women (CLES 2014). TUC desires "statutory rights . . . both in terms of training and facility time" for their environmental representatives (TUC 2015a). A women's committee meets five times a year (TUC 2015c), and a green economy conference occurred in 2014 (TUC 2014).

The Union of Construction, Allied Trades and Technicians (UCATT) is the only UK union specializing in the male-dominated sector of construction, claiming 80,000 members (UCATT 2015a). Their gendered work includes a 2014 survey on workplace harassment and a women's weekend the same year, and they have raised concerns about gender representation and unfair treatment within construction, where women constitute 11 per cent of the workforce, and 1 per cent of on-site workers (UCATT 2014). UCATT's GoingGreen@Work section addresses climate issues, including support for low carbon, sustainable industry. One demand is for government to "create a partnership agenda to support routes into the low-carbon construction industry for young people, women and black and ethnic minorities" (UCATT 2015b).

UNITE has 1.42 million members, covering industries such as energy, agriculture and forestry. Their gender-focused publications include a study on women's health, safety and well-being at work (Unite 2012b). UNITE's energy position argues "emissions targets will only be met by the expansion of nuclear and renewable generation and by the development of carbon capture and storage (CCS) technologies" (UNITE 2012a). They accuse world leaders of betraying their citizens due to failure to address climate change. UNITE also supported a 2015 climate march and transitioning to a low-carbon economy.

The Public and Commercial Services Union (PCS) organize around 220,000 civil service members and private sector workers, with women making up 60 per cent of its membership (PCS 2015a). A "Women's Equality Toolkit" is published, and positions are available for green representatives, allowing for a promotion of climate change concerns and its connection to workplace practices.

Overall, grey literature from UK trade unions typically presents climate change and gender as important issues, although as further examination shows, these issues are more separated than integrated in their analysis and activism. Exceptions include TUC's analysis on international development, which discusses gendered impacts of climate change in poorer countries, and the journal *PCS Women* publishing the short article "Climate Change and Women" (PCS 2015b, 6). A likely reason for this typical separation of issues is due to different, dedicated sectors of these organizations creating the literature and action on each topic. For example, union environmental committees produce literature devoted to climate change, while women's committees look at gender issues, with little overlap between the two. This raises the question of how and why there could not be more integration between these largely discrete sectors.

UK environmental groups' grey literature on gender and climate change

Specific UK environmental groups were selected for study on the basis of their addressing both climate change and gender concerns in their literature and activism. Although, as with trade unions, there is a tendency to address these issues separately, several groups studied indicate a notably greater integration of the two topics than is found within trade unions.

A key finding from study of the print material and other literature from UK environmental groups is the greater diversity in their internal structures compared to trade unions. This raises the question of the impact, if any, this greater organizational diversity has on their analysis and activities. For example, Earth First! Friends of the Earth, Plane Stupid and the Transition Network operate as hubs for more autonomous local, affiliate organizations. Friends of the Earth (FoE) (2014) does make explicit connections between sustainability, climate change, environmentalism and women. Nonetheless, FoE is typical among mainstream environmental (and trade union) organizations, in that the major focus connecting gender and climate change relates to poorer countries.

Women's Environmental Network (WEN) (2015) address impacts of climate change on gender, dividing discussion between women in the global north and south. It describes itself as "the only organization in the UK working consistently for women and the environment", noting particularly how women are less involved in decision-making roles that can influence climate change responses. Writing for WEN on "Gender and the climate change agenda", Haigh and Valley (2010) make links between the impoverishment of women, and their vulnerability to climate change's impacts. Poorer countries are given more emphasis than richer ones, given that climate-related health problems in wealthier states are "not of the same magnitude as developing countries" (WEN & National Federation of Women's Institutes 2007, 23).

Further examples from other UK grey literature suggests that some environmental groups may give greater consideration to the intersection of gender and climate change than currently observed in trade unions. Their organizational structures are also dissimilar to that typically found in trade unions, suggesting there may be a connection between these features. Transition Network, the hub for the transition towns movement, aims to deal with the climate crisis at a community level, ostensibly rethinking the ways in which people can sustainably live and work together. Their analysis includes climate change's impact on work, leading them to favour sustainable, localist and co-operative approaches to mitigate its effects. Their structures are less centralized than is common in trade unions, with autonomous, horizontally organized groups a common feature. Blog reports of meetings attended by activists show a consideration of how gendered patterns of behaviour can limit participation (Hopkins 2014). A sub-group, Women in Transition, meets to discuss gender concerns within transition groups (Ward 2014). This includes consideration of dominant voices within meetings and communities, and the impact of stereotypically gendered values and personality traits on group composition and skillsets.

Deep Green Resistance (DGR) is a radical feminist environmental group that combines a steering and an administrative committee with horizontal organizing throughout the rest of its membership and affiliated groups. This central committee model is maintained to ensure their policies and strategies "are not watered down as so often happens to radical campaigning groups" (Deep Green Resistance 2015b). DGR explicitly "aligns itself with feminists and others who seek to eradicate all social domination and to promote solidarity between oppressed peoples" (Deep Green Resistance 2015a). DGR's gender analysis is central to its position, stating that:

> Men as a class are waging a war against women . . . Gender is not natural, not a choice, and not a feeling: it is the structure of women's oppression. Attempts to create more "choice" within the sex-caste system only serve to reinforce the brutal realities of male power. As radicals, we intend to dismantle gender and the entire system of patriarchy which it embodies. (ibid.)

DGR recognizes this position has led to controversy, particularly with regard to its attitude to transgender people (Deep Green Resistance 2015b, 2015c). Nonetheless, it is a rare example of environmentalism in the UK centrally placing a gender critique in their analysis and activism.

Environmental groups' grey literature reveals a greater diversity of organizational structures than within trade unions. Hierarchical models are more ingrained within the trade union movement, whereas some more space is given to horizontal organizing among environmental groups. Gender is often given less explicit consideration in analysis and activism in environmental bodies, while climate change is, not unsurprisingly, a major focus. The noted exceptions to the integration of gender and climate change give rise to the question as to why some organizations are more successful in integrating these issues, to an arguably more detailed and nuanced extent than can be found within the trade unions, with the possibility that structural factors play an important role.

Academic literature on gender, climate change in UK trade unions and environmental groups

Drawing connections between gender and climate change is more common in the academic literature than in that of activist groups. For example, in discussing Transition Network, Barrineau (2011, 6) argues that despite the consideration the organization gives to gender, elements of a gendered division of labour among participants can be observed, including food preparation and gardening tasks dominated by women and physical labour dominated by men. Barrineau suggests "the cultural myth that associates women with nurturing tasks" (25) has not yet been fully confronted, while noting that "transition culture aims to create new stories about the way we should live, yet [it] still clings to old stories concerning gendered norms" (21).

The lack of consideration regarding gendered implications of climate change might be considered surprising given women's high level of climate change activism. Some analysts explain this by pointing out that women's more community-based activism, rather than a critical analysis perspective has been a more common approach to women's climate change activities (Buckingham-Hatfield 2000, 61). While women might be expected to have key roles in ENGOs because of their greater numbers in these organizations and in the voluntary sector more generally, the majority of senior and decision-making positions in these organizations are held by men. Within the environmental movement, a culture of machismo can work against women and their interests, and gender is often overlooked when considering environmental injustice, possibly because women are not a "geographically concentrated" group, and this limits their visibility (Buckingham & Külcür 2009, 661).

Many ENGOs can be criticized for undemocratic organizational structures, and for uncritically replicating gendered hierarchies found in wider society. Carter (2007) recognizes this lack of democracy in groups such as Greenpeace – "Its founders had a clear organizational blueprint of an elitist, hierarchical structure where control resided with full-time staff and professional activists" (150). Friends of the Earth UK has a majority of elected members on its board,

> and local groups can influence strategy through the annual conference, but with the continued growth and professionalism of the organization, it is a matter of some debate how democratic FoE is in practice. (150)

This critique could be extended to UK trade unions.

Buckingham and Külcür (2009) argue environmental campaigning groups can be "gender blind at best, masculinist, at worst". Pre-existing gender imbalances found within the household, work, and other institutions are likely "to have an impact on the decisions made" (667). This is "compounded by an environmental sector whose staffing profiles closely mimic those in the governments and industries they lobby and campaign against" (ibid.).

Addressing trade unions, Kirton and Healy (2013) note that despite moves towards better levels of gender equality, men still hold a disproportionate amount of power in these organizations. They argue there is a tendency in unionism to be suspicious of attempts to focus on "separate spheres" such as women within the working class, where these efforts may be read as patronizing, tokenistic, wrongly universalizing women's experiences or undermining worker solidarity. Ledwith (2012) and Stuart, Tomlinson and Lucio (2013) identify an enduring culture of male hegemony in trade unions that has proven difficult to overcome. Despite a growth in initiatives to meaningfully increase the involvement of women, the enduring masculinized culture ensures men's positions in influential roles, not least the tendency of existing male leaders to spot and encourage potential successors in their own image (Ledwith 2012, 193).

The academic literature on environmental groups and trade unions points to issues of (institutional, inherited) sexism, and problems with top-down organizing with regard to a diversity of viewpoints having influence over directions taken. This paucity of material led to the interview research carried out, where respondents were invited to consider barriers they perceived regarding the integration of gender and climate change.

Interviews with feminist activists in UK trade unions and environmental groups

Potential activists who were asked to participate in this study were selected from those whose relevant work and publications had been uncovered during the research process, although some emerged through snowballing recommendations by other contacts. Interviews were conducted online, and no one was paid for their involvement. Simon Fraser University's ethics and consent process was adhered to. Interviewees could discuss multiple organizations they had recent experience with.

Six interviews were conducted, and participants referred to in this essay are pseudonyms. Two of the interviewees – Amanda and Carolyn – were full-time public sector trade union activists, whose official work and personal activism included promoting environmental and climate issues within their own and other unions' membership, and environmental organizations, such as anti-fracking groups. Carolyn had held the union role of "green representative", an awareness-raising position also focused on practical implementation.

The other four interviewees were environmental activists. Michael had experience in volunteer-organized, locally based organizations. Until recently, his activism involved pressuring local government to make and follow through on environmental and climate promises. However, he noted this tactic has been "based around the idea that fundamentally politicians were in some way shameable into action. And they're not." His current activities mostly focused on writing letters and articles for newspapers, informed by material gathered through freedom of information requests. Kate had been a member of a group organizing around cycling promotion. Through their public cycling events, strong connections were made to the environmental, social and climate impacts of car culture both globally and locally, encouraging use of this greener method of transport. Tim worked at the national level for a large environmental group, recently campaigning around climate change legislation, factory pollution and "women's empowerment in environmental sustainability". Kira was a member of a small group, although part of a larger network, focused around climate change and anti-oil activism, in particular protesting "oil companies sponsoring art institutions and galleries".

Gender and green analysis and activism in UK trade unions

Typical hierarchical structures of UK trade unions include a variety of committees and boards, with explicit chains of communication and accountability, often with a central executive committee with connections to all other divisions. Union composition includes a mixture of appointments, with some roles elected by the membership, while others are non-elected employees of the union. Annual conferences are the typical venue for unions to vote on policy, often informed by committee and regional branch contributions.

Dedicated women's committees are common, frequently with the stated goal of providing a space to consider and represent specific interests of women members. The presence of union committees focused on environmental issues has also become more common. Public sector trade unionist Carolyn argued that "gender issues are very important to the trade union movement". She believed one intention and advantage of having designated women's representatives and fora – such as RMT's women members' section (RMT 2015) – was to enable women members to participate. These members are

> sometimes put off by certain formal structures that can be male dominated. Especially with childcare responsibilities and domestic responsibilities, women find it more difficult to attend evening meetings or weekend meetings, or even to make themselves heard in [what is still] a male environment. Parallel structures or supportive structures are very important to implement.

Democratic structures and accountability measures are frequently built into union structures. Members can contribute to composing and promoting union policies, and/or run for election to a variety of dedicated positions. However, the existence of such posts means a minority of members will hold them, while a majority will not. Union officers, committee members, representatives and full-time employees will be better placed to exert influence over the direction of the union and its policies than regular members. Nonetheless, Carolyn was happy to characterize her own union's decision-making process as "very democratic". She described the national women's forum as the executive's "first point of call to consult and involve in advising" on policies identified as having a gendered dimension. This raised the question of why a gendered analysis of climate change is not more common.

UK trade unions commonly raise green issues. Carolyn's national executive committee featured a sub-committee devoted to "climate and environmental issues". Carolyn and Amanda's public sector union activism involved collaborative work with different organizations on public ownership of energy, and its affordability in a time of increasing prices and sector oligopoly. Both she and Amanda had separately been involved in union activism relating to fuel poverty, noting how women's committees' engagement with the issue has led to greater recognition of the issue, where single mothers and elderly women were thought to be disproportionately affected. Carolyn judged that, within her union, "gender is addressed, climate change is addressed, but I think the crossroad between the two is actually quite new".

Trade union struggles to integrate gender and climate change

As jobs in many sectors have been under high levels of threat since 2008 and the growing effects of austerity cuts imposed by the government, unions have dedicated much of their activism to dealing with these economic and political issues, with other considerations often treated as secondary. Carolyn discussed a "climate jobs" initiative she is involved with across multiple trade unions, noting how "it's sometimes difficult to engage workers with these issues. They may see the priority as totally irrelevant to them". There had been some success in bringing together "the issue of climate and the issue of jobs" since the initiative's 2007 roll-out, including internationally, as climate-friendly jobs strategies could now be seen in the USA, Australia and France. But the impact of austerity has presented many barriers to its greater success. Given these difficulties for trade union activists who wish to place priority on climate issues, it helps to explain why a more detailed, gendered analysis is less common than it might otherwise be.

Another reason why UK trade unions may struggle to integrate gender and climate change is the tension between different sectors with regard to environmental issues. Amanda reported friction between unions representing workers in climate-sensitive sectors, such as energy and construction, and further promotion of "green" agendas, which can be perceived as placing particular jobs under greater threat – for example, those of North Sea oil workers. While their unions sought to prevent this, she noted, "they're working on a finite resource, and we need to be looking at alternatives." Amanda argued this tension between jobs and climate can continue when working with the TUC, due to their representing people in a wide variety of sectors. While it is fully understandable unions take their primary concern to be the protection of members' jobs, greater consideration needs to be given as to what the nature of that work is, and its climate impact. Competing interests can delay the process of unions working together. In such cases, significant barriers to union climate change activism can halt any further development, let alone that with a gendered analysis.

A further possible reason for the difficulty in integrating gender and climate change in trade unions is the lower status of those working on roles dedicated to these issues. Although Carolyn's public sector union officially recognizes green and women's representatives, a major barrier against their success was an absence of "legal provision for facilities time, which is paid time off work, paid by the employer" to organize around climate issues. However, the union was currently pushing to grant them this legal recognition with employers. Additionally, Carolyn reported how "before the Conservatives came to power . . . negotiating structures around climate and greening the workplace" were at least operational. But since Labour lost power, "those conversations are not taking place, [halting] a number of projects we were planning to run, such as sustainability assessment in the workplace". Carolyn felt there had been resistance to placing emphasis on climate change issues within trade unionism, as it was often regarded as an international question that did not directly affect members. To work towards countering this perception, it had been useful to

> establish a link between what is affecting people on an everyday basis, and the issue of climate, and gender, where relevant. For example, linking unemployment and how we can create climate jobs. Linking private energy, fuel poverty, the link with low pay, and the link with climate change . . . You have to make it relevant to the everyday concerns of your members, and show how climate is actually relevant to many spheres.

But this can be a difficult and time-consuming conversation, particularly when battling the perception of climate issues being in conflict with job security.

An issue related to the status of those dealing with climate change is the dominance of men within union hierarchical structures, and the difficulties women face in achieving positions of influence. For example, Amanda recognized that although her public sector union's membership numbers have a slight imbalance towards women, "the more higher up it goes, the less gender diverse it tends to be", with the imbalance switching more decisively towards men. Although there had been regular, purposeful attempts to encourage the involvement of women members – for example, through their presence on speakers' lists and attendance on panels – she was "not aware of any specific policy around it". In her experience of less gender-diverse unions, women can "feel very isolated and alienated and not able to bring their views", often being "shouted down in meetings". However, she felt there was reason to be optimistic with regard to a new wave of "younger, confident women" activists emerging from the universities.

Carolyn felt that "women are not excluded, but it will be more difficult for a woman to rise in positions of power or influence". It is still "mainly men who are leaders of trade unions". Furthermore, in her experience of negotiating climate issues with energy sector unions, those attending are predominantly men. This was not raised as being necessarily problematic, and a positive note was that the male-dominated union representing fire prevention workers is now involved in discussions around climate and jobs. This was due to the nature of their work having altered due to climate change, including greater instances of wildfire and flooding. As it stands, women's comparative lack of influence in the UK trade union movement suggests a further reason why analysis and activism connecting gender and climate change is not more widespread.

Gender in UK environmental groups

While organizational structures of UK trade unions frequently enable explicit consideration of gender issues – albeit with the limitations noted above – this is not always the case with environmental groups. Although, like trade unions, some organizations exhibit hierarchical structures and set roles for their workers and volunteers, others use more decentralized arrangements. It could be expected that less hierarchical organizations are better suited to include gender-focused activism, as they could be freer from the male-dominated structures common in trade unions. Furthermore, as they are environmentalist groups, they would not be affected by competing concerns between protecting jobs that contribute heavily to climate change and combating these very causes of it. Given this, it could be considered surprising that UK environmentalist groups do not include a more gendered analysis of climate change.

In Amanda's experience of locally-minded anti-fracking groups, it had been common to find "slacker structures, much more consensus decision-making", compared to trade unions. Tim described his national group as essentially run by the grassroots members, because they "vote in the majority of the board, and through that can influence the work of the organization". While "campaigns [were] predominantly chosen by staff . . . we are trying to improve our dialogue" between all members.

Proponents of a non-hierarchical approach frequently speak in favour of it being more democratic and inclusive, allowing all people involved in a decision to contribute to its outcomes. Given this, it appears possible greater gender inclusivity could be achievable, with a correspondingly more developed gender analysis. Kate's activist cycling group used non-hierarchical organizing, with the membership intentionally kept to around twenty people, a structure they felt was largely effective for them, and "fairly even" in terms of gender diversity. There was a purposeful intention to show women in promotional materials, often performing in "nonconventional roles, like pictures of women doing mechanics and teaching cycling training". Additionally, the group deliberately tried to have women mechanics working alongside men at their free bike-fixing events. Kate recognized how "a lot of the skills we were willing to take on were related to what we'd already been trained in, and that had obviously been shaped by what society is". This had gendered implications – for example, men were more inclined to take on mechanical and technical responsibilities. The group sought to overcome such barriers by "role-swapping, skill-sharing, consensus training, [and] trying to talk openly . . . if any issues came up."

In Tim's national ENGO, women composed around 60 per cent of staff, activists and supporters. However, this was not consistently reflected in upper management, where men were often in a greater majority. In recent years, "within the staff grouping, there's been an internal push . . . to take gender and broader diversity more seriously". Motivations for this included recognizing it being the moral thing to do, better diversity leading to better decision-making, and also the fact that "the people who are most impacted by climate change are going to be predominantly disadvantaged communities". Plans had been set to increase diversity, firstly in the more achievable staff grouping, followed in coming years by activists and then more general supporters of the organization. The establishment of a steering group on diversity had helped progression toward these goals. Tim argued the importance of "designing our campaigns so they reach out better to a broader section of society". Work had also been carried out in collaborating with writers on the connections between environment, sustainability and gender.

Kira's anti-oil activist group was largely gender-balanced, and composed of a core group of roughly ten, with several dozen more less regular members. The organization operated in a loose coalition with other groups which came together occasionally to jointly organize. Commenting on their general independence from larger ENGOs, Kira argued "one of the great things about grassroots groups [is] you don't have to be bound by where your funding comes from, or rules in terms of hierarchies and how you organize, and I think it is important to have that form of diversity within a movement." A key element of the group's public protests involved performance pieces, wherein gender was given consideration, such as having "gender-flipped roles . . . '*BP*' has always been played by different genders". Kira argued, "if you are using performance as your means of protest, it is something that needs to be considered", given potential for misrepresentation and problematic depictions of gender to the public.

One barrier to environmental activism identified by several interviewees was the negative influence of law enforcement, which had increased alongside the rise in austerity measures. Some identified this as having a gendered element, with public sector trade unionist Amanda commenting how "you read reports of, for example, mothers being interviewed by police about the activities of their children". Also of concern is the infiltration of activist groups in the environmental movement (and elsewhere) by undercover police, "taking advantage of women", with several known cases of their having been in sexual relationships with female activists. Amanda claimed these actions meant some of her own friends were hesitant to get further involved in activism. Kira's anti-oil group encouraged participation in action by emphasizing a range of roles people could take on, with varying levels of risk regarding arrest. This hostile environment as a result of tougher law enforcement, which can have additional impacts on women, can be seen as a deterrent to their further involvement. Importantly, this barrier affects trade union activists as much as environmentalists, and can be considered a possible factor in relation to further development and integration of gender and climate change issues.

With regard to the impact of austerity measures on environmental groups, environmental activist Michael judged that "one of the things that's happened to climate activism over the past five years is that people who would be active on this issue have got other, more immediate campaigning concerns, around defending the welfare state and the effects of austerity. And that causes a huge amount of displaced activity." Tim also identified austerity as "a massive challenge" to organizing around climate change, as many have taken the economy to be the "overriding priority before everything else". As shown, a prioritization of these concerns for UK activists goes towards explaining the slow progress in developing a gendered analysis of climate issues.

A gendered analysis of climate change does occur within some UK environmental groups, to a greater degree than that within trade unions. While this appears more common in environmentalist organizations which move away from the hierarchical structures seen in trade unions, it is not a uniform feature among groups of this type. This raises the question of what additional influences are affecting them.

Environmentalist groups are less restricted than trade unions when it comes to addressing climate change, but they are still part of a wider culture of sexism which marginalizes gendered issues. Adopting a "non-hierarchical" structure is not enough. As seen with groups such as Transition Network and Deep Green Resistance, the inclusion of a gendered approach to climate activism came with the conscious adoption of a gendered analysis to both the processes undertaken by the organization, and a critique of the gender expectations of the wider culture. None of these undertakings are simple to carry out, nor are they likely to be easily accepted by others.

Amanda noted the decentralized structures found in some environmentalist groups does not in itself eradicate the influence of less explicit hierarchies, which "definitely exist" in both hierarchical and nominally non-hierarchical organizations. Considering the relative benefits and drawbacks of the two broad approaches, she concluded, "There is no perfect system, and I think you still need that management facilitation in there, [to] allow for people to have confidence in those environments, to feel like they can participate."

This points toward a recognized area of concern regarding non-hierarchical groups. A key text is Jo Freeman's (1970) "The Tyranny of Structurelessness", which in Michael's activist experience was "occasionally invoked, but hardly ever actually read and considered". Freeman's concern was the ability of non-explicit hierarchies to limit the equitable involvement and influence of group members. For instance, charismatic members can sway the direction of a meeting regardless of whether they hold the position of chair or president. In his experience, Michael noted the "hidden power" present in these settings does not emerge from official positions people hold, but factors such as "confidence, education, credibility, free time, and – less charitably, I would say – willingness to work the process to get [their] own way."

Kira's anti-oil group had run strategy days on tactics and decision-making processes, including discussion of non-explicit hierarchies in the organization. One problem raised, common to many small activist groups, was the tendency for the same core group of people to take on the bulk of the organizing, which can fall along gendered lines, for example men taking on public speaking roles. Explicitly raising the issue had led the group to consider revolving positions of responsibility. Furthermore, a tendency of more involved activists nominating less active members for particular tasks helped to diversify responsibility. Kira noted this "shouldn't be done in a tokenizing manner, but if there are people you think would be genuinely good at something then quietly talking to individuals rather than just expecting people to speak up in a group situation can be effective in combating potentially gendered dynamics in terms of confidence and speaking up".

Tim and Michael spoke of a lack of momentum among activists following the perceived letdown of the 2009 Copenhagen Summit. As well as being dispiriting for campaigners, Tim felt the media had moved on to other topics following its coverage, further deescalating energy around climate activism. This also impacted on potential funders, a point exacerbated by the growing precedence given to austerity concerns. Michael argued that in his experience, social movements such as those working around climate activism were "incredibly 'small-c'conservative", and he had commonly encountered three types of response from people as to why they continued to organize in ways that were not gender-inclusive. Firstly, activists would describe themselves as lacking the necessary "confidence and charisma" they felt was needed to make internal changes. Secondly, they would maintain "our current system works really fine" (i.e. for them), a response most often voiced "in the aftermath of a successful event or a successful demonstration, and then they tend to shut up if there's been an unsuccessful one". Thirdly, and most frequently, the responses took the form of "'Well, we don't have time for that. This speaker's really important, and we need to hear from him' – usually him." This suggests a lack of imagination and desire to change existing approaches, which can further hold back the development of gendered considerations in climate activism. Michael also criticized climate groups for ineffectively policing "shitty male behaviour" in meetings, where women are frequently interrupted, not taken seriously, or, in its "sleazier manifestations", are treated in an "overly attentive" manner. "It's just a rampantly patriarchal, misogynist society, so it's unsurprising [these groups have] those pathologies." Tim's earlier experiences of environmental groups had been impacted by a "macho culture", though he felt this was gradually being confronted, at least in his own working environment. Kira felt men were more likely to speak up in meetings, although her group had helped to increase the diversity of ideas expressed by splitting members into smaller groups for brainstorming and discussion before feeding back to the wider membership.

Amanda added to this by saying,

> I feel in terms of gender we've gone back like thirty years . . . It just shocks me when expectations around women's roles and sexualization [have] just gone to an extreme now, which I frankly find very depressing . . . I don't agree so much with all this talk of "We need more women". [More importantly] we have to change . . . systemic inequalities.

Within climate activism, there also still existed the all-too-common situation where women looking to participate further get the confidence-sapping message, "Well, they can't have this position because they're not going to turn up, they'll have to be looking after kids."

These criticisms of non-hierarchical organizing help to show why this nominally more democratic and inclusive approach does not necessarily lead to everyone involved having an equal role and influence. Those able to exert influence due to being in a privileged position can have as much power as any designated official. Within a wider culture of gendered power imbalance, women's voices can be marginalized just as much as under explicitly male-dominated hierarchies.

While non-hierarchical structures and consensus models are more common in environmentalist groups than trade unions, a number of environmentalist organizations still use hierarchies. Groups like these are likely to find themselves affected by both explicit and non-explicit hierarchies which favour men, suggesting further why gendered consideration of climate change is less prevalent than it currently is. Furthermore, as Buckingham-Hatfield (2000), Buckingham and Külcür (2009), Carter (2007) and Murray (2011) argued, environmentalist groups and trade unions can experience problems with male dominance and a lack of internal democracy.

Conclusion

There is significant variety in the organizational structures of UK environmental groups and trade unions. Environmental organizations with hierarchical structures similar to, or even less democratic than, trade unions, show a similar level of integration regarding gender-focused climate activism. However, it is also infrequent for non-hierarchical environmentalist groups to consider this issue, a possible reason being the influence of a wider, sexist culture, and the considerable difficulty involved in organizations genuinely ridding themselves of non-explicit hierarchies – something which may be even harder to accomplish in larger groups.

Nonetheless, it should not be discounted that part of the problem with this lack of integration could be related to the tendency to treat gender and climate change as discrete issues. This could perhaps be overcome without a radical restructuring of these groups. A number of possible courses of action suggest themselves, which could not only contribute to a greater integration of these issues, but also have wider benefits for internal diversity, democracy and accountability. For example:

use of committees, publications and events dedicated to gender and green issues; discussion and skill-sharing between different organizational sections; ensuring gendered aspects of sustainability and austerity are on the agenda; emphasizing diversity in media produced; funding travel and childcare expenses wherever possible; challenging (male) voices that dominate meetings and decision-making processes – for example gender diverse lists of speakers, less reliance on people coming forward to speak by themselves; consideration of gendered and environmental impacts of austerity; and strategizing with regard to successful activism in the face of intervention from law enforcement – including ways people can participate that are less at risk of arrest.

These proposals may well generate resistance from some activists. As suggested by interviewees, such objections more often come from those whose – typically, male – position within the organization is among the most secure and comfortable. Indeed, for many of these points to be taken up, it is first necessary to recognize that established methods of organizing have their deficiencies. But if greater integration of these important issues is to occur, enactment of these measures may be an important step in bringing that about.

References

Barrineau, K 2011, "An alternative approach to food justice: A gendered analysis of transition towns in the UK", (https://www.transitionnetwork.org/resources/alternative-approach-foodjustice-gendered-analysis-transition-towns-uk).

Buckingham, S & Külcür, R 2009, "Gendered geographies of environmental injustice", *Antipode*, vol. 41, pp. 659–683.

Buckingham-Hatfield, S 2000, *Gender and the environment*, Routledge, Florence, KY.

Carter, N 2007, *The politics of the environment: Ideas, activism, policy*, Cambridge University Press, Cambridge.

CLES 2014, "Austerity uncovered", (https://www.tuc.org.uk/sites/default/files/TUC%20Final%20Report%20Dec%2714_1.pdf).

Deep Green Resistance 2015a, "About DGR-UK", (http://deepgreenresistance.uk/about-dgr-uk/).

Deep Green Resistance 2015b, "FAQs", (http://deepgreenresistance.uk/faqs/).

Deep Green Resistance 2015c, "Radical feminism frequently asked questions", (http://deepgreenresistance.org/en/who-we-are/faqs/radical-feminism-faqs).

Freeman, J 2012, "The tyranny of structurelessness", (http://flag.blackened.net/revolt/hist_texts/structurelessness.html).

Friends of the Earth 2014, "Why gender equality is necessary for environmental sustainability", (http://www.foe.co.uk/sites/default/files/downloads/gender-equality-environmental-sustainability-22099.pdf).

Haigh, C & Valley, B 2010, "Gender and the climate change agenda", (http://www.wen.org.uk/wp-content/uploads/Gender-and-the-climate-change-agenda-21.pdf).

Hopkins, R 2014, "Is gender an issue in transition?" (https://www.transitionnetwork.org/blogs/rob-hopkins/2014-10/gender-issue-transition).

Kirton, G & Healy, G 2013, "Strategies for union gender democracy: Views of trade union leaders in the UK and the US", *Travail, Genre et Sociétiés*, vol. 2, pp. 73–92.

Ledwith, S 2012, "Gender politics in trade unions: The representation of women between exclusion and inclusion", *Transfer: European Review of Labour and Research*, vol. 18, no. 2, pp. 185–199.

Manchester Climate Monthly 2014, "Meetings are institutionally sexist": Discuss (White-knighting by #Manchester #climate bloke), http://manchesterclimatemonthly.net/2014/05/13/meetings-are-institutionally-sexist-discuss-white-knighting-by-manchester-climate-bloke/.

Murray, R 2011, "Sometimes it's hard to be a woman", (http://www.greenpeace.org.uk/blog/what-you-can-do/sometimes-its-hard-be-woman-20110308).

PCS 2015a, "About PCS", (http://www.pcs.org.uk/en/about_pcs/about_pcs.cfm).

PCS 2015b, "Climate change and women", *PCS Women,* vol. 9, p. 6.

RMT 2015, "Women members", (https://www.rmt.org.uk/about/women-members/).

Stuart, M, Tomlinson, J & Lucio, MM 2013, "Women and the modernization of British Trade Unions: Meanings, dimensions and the challenge of change", *Journal of Industrial Relations*, vol. 55, no. 1, pp. 38–59.

Thiranagama, N 2008, "Closing the gender pay gap: An update report", (http://www.etuc.org/sites/www.etuc.org/files/TUC_Closing_the_Gender_Pay_Gap-2_1.pdf).

TUC 2013, "Research reveals gender stereotyping in apprenticeships", (https://www.tuc.org.uk/equality-issues/research-reveals-gender-stereotyping-apprenticeships).

TUC 2014, "Can we ever build a green economy?" (https://www.tuc.org.uk/events/can-we-ever-build-green-economy).

TUC 2015a, "Green workplaces", (https://www.tuc.org.uk/workplace-issues/green-workplaces).

TUC 2015b, "TUC directory 2015", (https://www.tuc.org.uk/sites/default/files/TUC_Directory_2015_Digital_Version.pdf).

TUC 2015c, "TUC women's committee", (https://www.tuc.org.uk/equality-issues/gender-equality/tuc-womens-committee).

UCATT 2014, "Women in construction newsletter", May, (https://www.ucatt.org.uk/files/publications/Women%27s%20Forum%20Newsletter%20May%20).

UCATT 2015a, "About UCATT", (http://www.ucatt.org.uk/about-ucatt).

UCATT 2015b, "Going Green@Work", (http://www.ucatt.org.uk/going-greenwork).

UNITE 2012a, "United position paper on energy", (http://www.unitetheunion.org/uploaded/documents/Energy201211-3509.pdf).

UNITE 2012b "Women's health, safety and well-being at work", (http://www.unitetheunion.org/uploaded/documents/Women%27s%20Health,%20Safety%20%26%20Well-being%20at%20Work%20%28Unite%20guide%2911-5062.pdf).

Ward, F 2014, "Is gender an issue in transition?" (https://www.transitionnetwork.org/blogs/rob-hopkins/2014-10/gender-issue-transition).

WEN 2015, "About WEN", (http://www.wen.org.uk/about-wen/).

WEN & National Federation of Women's Institutes, 2007, "Women's manifesto on climate change", (http://www.wen.org.uk/wp-content/uploads/manifesto.pdf).

7

TRANSPORTING DIFFERENCE AT WORK

Taking gendered intersectionality seriously in climate change agendas

Leonora C. Angeles

Introduction

Understanding gendered forms of mobility and the linkages between gender, work and transportation issues will help address climate change challenges so that better adaptation measures can be put in place. There has been an increased focused on people's mobility or so-called "mobility turn" or "new mobilities" paradigm that demands attention to combined analyses of mobile bodies, power constellations, identities creation and scalar analysis of micro-geographies that are clearly gendered (Cresswell 2010; Sheller & Urry 2006). Once limited to the grey literature of practitioners, the gender dimension of transport policy and planning has grown in the academic field (Grieco & McQuaid 2012). Yet, while the gender dimension in city urban environments has focused on housing, employment and welfare, transport, or how people move within the cities in the developed world is not given much attention (Schmucki 2012). Thus, this paper focuses mainly on land transport issues and how they intersect with gendered work and climate change at the macro-level in the developed industrialized world. We need to understand how gendered impacts of climate change could affect existing gendered relations, ideologies, lived experiences and division of labour at work in various employment sectors, including work within the transportation sector itself.

Gendered consumer demands, transport use patterns and car driving trends show gendered ecological footprint and contributions to climate change. Cross-fertilizing the hitherto separate bodies of literature gender and climate change, and gender, work and transportation issues entails examining linkages between gendered forms of work, gendered forms of mobility, and climate change challenges, and addressing what it would mean in policy and practice if we take gendered intersectionality seriously in climate change debates and agendas. This chapter demonstrates that while available data on these linkages are still piecemeal, limited and asymmetrical, thus preventing researchers from making robust comparative analyses across the developed, industrial countries, we could draw from multiple city- and country-level case studies to meaningfully incorporate gender analysis within work- and transportation-related climate change debates to guide alternative policy-making. It addresses the questions: Why does gender matter in mobility issues in the context of the gendered impacts of climate change on work? How can we address mobility and transportation issues related to work and employment enriched by gender and climate change analytical lens?

To answer these questions, materials on transportation, work and gender issues were taken from two databases, Academic Search Premier and ScienceDirect, and analysed through a gender and climate change analytical lens. Feminist geographers and planners have uncovered how axes of social differences based on gender, class, race, sexuality and ability shape our urban mobility experiences. Three aspects of gender analysis must be considered when discussing climate change-related issues and impacts on gendered mobility and work. First, we need to consider how and why gender is a relationship, and why studying only women cannot explain everything (Harding 1997). This means that women's mobility and transportation must be considered in relation to men's and children's communities; to work, employment and other aspects of our lived human experience; and to our built and natural environments that are all implicated in climate change impacts. Second, gender is a characteristic not only of people, but also a characteristic of our social structures and symbolic systems (ibid.), seen in how work and employment, transportation and climate change are lived and interpreted involves studying not only the gender of people, but also of things, of transportation means and the gendered discourses on climate change itself. Third, gender and transportation systems are part of our hierarchical social structures, institutions, policies and symbolic systems that interlock with class, caste, race, ethnicity and other hierarchical social relations. Gender could not be discussed in isolation as a stand-alone social dimension, but examined using an intersectionality lens.

The chapter is organized as follows. First, it discusses why transportation and gender matter in work and mobility issues when viewed through an intersectional lens. Second, it explains how can we enrich our understanding and address gendered work, mobility and public transportation issues using climate change and climate resilience as analytical lenses. The last part identifies some key policy and planning options and considerations when bringing gendered intersectional analysis and climate change together in work-related transportation issues.

Why transportation and gender matter in work and climate change debates

This section argues that gender matters in public transportation and mobility issues, given the gendered employment patterns within the transport sector, indicating gendered differential impacts of climate change on work within this sector and beyond. The gender imbalance in employment in the transport sector offers immense opportunities for climate change mitigation and adaptation strategies to incorporate gender analysis. But first, we need to examine its employment significance, establish the gendered profile of this sector and explain its climate change contribution.

Transportation access, a capability deemed important to human development and critical to work and employment, is an indicator of mobility and social inclusion (Nussbaum 2000; Sen 1999). While a significant direct and indirect employer, the transport sector is among the biggest contributors to greenhouse gas emissions, carbon fuel use, and climate change, contributing about 24.5 per cent in Annex I (industrialized) countries and 14.5 per cent in Non-Annex I (less industrialized) countries to worldwide greenhouse gases (GHG). Transport is the second largest sector of GHG emissions in both the European Union and North America. Land transportation accounts for 74 per cent of total transportation to GHG contribution, followed by shipping at 14 per cent, aviation at 11 per cent and rail transport at only 1 per cent. While rapid mass transportation facilities can help mitigate private automobile use, their building, operation and maintenance also require high-energy costs, but lower GHG emissions.

But how did we reach this state of dependence on gasoline-driven modes of transportation and how can gender analysis sharpen our revisiting this social history? Western culture's century-old obsession with cars and motorized vehicles peaked as post-war mass production lowered the cost of car ownership and was incentivized by social re-engineering to accommodate oil-dependent vehicles through public infrastructure of highways and arterial road networks (Schmucki 2012). Thanks in no small part to the alliance between car manufacturing and oil industries, public infrastructure experts, and policy-makers, explained in the documentary films *Who Killed the Electric Car* and *The End of Suburbia: Oil Depletion and the Collapse of the American Dream*. Historically, transport planning in the industrialized world in Europe and the Americas has evolved into four distinct phases with similar ideas emerging in many key cities: (1) the traffic-friendly city with its origins in 1945–55; (2) the automobile-friendly city realized in 1955–71, which showed its "ugly" side effects of congestion and pollution; (3) the continuities and changes in the city-friendly traffic from 1971–80; and (4) the human-friendly city post-1980. Each stage has implications for gender, safety and the urban planning of streets and other public spaces (Schmucki 2012). Road networks link downtown and uptown work areas to increasingly more distant residential home locations, suburban neighbourhoods, industrial-commercial centres, and big box stores, malls and other shopping areas. Public transportation systems became radial, with main routes connecting city centres and employment districts towards the suburbs, whereas intra-neighbourhood transportation routes are more disperse and sporadic (Blumenberg 2004; Dobbs 2007).

The commuter culture of modern industrial societies and post-war emergence of car-dependent suburbia propelled the burning of fossil fuels and the rise of the suburban housewives and their commuter husbands working downtown, and coincided with the social isolation and dissatisfaction of highly educated women's "problem with no name" (Friedan 1963). Some women who do not drive or could not afford private vehicles experience relative exclusion from mobility, or labour participation altogether, due to the pressures for women to return to or concentrate on their domestic roles. For others, particularly senior women, gendered forms of fear and safety concerns in public transport and public spaces encourage them to prolong their car ownership and driving privileges for as long as possible (Li et al. 2012), hence contributing to GHG emissions.

Daily long commutes are also increasingly stressful, contributing to subtle, slow creep, imperceptible transformations in our daily habits as commuters (Bissell 2014, 2015). Public officials and experts have raised major public health concerns over stress-related commutes that affect our bodies (Novaco & Gonzales 2009), particularly elevated blood pressures, and that jeopardize work–life balance, work productivity and overall economic-related risks (Lyons & Chaterjee 2008; Wener, Evans & Boately 2005). Commuting effects on mental well-being and psychological health are noted to have gendered differences in Britain (Roberts, Hodgson & Dolan 2011), Tel-Aviv (Prashker, Shiftan & Herschkovitch-Sarusi 2008) and Sydney (Bissell 2014, Golob & Hensher 2007). Women and working mothers experience more commuting-related stress and health issues, not so much because of their long-distance journey to work, but because of childcare and housework responsibilities (Roberts, Hodgson & Dolan 2011). Higher GHG emissions from the long commutes exacerbate climate change and negatively affect human and environmental health.

Transportation issues intersect with social divides, such as class and racial hierarchies, which interact with gender differences in transport use. Public transportation routes linking city centres to the main employment areas often do not consider the travel needs and patterns of working women looking for more affordable suburban housing. Abrahamson (2005) examined three low-income neighbourhoods – Mannheim in Germany, Nantes in France and York in the UK – and identified single and married low-income mothers' strategies to cope with work and family care-giving demands. In all cases, women strategize when caring for their own and other's children or help with each other's shopping needs and count on relatives' support to address the high costs of rent, transportation and especially childcare. Childcare reliability and quality shape mother's continued engagement in paid work, often within the same neighbourhood. Transportation costs and travel time also put pressure on working women's low-incomes and quality-of-life, but are mitigated by free full-time publicly-funded childcare, as in the case of Nantes, and family and social network support in childcare, as in Mannheim. Low-income mothers were most stressed in York, where there is no subsidized childcare and unregulated labour market wages barely cover childcare and transportation costs (Abrahamson). American, British and Canadian case studies consistently show similar results: women's transit trips are shorter but more numerous than men's, more dispersed destination-wise (counted as a number of trips) and positively correlated to two socioeconomic status measures (education and income) (Aguilar-Zeleny 2012, 28; Hanson & Hanson 1981, 343).

Work and employment related demographics within the transportation indicate gender imbalance and low female labour participation. In the 27-member European Union countries, only 21 per cent of workforce in transport service is female, compared to total 44 per cent share of all services, due to poor employment retention, partly owing to the prevalence of gendered violence against women transport workers. Thus, male over-representation and domination in top decision-making and leadership positions, and front-line service provision, such as bus driving, set the stage for marginalization of women, the poor and low-income seniors. This gendered labour participation suggests that should biophysical limits imposed by climate change cause drastic declines in employment opportunities within the transport sector, more men than women would suffer unemployment and ensuing social and economic dislocation.

Differences among women, and classes of women, may be as important as differences between women and men, and also divide along ages, sexualities, gender orientations and geographies. As Hanson (2010) argues:

> Mobility is not just about the individual . . . but about the individual as embedded in, and interacting with, the household, family, community and larger society. That is, it should be impossible to think about mobility without simultaneously considering social, cultural and geographic context – the specifics of place, time and people. (8)

Data show that more men hold a driver's license than women, who in the industrialized world, and elsewhere, experience greater barriers in transport options than men. The UK's National Travel Survey shows that between 1935 and 2010, two-fifths of women compared to only a quarter of men do not drive, although the proportion is narrowing. More women, regardless of work status, use the public transit more often than men. There are notable gender differences in transport means, distance travel, work trip lengths and complexity that can inform sustainability questions. Women work closer to home and do shorter but more numerous trips than men, who tend to work long-distance or farther away from home. Women also use public transportation and do more non-work travel and make trips with multiple stops, usually accompanied by other passengers, especially children. Women's trips are significantly more complex than those made by men, especially when employment activities are added to their family care activities and desire to include more passengers in their trips (McGuckin & Nakamoto 2005; Rosenbloom 2005). Travel and spatial mobility needs interact closely, not just with work, but gendered work–family arrangements, even for highly skilled immigrants and dual career couples, as shown in a German case study (Shinozaki 2014).

The geographic context of work and transport needs is significant, particularly for women and social groups that have access and availability issues. Women's mobility access and transport choices are limited, particularly in rural areas in both the developed and developing world, where public transport, walking, biking and hitch-hiking remain predominant transport modes among poor populations

(Porter, Blaufuss & Acheampong 2012) and older people (Ahern & Hine 2012). Low-income groups in industrialized countries are twice as likely to use public transport than higher income groups. Women comprise a disproportionate share of the poor in the industrialized world, thus raising intersecting health, safety and mobility concerns especially among racialized populations. The case of missing and murdered women, mostly Aboriginal, along the "Highway of Tears" in Canada's British Columbia Interior was largely linked to racialization and violence against Indigenous women, exacerbated by unreliable rural public transportation. An Australian study shows that female retired seniors, especially caregivers, dominate the "vulnerable" and "impaired" categories experiencing social exclusion in terms of safety and access to public transport (Delbosc & Currie 2011).

Life cycle position shapes gendered mobility, which has climate change implications. Studies on seniors' travel patterns suggest bi-modal tendencies: both women and men seniors are heavy public transport users, especially during off-peak hours, but more male than female seniors tend to be prolonged car owners and drivers. Studies in Norway (Hjorthol & Sagberg 2000), Finland (Siren & Hakamies-Blomqvist 2004), Australia (Golob & Hensher 2007) and the UK (Schmocker et al. 2008) noted seniors' prolonged desire to drive. However, driving cessation among seniors is a gendered phenomenon (Curro 2012; Davey 2007). A Vancouver and Toronto study shows that male seniors hang on to their driver licenses longer than women seniors, who use their social networks and public transport knowledge for continued mobility even when they become widowed or can no longer drive (Curro 2012). Men are less prepared to reduce their dependence on cars as they age while women seniors tend to walk and use public transportation for medical and health trips, even if their households have limited or actual access to a car, as shown in urban Norway (Norbakke 2013) and rural Ireland (Ahern & Hine 2012), where attempts to address differential rural transport deprivation can have contradictory effects (McDonagh 2006). Older drivers with health problems across genders in Scandinavia and Canada were found to experience higher accident risks (Brorsson 1989; Gresset & Meyer 1994). A longitudinal study of male and female seniors, aged 65 years and above, showed that driving cessation occurs with the reduction of social activities outside of the home, regardless of health and sociodemographic conditions (Marotolli et al. 2000). In a German case study, no correlation was found between car ownership, driving and out of home activities among seniors, which suggests public transportation efficiency and accessibility (Scheiner 2006). These studies suggest the lifelong car-driving habits of multiple generations of seniors and some compensatory benefits of driving when health and mobility problems prevent them from walking and taking public transportation (Nordbakke 2013, 167). GHG emissions could get higher the longer people drive during lengthening life expectancies.

City-level research can help provide more nuanced and detailed picture. Using Metro Vancouver's TransLink's Daily Travel Survey (2011), Aguilar-Zeleny (2012) confirms secondary literature findings that, on average, men travel longer distances per day (10.5 km in average) than women (8.3 km), but women take a greater number of trips (3.3) than men (3.04). The picture becomes more complicated when we consider age: female transit trips in the category 20–29 years old are considerably greater than men's due to household maintenance, childcare, and ride-sharing responsibilities. Women and men in the 30–39 age category make more or less the same number of trips but much fewer trips than the younger categories. Women do more active transit trips than men across all age categories. They undertake more but shorter trips by car than men but also use transit and walk more than men, but more men drive the car for longer trip distances. For personal business destinations related to household provisions and childcare, women depend on the car for more multi-purpose and chain trips and longer distances than men. Men, in contrast, use the car more for longer and more trips than women when they travel for work or post-secondary education. Despite increased women's labour market participation, unattached women and female-headed households, urban transportation systems in Metro Vancouver in British Columbia have not catered to gendered needs and travel patterns, particularly of low-income working women.

There are also noted gender differences in the use of active transportation such as walking and biking, which can contribute to lower GHG emissions. Biking in Metro Vancouver has slightly more male participation regardless of destination. An Australian case study shows that more school girls (44.3%) walk to school than boys (37.4%) but fewer bike to school (8.3% vs 22.4%), thus reducing their overall active transport participation (Leslie, Kremer & Toumbourou 2010). But gender difference is narrowing in other countries. Women's bicycle trip shares stand high at 45 per cent in Denmark, 49 per cent in Germany and 55 per cent in the Netherlands (Pucher & Buehler 2008, 504).

Most of these studies point to the hetero-normative and essentialist views of heterosexual women and men, married couples with children and households with working mothers, and do not take into account diverse ethnicities and sexualities, such as LGBTQ seniors or Indigenous homeless youth. Data limitations and neglect of diverse populations' differential situations in these studies point to the ever greater need to link gender, diversity, sustainability and climate change in transportation policy and planning, given the strong association between travel, motorized transport and carbon footprint.

Intersecting gender and transportation through a climate change lens

The challenge of addressing climate change through a gender analytical and intersectional lens in the transport sector is complicated by piecemeal data and lack of understanding of how various policy options and interventions to implement sustainable forms of transport compare with each other in terms of their gendered responses and impacts. As Hanson (2010) argues:

> [S]ustainable mobility will necessarily entail reducing vehicle miles traveled. Among the many and varied approaches to achieving this goal are to: increase vehicle occupancy, reduce number and/or length of trips, shift travel to transit or non-motorized modes, use information technology instead of traveling or change land use patterns to bring destinations closer together. How might each of these options affect understandings and practices of gender? How does social and geographic context affect these processes? And how might gender affect the viability of any of these options in various contexts? At present, we have no idea. (16)

However, there are data available on gendered GHG emissions, demonstrating gender differences in ecological footprints, evidenced by women's overall consumption patterns, their general lack of access to motor vehicles, shorter driving over the lifespan and higher participation rates in public and active transport use, in both historical and contemporary times (Cohen 2015; Hanson 2010; Schmucki 2012). Since gender matters in understanding the causes and impacts of GHG emissions in the transport sector, the question remains: how can gender analysis help us address not only gender specific needs and issues within transportation, but also connect gender-related transportation issues to climate change solutions?

Gender, transport-related mobility, work, social inclusion and climate change should be analysed in terms of their relationships rather than in isolation from each other. Gender, class and other social axes of difference shape social exclusion patterns in transportation issues. Some of the critical inter-related accessibility factors that shape these patterns are mobility, land use and centralization or decentralization of public services; time use, management and constraints; availability of and entitlements to public spaces; and the use of information and communication technologies (Lyons 2003, 341). In turn, these patterns can help us examine which forms of transport-related mobility take place and where, and what embodied mobility or mobile bodies are affected by or contribute to climate change the most, and why.

Climate change impacts exacerbate transport-related access to work and leisure opportunities. A "capability deprivation" is a form of social exclusion. Social inclusion and exclusion should not be confused with poverty, social or physical isolation. It is a multidimensional concept that has animated transport policy, planning and climate change debates due to its association with a variety of things that are significant to human settlements, such as work, livelihoods, employment, health, housing and land use patterns. As one summary of the literature on social exclusion and transportation puts it:

> The transport and land use system can reinforce social exclusion by increasing the *generalized use of costs of travel for persons at risk.* If persons find themselves in circumstances where their home location relative to the transport system (low speeds, large distances to stops and stations, large distances to possible places of work, service and consumption), and their available mobility tools (car ownership/access, public transport season ticket ownership), *increase*

> *these costs* (e.g. the psychologically weighted sum of travel times, out-of-pocket costs and comfort) relative to the population average, then they are less likely to engage in travel and are therefore less likely to participate fully in society. *While low generalized costs of travel do not assure social inclusion, they make it easier to achieve.* (Schönfelder & Axhausen 2003, 273, emphasis added)

The problem in most analyses and policy approaches however is that "transport-related social exclusion is not always a socially and spatially concentrated process", and thus requires a schematic analysis that can create more socially and spatially differentiated conceptualizations of and inter-relationships between accessibility, mobility and social exclusion patterns (Preston & Raje 2007, 51), and by extension, more contextual and textured analyses of climate justice patterns.

Secondary data sources from developed countries demonstrate this differentiation and potential contribution to climate change debates. For example, a Leeds case study shows that not all car owners are affluent and that low-income people, including most working women in minimum wage jobs, require cars to get to work and are most negatively affected by congestion rates, or toll fees charged to drivers (Bonsall & Kelly 2005). A case study in four British locations (Nottingham, Bristol, Barton and Oxfordshire), found areas and population groups (e.g. Nottingham university students, working recent graduates) with high area mobility, high personal mobility and high accessibility that also demonstrate the most deep and negative effects of gentrification, congestion, transport-related or pedestrian accidents, and pollution (Preston & Raje 2007, 156). In contrast, Asian minorities in Bristol have low personal mobility (half of the white population) within high area mobility and accessibility (ibid., 157–158). High-income car owners, especially in Barton and Oxford gated communities, have high personal mobility and accessibility in areas with low mobility due to lack of reliable transportation networks. Seniors in Clifton, Nottingham register low in all three counts of personal mobility, area mobility and accessibility in contrast to teenage boys in Hartcliffe, Bristol, who despite their peripheral location in an area with low area mobility and accessibility, experience high personal mobility due to moped ownership, enabling them to travel to their working class jobs. Young people tend to engage in active transportation such as biking and walking for increased physical activity, given the right environmental and weather conditions (Panter, Jones & Van Shulijs 2008; Schofield, Schofield & Mummery 2005).

Car driving and car purchases preferences are socially differentiated by age, class and gender. Women are increasingly a significant car-buying segment as incomes and work opportunities grow. Women's car buying preferences tend to focus more on safety features than other aspects of vehicular design and performance (Koppel et al. 2008). A Canadian national survey revealed how gender differences in safety and performance priorities are dependent on the driver's age: women across the lifespan and all age groups highly rate safety as important, and this importance increased steadily for male drivers, but decreased slightly for the 55–64 male age group and increased in the 65 above male age group

(Vrkljan & Anaby 2011, 63). A parallel UK study revealed gender differences in considering price, car performance and self-image:

> The price of a car was shown to be important for both male and female buyers, but for different reasons. For male buyers, paying a higher price for a car meant that they could have higher expectations and impress others more, whereas for female buyers a higher price was more important in assuring them that their car would perform as it should. A high level of price consciousness tended to reduce satisfaction and loyalty for both male and female buyers, but to a much larger extent for male buyers. (Moutinho, Davies & Curry 1996, 144)

Significant gender differences exist in transport pattern use, and car-driving preferences have consequent gender differential GHG contributions. Due to men's intense travel patterns and frequent use of single-occupied cars, male energy consumption from transport – and thus, carbon emissions – is much higher than their female counterparts. While we cannot conclusively determine if men in general have bigger eco-footprints than women due to their differential daily caloric requirements and other consumer habits, we can draw some environmental and climate change implications of some of these gender differences.

Gendered transportation choices and transport use patterns demonstrate why transportation-related climate change impacts need gender analysis. For example, more men than women prefer driving pick-up trucks and SUVs, evidenced by advertisements targeting male audiences. Women tend to buy lower-priced cars in the compact and sub-compact segments (Candle 1991). This is perhaps because of their price consciousness (Moutinho, Davies & Curry 1996) and concerns over compact cars' performance in managing multiple but shorter trips on a daily basis. While women's environmental consciousness has not been factored into car buying preferences studies, it is encouraging that women buy compact cars despite their popularly cited reduced familiarity with cars and car technology, which Van Rijnsoever, Farla & Dijst (2009) consider significant in addressing the attitude-behaviour gap in car purchase decisions. Thus, gendered transport consumer choices, particularly car preferences, have implications for popular education on how consumer choices affect environmental and climate change.

While citizens in developed countries are increasingly aware of how their daily consumption habits affect the environment, there is evidence that transport choices and transport-related consumer preferences and behaviours are harder to change. For example, eight out of ten European Union citizens, or 80 per cent, consider that the type of cars people use, and how they use them, have significant local environmental impacts (European Commission 2007), but only 17 per cent are willing to change their car-driving habits (European Commission 2008). The barriers to addressing this attitude–behaviour gap in environmental and climate change concerns require interventions at the critical point of car buying, combining information provision as well as people's actual involvement with cars and car technology to translate positive environmental consciousness and attitudes into consumer behavioural and lifestyle change (Van Rijnsoever, Dijst & Castaldi 2009; Van Rijnsoever, Farla & Dijst, 2009).

Conclusions: Policy and planning considerations

Climate change mitigation through sustainable transport choices and promotion of active transportation means (such as walking and cycling) can only be addressed by addressing urban sprawl and providing affordable housing closer to work places. People's knowledge of local transport systems and travel patterns that can save time and money revolve around their primary residence, suggesting "home location as the *major hub* for daily life travel acting as a central node in the network" (Schönfelder & Axhausen 2003, 279, original emphasis). As significant household decision-makers, women's micro-social practices, competencies and strategies, such as those used in cost-sharing, escorting, negotiating travel transactions, scheduling trips among household members and between households (Hodgson 2012), and intra-household negotiations in car-deficient households (Scheiner & Holz-Rau 2012) are significant to consider in transport decisions' implications to climate change debates. Women remain critical subjects in any campaign and policy effort to maximize public transport use for carbon use reduction. It is the local, temporal-spatial and micro-level scale of our daily transport and consumption practices that will make a difference to our long-term climate change solutions.

Since gender relations have a *material base* that demonstrates critical aspects of access to and control over resources (Harding 1997), changing our cultural ideas and beliefs around the environment and transportation choices is not enough to address climate change. Some might even argue that climate change mitigation is a lost battle, requiring shifts to adaptation measures. While many developing countries in the South are addressing climate change through adaptation plans, industrialized countries have taken on the combined path of mitigation (i.e. reducing GHG emissions), resilience and adaptation (i.e. adjusting to new climate change realities).

Do we know enough about gender and transportation linkages to design good policy on climate change that takes gender into consideration? The chapter demonstrates that using gender lens to reduce GHG emissions requires creating gender-aware and integrated community development, employment, public transportation, housing and land use policies. There are enough conceptual and empirical studies that point in this direction (see Hanson 2010; Wekerle 2005). For example, since reliable, quality childcare is an important factor in mothers' decision to work, cities need to plan more mixed use, walkable neighborhoods with family-friendly amenities linked to places of recreation, worship, work and social services. In the Netherlands, the Groningen City Plan requires childcare facilities to be located near school buildings near housing and public transport systems, a policy enabled by Dutch central government guidance in urban planning (Greed 2005, 744). Since transportation cost and time impose serious resource constraints and affect people's quality-of-life (as time spent travelling), adequate policies regarding public transportation and land use should deal with better accessibility to jobs, affordable housing and reduction of travel time and costs. As Hanson (2010) argues:

> Moving toward sustainable mobility will require improved understanding of how fully fleshed-out gender and fully fleshed-out mobility connect together differently in different contexts. The centrality of context is especially apparent in advancing a sustainability agenda, as the kinds of changes that will be required to move toward sustainability will be different for different places and times, depending, as they will, on particular context-specific knowledge bases and practices. (16)

Increasing rapid transit service and the number of routes to low-income housing areas and between area neighbourhoods would be most welcomed not just by women, but also working men, seniors and youth who deal not just with gender but also other forms of social inequality.

There is a need to examine planning and governance arrangements in addressing social exclusion patterns in public transport (Stanley & Lucas 2008). More than 20 years ago, it was already noted, "transportation policy makers and planners must consider alternative ways of delivering transport services now and in the future. Conventional fixed route transportation services are not responding to the needs of working parents whose day-today travel patterns differ markedly" (Rosenbloom 1989, 87). Planners and policy makers can also address not only inequities in public transport pricing, but also the indirect effects of car congestion charges on road diversion, land use changes due to new transport use patterns that particularly affect environmental degradation in low-income neighborhoods Bonsall & Kelly (2005) and other policies aimed at creating more just cities (Fainstein 2010). Overall, the use of market-based mechanisms, particularly the use of regressive tax policies, to mitigate climate change impacts in the transport sector can have gender differential effects on the poor and working class. This is a huge topic that merits closer examination beyond the scope of this study.

While considerations of gender and transport issues are well developed compared to other policy areas in the industrialized world, there is little systematic implementation of that knowledge in climate-related policies, programmes and other measures within an integrated sustainable transportation, renewable energy, walkable neighbourhoods, affordable housing and green job creation strategy. Our growing cities must stop creating unsustainable, fossil fuel-driven transportation systems and networks that will further increase work–home travel distances and locate human settlement in more remote areas that are geographically separated from many of the daily needs and activity fields of people, particularly women, whose patterns of mobility are much more complex than those of men due to their different gender roles within and outside of the household. There have to be more systematic attempts by governments, civil society organizations, bilateral agencies and multilateral development agencies to incorporate gendered intersectional analysis in transportation issues and in climate change negotiations. Sustainable, publicly-funded mass transportation systems in both urban, peri-urban and rural areas must consider not just gendered and socially differential needs and transport requirements for the support of well-being, health and sustainable livelihoods, but also in terms of their impacts on the environment and climate change. Policy and planning agencies need better interaction and coordination of their work for greater consistency, better implementation, monitoring and evaluation of urban infrastructure development (e.g. public transport); land use (e.g. location of social housing complex, employment centres etc.); social policies (e.g. income assistance, health, unemployment security) and economic policies (e.g. job creation, small business incentives etc.), using gender analysis alongside a climate change mitigation and adaptation lens.

References

Abrahamson, P 2005, "Coping with urban poverty: Changing citizenship in Europe", *International Journal of Urban and Regional Research*, vol. 29, no. 3, pp. 608–621.

Aguilar-Zeleny, P 2012, "Women's mobility and public transportation: Implications for social and urban policy for low-income working class women in Metro Vancouver", Master's Thesis, University of British Columbia, British Columbia.

Ahern, A & Hine, J 2012, "Rural transport: Valuing the mobility of older people", *Research in Transportation Economics*, vol. 34, pp. 27–34.

Bissell, D 2014, "Encountering stressed bodies: Slow creep transformations and tipping points of commuting mobilities", *Geoforum*, vol. 51, pp. 191–201.

Bissell, D 2015, "Virtual infrastructures of habit: The changing intensities of habit through gracefulness, restlessness and clumsiness", *Cultural Geographies*, vol. 22, no. 1. pp. 127–146.

Blumenberg E 2004, "En-gendering effective planning spatial mismatch, low income women and transportation policy", *Journal of the American Planning Association*, vol. 70, no. 3, pp. 269–281.

Bonsall, P & Kelly, C 2005, "Road user charging and social exclusion: The impact of congestion charges on at-risk groups", *Transport Policy*, vol. 12, pp. 406–418.

Brorsson, B 1989, "The risk of accidents among older drivers", *Scandinavian Journal of Social Medicine*, vol. 17, no. 3, pp. 253–256.

Candle, J 1991, "Woman car buyer: Don't call her a niche any more", *Advertising Age*, vol. 62, no. 3, pp. 58–59.

Cohen, MG 2015, "Gendered emissions: Counting greenhouse gas emissions by gender and why it matters", in C Lipsig-Mummé & S McBride (eds), *Work and the challenge of climate change: Canadian and international perspectives*, pp. 59–81, McGill-Queen's Press, Montreal.

Cresswell T 2010, "Mobilities I: Catching up", *Progress in Human Geography*, vol. 35, no. 4, pp. 550–558.

Curro, M 2012, "Seniors' perceptions around driving cessation: A multi-ethnic, multi-cultural perspective", Master's Thesis, University of British Columbia, Vancouver.

Davey, J 2007, "Older people and transport: Coping without the car", *Ageing and Society*, vol. 27, no. 1, pp. 49–65.

Delbosc, A & Currie, G 2011, "Transport problems that matter: Social and psychological links to transport disadvantage", *Journal of Transport Geography*, vol. 19, pp. 170–178.

Dobbs L 2007, "Stuck in the slow lane: Reconceptualizing the links between gender, transport and employment", *Gender, Work and Organization*, vol. 14, no. 2, pp. 85–108.

European Commission 2007, *Attitudes on issues related to EU transport policy*, Directorate-General for Energy and Transport, Brussels.

European Commission 2008, *Attitudes of European citizens towards the environment*, Directorate-General for the Environment, Brussels.

Fainstein, S 2010, *The just city*, Cornell University Press, Ithaca and London.

Friedan, B 1963, *The feminine mystique*, W.W. Norton, New York.

Golob, TF & Hensher, DA 2007, "The trip chaining activity of Sydney residents: A cross-section assessment by age group with a focus on seniors", *Journal of Transport Geography*, vol. 15, no. 4, pp. 298–312.

Greed, C 2005, "Overcoming the factors inhibiting the mainstreaming of gender into spatial planning policy in the United Kingdom", *Urban Studies*, vol. 42, no. 4, pp. 719–749.

Gresset, J & Meyer, F 1994, "Risk of automobile accidents among elderly drivers with impairments or chronic diseases", *Canadian Journal of Public Health*, vol. 85, no. 4, pp. 282–285.

Grieco, M & McQuaid, R (eds) 2012, "Gender and transport: An editorial introduction", *Research in Transportation Economics*, vol. 34, pp. 1–2.

Hanson, S 2010, "Gender and mobility: New approaches for informing sustainability", *Gender, Place and Culture*, vol. 17, pp. 5–23.

Hanson, S & Hanson, P 1981, "The travel-activity patterns of urban residents: Dimension and relationships to sociodemographic charateristics", *Economic Geography*, vol. 57, no. 4, pp. 332–347.

Harding S 1997, "Just add women and stir?" in Gender Working Group (eds), *Missing links: Gender equity in science and technology for development*, pp. 295–308, UN Commission on Science and Technology for Development, Ottawa.

Hjorthol, R & Sagberg, F 2000, "Introductory report: Norway", in B Rosenbloom (ed.), *Transport and ageing of the population*, p. 177, Economic Research Centre, Paris.

Hodgson, F 2012, "Escorting economies: Networked journeys, household strategies and resistance", *Research in Transportation Economics*, vol. 34, pp. 3–10.

Koppel, S, Charlton, J, Fildes, B & Fitzharris, M 2008, "How important is vehicle safety in the new vehicle purchase process?" *Accident Analysis and Prevention*, vol. 40, no. 3, pp. 994–1004.

Leslie, E, Kremer, P & Toumbourou, J 2010, "Gender differences in personal, social and environmental influences on active travel to and from school adolescents in Australia", *Journal of Science and Medicine in Sport*, vol. 13, no. 6, pp. 195–201.

Li, H, Raeside, R, Chen, T & McQuaid, R 2012. "Population ageing, gender and the transport system", *Research in Transportation Economics*, vol. 34, pp. 39–47.

Lyons, G 2003, "The introduction of social exclusion in the field of travel behavior", *Transport Policy*, vol. 10, pp. 339–342.

Lyons, G & Chaterjee, K 2008, "A human perspective on the daily commute: Costs, benefits and trade-offs", *Transport Reviews: A Transnational Transdisciplinary Journal*, vol. 28, no. 2, pp. 181–198.

McDonagh, J 2006, "Transport policy instruments and transport related social exclusions in the rural Republic of Ireland", *Journal of Transport Geography*, vol. 14, pp. 355–366.

McGuckin, N & Nakamoto, Y 2005, "Differences in trip chaining by men and women", *Research on women's issues in transportation*, vol. 2, pp. 49–56, Transportation Research Board, Washington, D.C.

Marottoli, RA, Mendes de Leon, CF, Glass, TA, Williams, CS, Cooney, J, LM & Berkman, LF 2000, "Consequences of driving cessation: Decreased out-of-home activity levels", *Journal of Gerontology*, vol. 558, pp. 334–340.

Moutinho, L, Davies, F & Curry, B 1996, "The impact of gender on car buyer satisfaction and loyalty", *Journal of Retailing and Consumer Services*, vol. 3, no. 3, pp. 135–144.

Nordbakke, S 2013, "Capabilities for mobility among urban, older women: Strategies, barriers, options", *Journal of Transport Geography*, vol. 26, pp. 166–174.

Novaco, R & Gonzales, O 2009, "Commuting and well-being", in Y Amichai-Hamburger (ed.), *Technology and well-being*, Cambridge University Press, Cambridge.

Nussbaum, MC 2000, *Women and human development: The capabilities approach*, Cambridge University Press, Cambridge.

Panter, J, Jones, AR & Van Shulijs E 2008, "Environmental determinants of active travel in youth: A review and framework for research", *The International Journal of Behavioral Nutrition and Physical Activity*, vol. 5, no. 1, pp. 34–43.

Porter, G, Blaufuss, K & Acheampong, FO 2012, "Gendered patterns of IMT adoption and research: Learning from action research", *Research in Transportation Economics*, vol. 34, pp. 11–15.

Prashker, J, Shiftan, Y & Herschkovitch-Sarusi, P 2008, "Residential choice location, gender and the commute trip to work in Tel Aviv", *Journal of Transport Geography*, vol. 16, no. 5, pp. 332–341.

Preston, J & Raje, F 2007, "Accessibility, mobility and transport-related social exclusion", *Journal of Transport Geography*, vol. 15, no. 3, pp. 151–160.

Pucher, J & Buehler, R 2008, "Making cycling irresistible: Lessons from the Netherlands, Denmark and Germany", *Transport Reviews*, vol. 28, no. 4, pp. 495–528.

Roberts, J, Hodgson, R & Dolan, P 2011, "'It's driving her mad': Gender differences – the effects of commuting on psychological health", *Journal of Health Economics*, vol. 30, no. 5, pp. 1064–1976.

Rosenbloom, S 1989, "Trip chaining behaviour: A comparative and cross cultural analysis of the travel patterns of working mothers", in M Grieco, L Pickup & R Whipp (eds), *Gender, transport and employment: The impact of travel constraints*, pp. 75–87, Gower, Aldershot, UK.

Rosenbloom, S 2005, "Women's travel issues: The research and policy environment", in SS Fainstein & LJ Servno (eds), *Gender and planning: A reader*, pp. 235–255, Rutgers University Press, New Brunswick, NJ.

Scheiner, J 2006, "Does the car make elderly people happy and mobile? Settlement structures, car availability and leisure mobility of the elderly", *European Journal of Transport and Infrastructure Research*, vol. 6, no. 2, pp. 151–172.

Scheiner, J & Holz-Rau, C 2012, "Gender structures in car availability in car deficient households", *Research in Transportation Economics*, vol. 34, pp. 16–26.

Schmocker, JD, Quddus, MA, Noland, RB & Bell, MGH 2008, "Mode choice of older and disabled people: A case study of shopping trips in London", *Journal of Transport Geography*, vol. 16, 4, pp. 257–267.

Schmucki, B 2012, "'If I walked on my own at night I stuck to well lit areas': Gendered spaces and urban transport in 20th century Britain", *Research in Transportation Economics*, vol. 34, no 1, pp. 74–85.

Schofield, G, Schofield, L & Mummery, K 2005, "Active transportation: An important part of adolescent physical activity", *Youth Studies Australia*, vol. 24, no. 1, pp. 43–47.

Schönfelder, S & Axhausen, KW 2003, "Activity spaces: Measures of social exclusion", *Transport Policy*, vol. 10, pp. 273–286.

Sen, AK 1999, *Development as freedom*, Oxford University Press, Oxford.

Sheller, M & Urry, J 2006, "The new mobilities paradigm", *Environment and Planning A*, vol. 38, pp. 207–226.

Shinozaki, K 2014, "Career strategies and spatial mobility among skilled migrants in Germany: The role of gender in the work-family interaction", *Tijdschrift voor Economische en Sociale Geographie*, vol. 105, no. 5, pp. 526–541.

Siren, A & Hakamies-Blomqvist, L 2004, "Private car as the grand equalizer? Demographic factors and mobility in Finnish men and women aged 65+", *Transportation Research Part F: Traffic Psychology and Behavior*, vol. 7, no. 2, pp. 107–112.

Stanley, J & Lucas, K 2008, "Social exclusion: What can public transport offer?" *Research in Transportation Economics*, vol. 22, pp. 36–40.

TransLink 2011, "Daily travel survey 2010, Preliminary data, Planning Department", Vancouver, British Columbia.

Urry J 2000, *Sociology beyond societies: Mobilities for the twenty-first century*, Routledge, London.

Urry J 2007, *Mobilities*, Polity Press, Cambridge.

Van Rijnsoever, FJ, Dijst MJ & Castaldi, C 2009, "Leaving no stone unturned: The determinants of information channel use during the automobile purchase process", ISU Working Paper Series No. 09–06, Utrecht University, The Netherlands.

Van Rijnsoever, FJ, Farla, J & Dijst, MJ 2009, "Consumer car preferences and information search channels", *Transportation Research Part D*, vol. 14, pp. 334–342.

Vrkljan, B & Anaby, D 2011, "'What vehicle features are considered important when buying an automobile?' An examination of driver preferences by age and gender", *Journal of Safety Research*, vol. 42, pp. 61–65.

Wener, R, Evans, G & Boately, P 2005, "Commuting stress: Psychophysiological effects of a trip and spillover into the workplace", *Transportation Research Record*, vol. 124, pp. 112–117.

Wekerle, GR 2005, "Gender planning in public transit: Institutionalizing feminist policies, changing discourse, and practices", in SS Fainstein & LJ Servon (eds), *Gender and planning: A reader*, pp. 275–295, Rutgers University Press, New Brunswick, NJ.

8

THE US EXAMPLE OF INTEGRATING GENDER AND CLIMATE CHANGE IN TRAINING

Response to the 2008–9 recession

Marjorie Griffin Cohen

Green job initiatives instituted by governments, either in training or green job creation, that specifically include women are not prominent in practice and are highly contingent on policies geared towards the creation of green jobs in the first place. But their prevalence is also related both to the extent that gender issues are a normal part of government policy and to a government's willingness to make "gender mainstreaming" something that is effective.[1] Some developed counties specifically include women in programmes that target labour when green jobs are the focus of policy.[2] The best of these gender-inclusive policies include training for sustained work, but the focus is too often predominantly informational or centres on giving women only a taste of what work in a particular trade might mean. Usually multiple objectives are attempted, such as concentrating on women at risk or with multiple barriers to workforce entry. Rarely are the programmes tied to work initiatives that lead to continuing employment and the programmes are of short duration, funded more modestly than programmes for males, and are not strongly related to action on climate change.

Part of the problem stems from inertia on green jobs altogether. Job creation of this kind by governments is seldom a priority, except when major projects to build infrastructure are undertaken. The problem is that the "green" components of these jobs are not strong, so the effects are rather limited. One potential exception to this was related to the massive amounts of money governments used to rescue economies during the Great Recession of 2008–9. As this chapter will show, although unprecedented government "stimulus" actions were undertaken, the impact on both green jobs and the integration of women into a green workforce were exceptionally modest. Funding required to stimulate the economy through "green" economic initiatives, in both Canada and the US was primarily directed at "green" corporate activity, with a small proportion focused on job-related programmes for green employment. Table 8.1 shows what proportion of the rescue funding in select countries went towards green initiatives of any kind.

TABLE 8.1 Examples of economic recovery programmes' funds for green activities

Country Programme	*% of total stimulus funds for Green Initiatives*
France Economic Revival Plan	21.2%
American Recovery and Investment Plan	12.0%
Canada Economic Action Plan	8.7%
UK Recovery Plan	6.9%
Australia Nation Building and Jobs Plan	9.3%

Source: ILO (2011).

In some respects, the amounts indicated are misleading. For example, in Canada the Economic Action Plan provided $1 billion (only three-quarters of which was actually spent) for a Green Infrastructure Fund that operated on a cost-share basis with the provinces. In British Columbia, for example, this went towards building a large electrical transmission line to support a new mine in a remote area. Because electricity is hydro-based, this was considered a "green" initiative. There was no plan for including women in the jobs created and no green job training involved.[3]

The situation was somewhat different in the US, where the *American Economic Recovery and Reinvestment Act* (*ARRA*) 2009 supported projects aimed at the employment of women in green jobs. Of the total expenditure of about $831 billion, about $27.2 billion was devoted to energy efficiency and renewable energy research and investments, and $3.95 billion towards jobs and training. Of this, a small proportion, $500 million, went to projects to prepare workers for careers in renewable energy or energy efficiency-related jobs. A $100 million fund was allocated to the US Department of Labor (2010a) for 25 special projects (Energy Training Partnership Grants), including $28 million going to communities impacted by auto industry restructuring and $5 million to fund the BlueGreen Alliance whose focus was unemployed steel workers. While the bulk of the training in the 25 projects was for males, the US Department of Labor's Women's Bureau (WB) (2010b) developed ten projects to train women for green jobs.

Most of the grants awarded related to apprenticeship training and employment in non-traditional occupations. This area has been a focus of the WB since the US passed *The Women in Apprenticeship Occupations and Nontraditional Occupations Act* (WANTO) in 1992 and provided funding intended to help women access high-paying jobs in fields that are traditionally male-dominated (*US H.R.3475* 1991–1992). The assumption in pursuing activities primarily in the area of non-traditional jobs for women is that the emergence of a "green sector" in the economy, particularly as a result of the efforts of the ARRA's influx of new funds for green jobs, would see a surge in demand for labour in areas that are outside normal employment for women. These were expected to be primarily in the skilled trades in the building sector and in the burgeoning clean energy sector. The major roadblock to women's participation in the "green sector" is that most of the jobs would be related to the skills trades, an area that over the years has been notorious difficult for gender integration. In both the US and Canada the proportion of women in this labour force has been stuck at about 4 per cent. Similarly, the energy sector is primarily a male sector of employment with women's jobs in the sector mostly confined to support activities (Cohen & Calvert 2013). Despite the record of poor female employment in these male dominated sectors, very specific programmes in the US, like WANTO, recognized that both labour shortages in the skilled trades and women's need for decent well-paid employment should provide opportunities for women (Hegewisch et al. 2013). As will be seen in a later section, some excellent and effective programmes have succeeded for a short period of time, but are frequently abandoned when the short-term objective has been met or there is a change of government and gender integration of the skilled trades is no longer an objective.

Women in green sectors and jobs

The US Bureau of Labor Statistics defines green jobs as including two different types of employment. One part of the definition counts green jobs as "jobs in businesses that produce goods or services that benefit the environment or conserves natural resources", and the other focuses more on the work itself by including "jobs in which workers' duties involve making their establishment's production processes more environmentally friendly or use fewer resources" (US Department of Labor 2010b). These are broad inclusive definitions which would include women in green employment while doing typically female-employment-type work, or which include any workers (even in dirty industries) who are engaged in some kind of work to improve the company's environmental record. This could, for example, include people working in public relations who "green wash" the company. But the usual understanding of green jobs is blue-collar jobs that would involve manual labour in businesses that improve environmental quality. The assumption is that this new category of work would be in areas that were poised for dramatic growth, some in areas requiring skilled labour, but for the most part would provide high quality jobs with relatively low barriers to entry (Stevens 2009). As a result, they would be a good fit for training low-income men and women and give them access to meaningful work with living wages and advancement opportunities (Pinderhughes 2007, 3).

Many of the jobs associated with this type of employment would, indeed, require fairly little training. These have been identified in areas such as bicycle repair and bike delivery services, organic food production, hauling and reuse of construction and demolition materials, hazardous material clean up, green landscaping, non-toxic household and industrial cleaning, parks maintenance, public transit jobs, recycling tree cutting and pruning, peri-urban/urgan agriculture, and weatherization.[4] Women's training for green jobs tends to be in these areas while male training is for the jobs requiring more skills. These would include energy retrofits, furniture making from certified wood, green building, manufacturing jobs related to large scale production of appropriate technologies, water retrofits and whole home performance (USBLS 2012).

Programmes for women

Many of the green training programmes provided through the Department of Labor Employment and Training Administration (ETA) and the Women's Bureau included women in one of their wide range of target groups. Ten of the many programmes were specifically designed to inform women about green jobs, facilitate their access to green jobs, or actually give them training for employment. All of these women-focused programmes were designed to be short-term, but often built upon existing programmes that had a track record in either educating women for work or providing training programmes.

Since this US experience was one of the few concentrated efforts to include females in green jobs initiatives, it is an important project to examine in order to understand what did and could work to get women employed in green jobs. The information used in this chapter about the programmes in the US was accessed through documents, both published and unpublished, and through interviews with key players in individual programmes. These interviews with key informants involved in five different programmes (out of ten) took place between January and May 2013. While clearly the programmes were intended to be limited in both scope and timing, the point of this chapter is not simply to state the obvious, that well-funded, full-scale programmes that have strong job placement aspects are what is needed, but rather to focus on what insights can be gained about what worked best, what needs to be flagged as problems, and what types of programmes have the potential to actually integrate women into areas that have been traditionally male-dominated. This is especially important because these programmes did tend to have as their target women who had multiple barriers to labour force participation. These programmes are assessed in this chapter both through the programme objectives and their ability to relate their programme design to actual employment. It needs to be noted that no programmes have had follow up analyses of participants, so this aspect of the assessment simply does not exist.

Types of programme

The grants to specific groups in ten different regions of the country centred on activities related to information dissemination, network mobilization, outreach and training programmes for women. The list below describes the ten pilot projects that became operational through the Women's Bureau. The funding was on average about $60,000 a year for each group, and often those groups receiving the funding were expected to match it. These were small-scale projects of limited duration that primarily focused on reaching a small number of unemployed and dislocated workers and those with other barriers to employment. They also tended to be related to employment in non-traditional types of employment in construction.

The actual training programmes were for jobs in industries where training could be accomplished rather quickly. So, for example, the "Step up to green carpentry" programme offered through Vermont Works for Women provided unemployed or

TABLE 8.2 Think women in green jobs: US Women's Bureau projects

Region I – Boston: Vermont Works for Women in Burlington, Vermont, developed an on-the-job training programme for women in the fields of green construction, renewable energy and energy efficiency. The programme provided unemployed and underemployed women with skills related to the installation of solar tracking systems, weatherization, window and door replacement, equipment operation, and energy auditing.
Region II – New York City: Sustainable South Bronx (SSBx) is a community organization dedicated to Environmental Justice solutions that are informed by community needs. SSBx created a full-time, hands-on job training programme for women in the fields of green roofing and urban agriculture and horticulture which trained and certified women in green roof design, installation and maintenance; landscaping; hazardous waste cleanup; and related specialties. The training programme included a six-week internship, as well as formal mentoring.
Region IV – Atlanta: The "Women Going Green" project, contracted with 3D Management Enterprise, Inc., to design and implement a 36-week training programme to educate women on the diversity of career paths available in green industries, including opportunities in green entrepreneurship. This included a curriculum covering energy efficiency and renewable energy industries, developing a green business plan, and financial literacy and small business skills.
Region V – Chicago: Detroiters Working for Environmental Justice (DWEJ) in Detroit, Michigan, increased the participation of, and support for, women in its existing green jobs training programme through a 14-week training programme in green careers. DWEJ recruited and provided women with training in lead, asbestos and mould remediation; energy audits and retrofitting; deconstruction; and environmental site assessment; as well as training for HAZWOPER certification and OSHA10-hour construction certification.
Region VI – Dallas: Austin Community College (ACC) focused on increasing the participation of women in its energy efficiency and renewable energy industry training programmes, including outreach strategies recruitment materials designed expressly to target women. ACC offered two sections of its entry-level solar photovoltaic installer course taught by women instructors, for women, in spring and summer of 2010.
Region VII – Kansas City: The YWCA of Greater Kansas City, YWomen CAN (Career Action Network) and Employ Direct developed a training programme to increase women's knowledge of the types of green jobs available and the skills required for those jobs. The training programme offered energy auditor training, environmental abatement training and environmental remediation training. Participants who received training or who were looking at starting a business were connected with mentors.
Region VIII – Denver: The Alliance for Sustainable Colorado created a "Green Jobs Pipeline for Women in Colorado" by designing an outreach and recruitment model to increase the number of women aware of green jobs and the skills needed to prepare for a green career. The Pipeline included a consortium of organizations – businesses and professional associations, environmental organizations, government agencies and educational institutions – that had access to training programmes in cities and rural areas throughout Colorado. The Pipeline model centred on technical assistance for the recruitment and retention of women (such as linkages to workforce systems, mentoring, changes in the training programmes etc).

(continued)

TABLE 8.2 *(continued)*

Region IX – San Francisco: Women in Nontraditional Employment Roles (WINTER) is a non-profit organization designed to encourage and support women and youth's training, education, employment and retention in high-wage, high-skill jobs. The "WINTERGreen" training component was designed to assist women in entering a pre-apprenticeship and environmental education training programme that would lead to a Green Building Certificate and other industry-recognized certificates. The 11-week training programme included instruction on safety, HAZWOPER, asbestos abatement and green building components.

Region X – Seattle: Oregon Tradeswomen, Inc. (OTI), in Portland, Oregon, recruited and trained women to earn a green industry-recognized credential or certification, and assisted the women in identifying employment and apprenticeship opportunities. The training included a pre-apprenticeship program, HAZWOPER classes and construction safety classes. The green building curriculum was made available for other community-based organizations to use, free of charge.

Source: Adapted from Women's Bureau (USDL-WB, n.d.).

under-employed women with a six-week training course in carpentry "soft skills". These were acquired through on-the-job training in green construction where the trainees would participate as "job shadows", mostly in residential construction. The training programme included work to acquire skills in weatherization, window and door replacement, equipment operation, energy auditing and installation of solar tracing systems. While 80 per cent of the women who went through the programme were placed in jobs when they initially left the programme, when the federal funding ran out the programmes were not replaced. This was one of several programmes provided by Vermont Works in conjunction with other partners, including businesses and other non-profits. The main problem, according to those working with this organization, is that while there was a great deal of funding both through the state and federal governments for weatherization training, there was not sufficient demand for workers. The programme was initiated at a time when households were experiencing job losses, so without specific incentives from the government to create demand for weatherization, there was simply no business model for firms to hire workers. In this programme the problem was not lack of interest from women, but lack of jobs to go to once the training was complete.

The training initiatives through existing educational institutions were different in that they were intended to increase women's participation in already existing trades-related programmes. The project at Austin Community College (ACC) received a $60,000 grant to double female enrolment in its training for green jobs. Males normally dominated these training programmes, with women accounting for only 10 per cent of the participants. The main project for women at ACC was a solar programme that trained people in installation and how to price and configure a system. It trained 22 people in three separate courses that lasted six or twelve weeks. While the course was called "solar for women", men were able to join, and women accounted for 75 per cent of enrolment. Most of these were older women in their forties and fifties. About one-half of the women made it through the programme and several sat for the national examination. Overall the ACC programme was considered successful and ultimately as a result of this overture to women there was an overall increase of 2 per cent in female participation in ACC's training programmes. Two issues were cited by key people involved in the programme as accounting for the programme's success. One was the need for appropriate marketing to get women's attention. Images such as "Green Rosie" were important, but so too were "bling" handouts (small bits of jewelry) on recruitment days.[5] Also very significant was the use of female instructors in a team for the entry-level course. The programme was considered so innovative that it was proposed as one of those focusing on women to be highlighted in a State of the Union address by the President. Ultimately this did not happen, primarily because those running the programme felt the honour was premature.

The main problem with the programme was that those completing it struggled to find jobs. The local electrical company had instituted a stimulus programme that included incentives for solar energy, but this was eliminated just as the women's programme was beginning. As was noted by the programme's organizers, timing is everything, and signals to consumers are crucial. The price of solar installations decreased considerably, so the utility felt the subsidies were no longer necessary. But this sent a negative signal to consumers and demand plummeted. Another issue that is important was the distinction between how male and female training was funded. The programmes for males at ACC were funded through a $4.4 million Department of Energy grant through the International Brotherhood of Electrical Workers (the BlueGreen Alliance) for solar school training and there was no cost to the males attending the programme. In contrast, the programme for women funded through the USDL-WB required each participant to pay $625 for a course.

Oregon Tradeswomen, like Vermont Works for Women, has a long history of providing programmes for women's employment in the skilled building trades. The infusion of cash from the USDL-WB from the economic recovery funding enabled them to include a green jobs component in their pre-apprenticeship programme. The programme ran for seven weeks and its green component included aspects of weatherization such as solar installation, caulking and window replacement. Each year there were four classes of 24 students in the programme, and about 75 per cent graduated and 64 per cent were placed in jobs after completion. This is particularly significant when about 95 per cent of the participants could be considered disadvantaged, in that they typically had incomes of between 50 and 80 per cent of the median family income. The programme worked very hard at placing students in apprenticeship programmes in both the private and public sector and the success of the programme is clearly related to getting women into full apprenticeship programmes. But specifically targeting women and integrating them into the training programmes is especially important. As the programme administrators pointed out, a critical mass of women is absolutely crucial for any programme in the trades and it just does not work if there are token women in them. The main problem with the green jobs component, as was evident elsewhere as well, is that there was little demand for workers in this area, so continuous employment for the participants was unlikely.

Another type of programme related to the skilled building trades focused primarily on outreach and recruitment. This was the "Green Jobs Pipeline for Women in Colorado" that was created by the Alliance for Sustainable Colorado. Its main goal was to increase the number of women aware of green jobs and the skills needed for a green career and to focus on recruitment and retention of women. The "pipeline" itself was an infrastructure network that connected a wide variety of organizations that could lead women into green jobs. These included organizations with information about job openings, education and training, developing mentors for women, potential employers and career placement services. The organization itself did not train women, and once the funding ended, so did the programme. This group, like others, found that the main problem was that there were few jobs for women once they were trained for "green" employment. Again, the issue of how the government stimulates demand for environmental upgrades is crucial for employment. One person associated with this project said that if every house sold was required to have retrofits that are up to "green" code, there would be no lack of available jobs.

The Sustainable South Bronx programme was slightly different from other programmes in that it focused more heavily on urban agriculture and horticulture, although there was some building-related training for green roofing. The programme had eight females participating in a 12-week programme, a pilot programme that was not repeated. The participants were recruited through another programme this group had undertaken (Best-Eco programme), which gave training for gardening on roofs, river restoration and culling invasive species. About 80 per cent of the women in the programme were living on welfare and were given $150 a week to participate. The main training was in green roof installation, tree care, tree identification and community garden work. All but two of the women found work after the training and a great deal of attention was paid to job placement. It is not clear how long they remained employed or whether employment continued after the first placement, but the Sustainable South Bronx programme itself devotes considerable energy to job placement, with two of its total staff of ten devoted to this. When the federal funding for the women-only programme ended, the training also ended.

Best practices or just practices

Several assumptions about green jobs seemed to guide the selection of which programmes for women would receive the economic recovery grants. One was that environmental issues would be a major driver in creating jobs in the future, and that programmes with a focus on women could encourage women's participation in green jobs creation. The areas given priority as green jobs were primarily blue-collar jobs in businesses that had some focus on green practices. The funders also wanted training in areas that would have relatively low barriers to entry, could potentially lead to relatively well-paying work, and where skills could be acquired inexpensively in a fairly short period of time. These requirements seemed to fit training for secondary industries of construction, manufacturing and energy production; however in the ten projects associated with the WB funding, the emphasis was most decidedly on construction jobs. The major question is whether these pilot projects are an approach that could or should be reproduced on a larger scale, and whether they are the most appropriate type of work for the populations of women that were targeted for training.

The target population was women, but in most projects there were also attempts to focus on women who had multiple barriers to labour force participation and in some instances this was the population of the entire training group. These were women who exhibited one or more of the following characteristics: they were chronically unemployed, on welfare, had dependents so needed regular work times, were visible minorities or immigrants, had criminal records and/or were older women. This focus on disadvantaged groups is fairly typical in government-funded job-related programmes, usually because these groups are especially affected by economic recessions and it is felt that when public funding is available the most disadvantaged should have priority. The problem is expecting women who are already disadvantaged in some way to also cope with the barriers that exist in the male-dominated sectors.

There are multiple conclusions that can be drawn from the US initiatives in integrating women into green jobs through short-term programmes. Two major problems are evident in training women for green jobs as these are currently defined, that is, as jobs primarily in construction industries. The first is related to the chronic discrimination women historically have experienced in this sector. Programmes to get women with multiple barriers to any type of employment into a sector that exhibits a high degree of discrimination based on race and gender are not likely to meet with success. The second major problem is that for most of the programmes the skills acquired were fairly minimal. As one women interviewed for this study who ran a programme said, "green training doesn't help women get a construction job". This certainly does not mean that disadvantaged women should not be the focus for green jobs, or that major efforts to integrate women into construction industries should not occur. But there is a big difference between training for skills trades for women (which absolutely must be done) and thinking that the minimal training for green jobs is the route through which this can be successful.

Related to these two major issues was the economic climate in which the limited training occurred. Every programme that trained for weatherization or solar installation jobs found the training aspects successful. Women learnt what to do, but the major problem was the failure to find ongoing employment when the training was complete. Because of the recession and in many cases the cancellation of energy efficiency incentives for households, the supply of workers outstripped demand for household retrofits.

There were some consistencies in the experiences of the projects for women across the country. Almost all reported "success", in that women were attracted to the project, and were eager to participate. The training seemed to be strongest when it provided training for a demonstrable skill and when those doing the training were either women themselves, or were specifically trained to deal with women. In some projects the trade unions were helpful in helping place women in work. But generally the overall likelihood of a programme leading to employment was tied to the equity language in the state where the training occurred and whether the government itself was stable. Virtually all programmes recognized the necessity of a strong buy-in by government and a willingness to insist on equity in hiring for government sponsored construction projects. As has been noted above, the willingness to link public policy to stimulate demand for workers who are being trained is especially important during an economic downturn. But in virtually all cases the programmes that were specifically designed to focus on job placement and follow-up were more likely to report good outcomes for the people who were trained or mentored.

Given the limitations of the programmes, the expectations that the training women were given would lead to green jobs and perhaps even encourage women to participate in more ambitious apprenticeship programmes were probably unrealistic. But, once again, the willingness of women to try to break the barriers to entry in construction-type jobs needs to be encouraged. A more serious consideration, however, is what would be considered a more realistic type of training for women who face multiple barriers to job placement. Ultimately the idea of what constitutes a "green job" needs to be expanded to include the type of work that women typically do. Much of it is inherently "green", and requires skilled workers who could be trained through government funded green projects.

Conclusions

Public policy potential for success is high when green jobs are defined broadly and resources are committed to gender inclusion. Policy failure for gender inclusion is more likely when green jobs are defined narrowly to focus primarily on greening "dirty" industries. This does not mean that programmes should not exist in this area, but that their likelihood of "success", in that the pilot projects will lead to sustained employment and increased diversity in the labour force, does not occur unless a broad ongoing commitment is made. The point is not to negate the significance of gender consideration in all types of public policy programmes, but to consider what works, what does not and how gender considerations in climate change labour programmes could provide an impetus for a truly green economy (Cohen 2003).

When a US government committee held hearings about the effectiveness of the green jobs programmes, Labor and Energy Department officials indicated that a large proportion of the funding intended for green jobs (45%) had not been spent, primarily because so few projects were "shovel ready". Of the money earmarked for training ($500 million) only about half of had been spent three years after the ARRA was passed. Among a total of about 47,000 men and women enrolled in training, 26,000 completed it and 8,000 found jobs. Of that 8,000 only 5,400 were employed six months later (Furchtgott-Roth 2012).

Although perceived as a new type of employment, the actual nature of "green" work tends to be caught in the history and inertia of unequal gendered divisions of labour. Some jobs related to a green economy are new, but most rely heavily on traditional types of work. This means that the barriers that currently exist for women to participate in male-dominated industries need to be acknowledged in the planning that occurs for green production and distribution.

Notes

1 "Mainstreaming involves ensuring that gender perspectives and attention to the goal of gender equality are central to all activities – policy development, research, advocacy/dialogue, legislation, resource allocation, and planning, implementation and monitoring of programmes and projects" (UN 2013).
2 This is especially interesting in Finland and Spain and is related to the support these countries have received from the European Social Fund (ESF) (European Commission 2007).
3 Little has been done on gender and green jobs in Canada. One exception is Joan McFarland's research on Nova Scotia (McFarland 2013).
4 The US Bureau of Labor Statistics (USBLS) lists 333 detailed industry groups where green goods or services producing industries exist. It is from among these industries that green jobs are identified (US Department of Labor Bureau of Labor Statistics 2012).
5 "Green Rosie" is a take on the World War II Rosie the Riveter posters.

References

Cohen, MG (ed.) 2003, *Training the excluded for work: Access and equity for women, immigrants, First Nations, youth and people with low income*, UBC Press, Vancouver.

Cohen, MG & Calvert, J 2013, "Climate change and labour in the energy sector", in C Lipsig-Mumme (ed.), *Climate @ Work*, pp. 76–104, Fernwood, Halifax.

European Commission 2007, "The European social fund in Spain 2007–2013", (http://ec.europa.eu/employment_social/esf/docs/es_country_profile_en.pdf).

Furchgott-Roth, D 2012, "The elusive and expensive green job", *Energy Economics*, vol. 34, pp. 543–552.

Hegewisch, A, Hayes, J, Bui, T & Zhang, A 2013, *Quality employment for women in the green economy: Industry, occupation, and state-by-state job estimates*, Institute for Women's Policy Research, Washington, DC.

ILO 2011, *Skills for green jobs: A global view*, International Labour Organization, (http://www.ilo.org/skills/projects/WCMS_115959/lang—en/).

McFarland, J 2013, "The gender impact of green job creation", Work in a Warming World Research Project, no. 2013–03, (http://warming.apps01.yorku.ca/wp-content/uploads/WP_2013_03_McFarland_Gender-Impact-of-Green-Job-Creation).

Pinderhughes, R 2007, "Green collar jobs: An analysis of the capacity of green businesses to provide high quality jobs for men and women with barriers to employment", City of Berkeley Office of Energy and Sustainable Development, (http://www.michigan.gov/documents/nwlb/Green_Collar_Jobs_236013_7.pdf).

Stevens, C 2009, "Green jobs and women workers: Employment, equity, equality", Draft Report, September, International Labour Foundation for Sustainable Development (Sustainlabour), (http://www.sustainlabour.org/IMG/pdf/women.en.pdf).

UN 2013, "Gender mainstreaming", (http://www.un.org/womenwatch/osagi/gender mainstreaming.htm).

US Bureau of Labor Statistics 2012, "News Release: Occupational employment and wages in green goods and services – November 2011", US Department of Labor, (http://www.bls.gov/news.release/archives/ggsocc_09282012.pdf).

US Department of Labor 2010a, "News Release: US Department of Labor announces $100 million in green jobs training grants through Recovery Act", (http://www.dol.gov/opa/media/press/eta/eta20091526.htm).

US Department of Labor 2010b, "Fact Sheet: Green jobs training. Funding, implementing, collaborating: The women's guide to great jobs", (http://www.dol.gov/wb/media/Gr%20Jobs%20Training_Fact%20Sheet_102810%20FINAL.pdf).

US H.R.3475 – Women in Apprenticeship and Nontraditional Occupations Act, 102nd Congress (1991–1992).

Women's Bureau (USDP-WB) n.d., *Thinking women in green jobs: Women's Bureau projects*, US Department of Labor, (https://www.dol.gov/wb/programs/greenskills_2.htm).

PART III

Vulnerability, insecurity and work

9

GENDERED OUTCOMES IN POST-DISASTER SITES

Public policy and resource distribution

Margaret Alston

Introduction

Australia has experienced a number of catastrophic climate-induced environmental disasters in the last two decades. These include major flood events, bushfires, cyclones and droughts. Climate experts suggest that these are the result of global warming and that the changes will accelerate and become more intense this century – temperatures will rise, seasons will vary and catastrophic events will become more frequent (Garnaut 2011). This will have significant impacts on people and communities across the country. Despite the destruction of life, livelihoods, property and landscape caused by major disasters, there has been a critical lack of detailed, systematic analysis of social impacts and far less of the gendered impacts (Alston 2015a).

This lack of attention to gender within Australia stands in stark contrast to the focus adopted in our foreign aid policies, where gender equality lies at the heart of most programmes delivered into developing nations (Bishop 2014). Further, Australia is a signatory to the Hyogo Framework for Action 2005–2015 aimed at building resilience to disasters. In this capacity, at least in the international arena, it has endorsed a "gender perspective" as a critical issue for action. What appears self-evident and a necessary factor in the developing world and global context is often overlooked within our own country (Tyler & Fairbrother 2013). Notwithstanding Australia's lack of focus on gender in this space, governments across the developed world continue to draft policies, programmes and practices that are almost universally what Enarson (2012, xix) refers to as "stubbornly gender-blind". In fact prior to the bushfires discussed below, Australia's emergency recovery plans are described as having a "pervasive gender-blindness", demonstrated in statements on diversity within these plans that focused on factors such as ethnicity and age but disregarded gender (Hazeleger 2013, 41). This failure to isolate gender suggests that those drafting such documents assumed gender is not a defining factor of disaster experience in the developed world context of Australia.

In this chapter, to disprove this notion, I draw on the experiences of people who lived through a catastrophic bushfire disaster, later termed the Black Saturday fires (Alston, Hazeleger & Hargraves 2014). These fires blanketed the state of Victoria in 2009. Drawing on this research I argue that gender defines vulnerability in disaster situations in both the developed and developing world and plays a significant role in how people cope with, and adapt to climate crises. Therefore, a lack of attention to gender in policy and practice responses will have significant social consequences for individuals and communities and their capacity to respond effectively. Following Black Saturday, gender was recognized as a significant factor before, during and after the event by institutions dealing closely with survivors (Municipal Association of Victoria 2011; Parkinson 2011; Parkinson & Zara 2011). A failure to recognize gender-blindness in policy formulation and disaster practices may enhance vulnerability and social exclusion, and affect adaptive capacity and resilience.

Gender: What is it and why is it important in this context?

Gender refers to practices and customs that shape what it is to be female and male in particular societies at particular times. For the purposes of this chapter it is useful to reflect that gender is relational, challengeable and malleable and constantly negotiated in the intimate sphere of the family and the wider society. Connell's (1995) work in defining and differentiating between the "gender order" of a society – that is the wider understanding of normalized gender relations reinforced by religion, legislation and institutional practices – and the "gender regime" – or the intimate, more localized experiences of individual women – is very useful in analysing gender and disasters and the gender-blindness of developed world disaster responses. This is because the existing order of a society frames how gender equality is viewed and acted upon at national levels and therefore how much space individual women have to negotiate and challenge their own situations.

Critically in Australia the gender order frames an ostensibly equal society – one where discrimination based on sex is legislated against, where educational attainment and access to employment is available to all and where our first female Prime Minister, Julia Gillard, has recently broken one of the last glass ceilings in the country. However it is also a gender order where more insidious discriminations exist – where women are treated as secondary by religious authorities, where they are denied equal pay for equal work, where the number of board positions in major companies held by women remains low, where male sporting events dominate popular culture and where violence against women is escalating and brutal. These insidious practices frame and normalize the positioning of Australian women as secondary, even shadowy, figures and limit the capacity for individual women to articulate more equitable conditions and responses in their own circumstances. Power relations are central to the positioning of Australian women and men in the gender order and gender regimes in which they operate.

In the disaster realm, Owen (2013) describes a toxic, valorized form of masculinity that overwhelms the disaster space – inflated confidence, bravado and glorified heroism (as well as genuine bravery) – that dominates prior to and during an event such as a bushfire. Emergency services management is very much a male-dominated area in Australia, operating along what Tyler and Fairbrother (2013) term a "command and control" military style, "culturally masculinized" environment, also noted in other post-disaster sites such as the response following the 9/11 attacks on New York (Tierney 2012). This is what Connell (1987) describes as a valorized form of hegemonic masculinity and this particular version is lauded by media depictions and harrowing stories of bravery. Arguably as Tyler and Fairbrother (2013) note this unremarked or invisible, but dominant, masculinity and the depiction of women as "other" frames gender power relations during and after a disaster.

Elsewhere (Alston 2015b) I have argued that disaster sites are spaces where gender inequalities may be challenged. Because social relations change, landscapes are damaged and livelihoods may be shattered, new ways of living must be established. It is in these spaces that more gender equitable relations can be negotiated and social contracts reframed. However in the immediacy of a disaster – and if gender sensitivity is lacking – policies and practices act to re-establish and reinforce gender normative behaviours. The opportunity to undertake gendered assessments and to reframe and rebuild with gendered awareness is often lost. In the hegemonic masculine environment that pervades disaster spaces, the prevailing gender order is quickly re-established and resource distribution and relief processes reinforce gender power relations.

Enarson (2012, 46–47), writing about Hurricane Katrina, notes that gender assessments are not given priority in disaster relief processes in the US because of "gender blinders" that include the lack of differentiation of the category "women"; the belief that women are emotionally stronger; that women and men are considered equal; that vulnerabilities are viewed as the result of factors such as poverty rather than gender; and that gender is less important to men, who benefit from gender relations, and it is men who dominate leadership and policy determination roles. I would add from an Australian context that the view that women and men are considered "equal" in the national imagination overrides the complex intersections of disadvantage and discrimination against women and their status as "other" that are so often overlooked in national policies.

Gender, vulnerability and resilience

Yet disaster research overwhelmingly reveals that gender is critical to life chances and choices and frames vulnerability (Lane & McNaught 2009). Vulnerability refers to the combination of factors that reduce one's capacity to respond effectively in a crisis and is characterized by gender, poverty, race, class and life stage. Research from across the globe reveals that women are far more likely to be living in poverty, to be food and water insecure, to have limited access to land and productive assets, to be time poor, to have less access to services, information and support, to hold fewer decision-making positions, and to be more constrained by care work (Dankelman & Jansen 2010).

While Neumayer and Plümper (2007) report that women are more likely to die in disasters, evidence from Black Saturday and other fires suggests that men are more likely to perish in fires (Hazeleger 2013), and that this is related to women being more likely to want to leave the fire situation and men to stay and defend (Tyler & Fairbrother 2013). In fact, reporting to the Victorian Bushfires Royal Commission, Professor Handmer and his research team (VBRC 2010) suggest that disagreements amongst couples on Black Saturday over whether they should stay and fight the fire (more likely to be men's preference) or leave (women's preference) may have led to increased numbers of deaths. Research from across the globe and from a diverse range of disaster sites reveals that violence against women escalates following disasters (Enarson 2012; Lynch 2011; Parkinson 2011; Whittenbury 2011).

Disaster policies are designed to build resilience by assisting people through change processes and restoring community. Yet gender shapes the level of vulnerability experienced in a disaster and limits the capacity of an individual to adapt and to build resilience. Thus, without a gendered assessment of disaster vulnerability and a gendered analysis of the types of policies and strategic actions required to build resilience, these policies may fail. As Enarson (2012, 197) so aptly notes "without paying attention to gender relations, as one of the defining characteristics of private and public life, we will not build an inclusive and gender-responsive approach to emergency management and disaster risk reduction". Gender mainstreaming of disaster policies and actions is critical to achieving such an approach.

Gender mainstreaming

Gender mainstreaming is a policy framework designed to bring about equality, by exposing gender as socially constructed and facilitating more gender-sensitive strategies, policies and actions. Gender mainstreaming was widely adopted by a number of countries following the Beijing Women's Conference in 1995 due largely to the work of women's transnational networks and non-government organizations (True & Mintrom 2002).

In the context of disasters, gender mainstreaming of actions, policies and programmes has the capacity to expose and address gender vulnerabilities. To illustrate gendered vulnerabilities in a developed world disaster context, I draw on research conducted in Australia to indicate the critical importance of gender-responsive disaster policies.

Examples from the Black Saturday fires

On 7 February 2009, after a lengthy drought and extremely dry conditions across the state, Victorians experienced the worst bushfire in Australia's recorded history. During the day the temperature climbed to over 46 degrees Celsius and hot dry winds reached gale force. Before the day was out 173 people had lost their lives, 5,000 people were injured, 2,029 homes had been destroyed, 78 towns were affected, three schools were totally destroyed and 47 others partially destroyed and 4,500 hectares of land had been burnt. The dimensions of this disaster were beyond what had ever seemed possible. As Australians woke on 8 February to the scale of this disaster, there was a huge outpouring of grief and bewilderment, followed by community contributions on a scale never seen before. At this point there were still many people unaccounted for and many areas remained closed to the public. For several days many people were unable to ascertain whether their relatives were alive, whether their homes had been destroyed and how their communities had fared. In one case a memorial service was held for a woman who later walked out of the disaster zone. In these days of confusion and anguish many questions were asked of the emergency services, including why there had been insufficient warnings and why services were so slow to respond.

In 2014 the writer and colleagues returned to one of the communities that had been particularly affected (Alston, Hazeleger & Hargraves 2015). This community (population approximately 1,300) was particularly affected by the fires, with 120 of those who died coming from this small town. The town, nestled along a high ridgeline, was directly in the path of fires that came up the valley from two sides. The one main road out of the town was cut off early, and thick smoke reduced visibility. The town is renowned for its picturesque mountain scenery, and residents included a large number of people who commute daily to the nearby city, as well as a number of people who have holiday houses in the area. It is a typical Australian small town with a mixture of people from diverse class, age and socio-economic backgrounds.

Research was undertaken in 2014 over a period of approximately nine months. Key informant interviews were held with 12 people (approximately equal numbers of women and men) who occupy state and local level government and non-government organization positions, all of whom had had a significant role to play in this community during and/or after the fires. In addition the research team visited the community on two occasions, staying for short periods of time, observing the general changes in the community. During these visits two focus groups (conducted by all members of the research team) were held with a group of five local women and one with a group of eight local men in the community itself. Although conducted five years after the event, this research provided significant insights into the gendered impacts of the fire over an extended period of time. Those in the focus groups were predominantly forty or older, and many had experienced relationship breakdowns since the fires.

Immediate crisis phase

Immediately after the fires several actions were taken. The Prime Minister and State Premier met on Sunday, 8 February 2009 to discuss actions to address the scale of the crisis. One of the outcomes of this meeting was the establishment of the Victorian Bushfires Recovery and Reconstruction Authority (VBRRA), a new state authority set up solely to deal with the people and communities affected by the fires. Within weeks VBRRA developed from an idea to a complete authority with new premises and 120 staff. VBRRA's aims were to clean up the areas; set up temporary housing for the displaced; establish community recovery committees in each community; and manage donations from the public. The meeting also led to the Prime Minister announcing that all affected people/families would have access to a case manager who would be provided by the state to support people through the crisis period. As a result of this decision over 600 case managers entered the communities in the following six months and over 5,500 households used the service (key informant).

Our research interviews and focus groups undertaken five years after the event allowed key policy makers, government employees and community participants to reflect on the experiences of the fires, the response and how effectively this had been done. For the purposes of this chapter a focus on gender factors is highlighted.

Gender factors: Responding to the fire

The research findings support the notion that there were significant gendered differences in the way people reacted as the fires approached. As it became clear that this was a major event, people had very little time to decide what actions to take. This is a critical point because respondents noted that gender differences at this point had a significant bearing on what occurred in the months and years that followed. Critically family/couple discussions centred around whether they should stay and fight the fires and try to protect their homes or leave early. Many men were in the fire brigades and absent from home. Those who were at home were more likely to want to stay and defend while women were more likely to want to leave, and many delayed their leaving because of differences of opinion. It is critical to see this flashpoint as much more than a gender divide between men protecting home and family and women fleeing. Parkinson and Zara (2011) have documented the experiences of 21 women on the day of the fire. They note in their introduction that these accounts are important because they give life to the experiences of women and expose them as active responders seeking to protect their family members. It is equally important to recognize the actions of (predominantly) men fighting the fires and protecting people and property. Five years after the event men in our focus group had still not come to terms with losing the battle against the fire. To them it was a living, breathing thing that had defeated them. In explaining their behaviour on the day of the fire, many men commented on the need to stay and protect.

> [I thought] this isn't going to beat me . . . I think it's a male thing. . . . [but] I failed my family . . . because I let the fire take my family's home. (participant in male focus group)

Men viewed their need to stay as a personal decision whilst also trying not to expose their family to risk.

> I packed the car and I sent my wife off but I wouldn't leave. (participant in male focus group)

Another noted that even five years after the event he still felt crippled by his inability on the day to save people.

> [I have a] lot of survivor guilt – I couldn't save people on the day. (participant in male focus group)

These gender differences on the day of the fire show the palpable reality of the decisions that had to be made – women focused on saving people, men on protecting their homes and "beating" the fire. Although it was clear that gender stereotypes fragmented on the day of the fire, many government employees commented on the hyper-masculinity exhibited.

> Males think they have to be capable and competent, and [think] "I will defend my house at all costs. I can overcome this massive disaster that's unfolding. I'm not going to leave my house". Generally women are more focused on safety and leaving and making sure that everybody is safe, whereas men want to defeat the threat. (male government employee)

Ongoing gender tensions resulting from decisions on the day

What is significant about the gender differences in decisions taken on the day of the fire is that they shaped subsequent tensions, influenced future planning and defined the ongoing conflict between many partners. Several informants noted the link between ongoing relationship problems and the different decisions taken.

> This disagreement on the day of the fire is linked to ongoing fights and separations . . . and often really nasty separations. (female NGO welfare case manager)

> Separations were inevitable. (male state employee)

> Women deferred to partners on the day and in the long-term this is now viewed as the cause of family dysfunction. (female local community worker)

While the figure is unsubstantiated, the five women in the focus group noted with some exaggeration that this ongoing tension had resulted in "75 percent of marriages breaking down". Critically women reported that the rise in family violence is related to this breakdown in relations. Parkinson (2011) published the results of a study undertaken with women in the Goulburn North East region. Nine of the sixteen women in their study experienced violence, many for the first time.

Ongoing tensions in the post-disaster period also related to decisions about the future, particularly whether the family should leave the area entirely or go back. Men in the focus group noted that many wives were too terrified to return and had stayed away, thereby effectively separating. Men articulated a need to rebuild as a form of resistance to demonstrate the fire had not beaten them. One female government employee noted:

> I resolved a lot of tension and conflict between couples. . . . men – "I want to go back" . . . women – "I don't think I'll ever feel safe there". . .

In the past, emergency management policies urged families to have a fire plan and to be prepared to either stay or go, but to make the decision early (Bushfires CRC 2010). The Bushfires Royal Commission held after the fires found that this advice was inadequate as many people had a poor quality plan, were unprepared for the scale of the fire, had no contingency plan and were waiting for official advice that never came (Bushfires CRC 2010). The results outlined here suggest that both women and men should have their own plans prepared beforehand and be ready to act on them as individuals.

The immediate aftermath of the fire

The loss of livelihoods following the fires was extensive and this affected women and men across the region. Businesses and agricultural activities across the region were destroyed or critically damaged and this had a significant impact on the resilience of people. Further the loss or destruction of schools meant that many children were unable to return to the routine of their daily lives for months.

Because the fire had been so catastrophic, the immediate aftermath represented new territory for workers, policy makers and community members. In hindsight many decisions were made hastily and that had adverse consequences for local people. One that had a critical impact on local men was Victorian Bushfires Recovery and Reconstruction Authority's (VBRRA) appointment of a large national contractor to come into the areas and clean up. Up until this point, local men and emergency services volunteers, together with the army had begun the task of cleaning their communities and this had given a sense of purposeful activity in the immediate aftermath of the fire. The appointment of a contractor and the direction to locals to stay away had a significant impact on the coping ability of men.

> So many men were shut down – [they] just withdrew home. (women's focus group)

> The way they (men and women) were treated post-disaster fundamentally can do greater harm than the disaster itself. (female government employee)

> [There was a] lot of anger . . . It was chaotic. (female government employee)

By contrast, women took on the role of nurturers and focused on the well being of their families and the community. They formed groups, facilitated community services such as community dining, childcare, information sharing, distribution of donated goods and generally checking on others. This created a central community hub where meals provided by community organizations were served to locals. These hubs also provided a meeting place where information could be shared, donations distributed and where people could come together to share their experiences and their grief.

> Early on community dinners were successful. It was an opportunity for the community, at a time when we were very fragmented to come together in a very social and relaxed environment and be able to connect, swap information, what's happening, what's working for you, what's available. And it was run by two amazing women who seemed to understand what were the barriers for women, and they arranged child care. (local male community worker)

Key informants commented on the activities undertaken by women, noting the significance of these to community resilience building.

> Women want to nurture – men want to fix things the way they were. (women's focus group)

> Community women were outstanding. (male government employee)

The emergence of the Firefoxes, and its development by two young women into a significant women's community group and voice for women, is a critical example of successful organizing by women. This group continues, has its own website and has produced a film – *Creating a New Normal* (Firefoxes 2014) – to wide acclaim. They have received a number of awards in the years since including Community Group of the Year 2013, State and National Winner of the Resilient Australia Award and Women of Fire Awareness Award.

The critical nature of the work of local women in building community in the immediate crisis phase is recognized as a key factor in community healing. One male government employee noted:

> The sheer loss of life . . . the exposure of trauma, the exposure of grief, as well as loss of material possessions had a much more profound effect for people where they needed that nucleus of community, they needed to come together to survive.

In commenting on the film, *Creating a New Normal*, international researcher Elaine Enarson commented:

> This film clearly shows that connecting as women, with women, and for women is a powerful strategy for meaningful recovery both at the individual level and for the family and community. (Firefoxes 2014)

These insights suggest that disaster policies must recognize, facilitate and resource women's nurturing activities and men's active engagement in rebuilding tasks. These empowering activities are critically important for individual and community resilience and require recognition and planning.

Medium term

A few months after the fires several services had been established, the clean up was largely complete, people had been able to go back to their properties and sift through the rubble for possessions and many were deciding whether to rebuild or move on. Four "villages" had been established for displaced people. These were largely the result of a significant donation of "dongas" or huts sent over from Western Australia by a mining company. These villages were supposedly temporary while the areas were rebuilt. Case managers had been appointed and those in the fire zones were being assisted to receive support to rebuild. While the process seemed like an orderly progression, in fact locals felt completely disempowered as it felt as if their lives and property were taken over by bureaucrats and complex processes. Some expressed anger at the way the "recovery" process was put in place with indecent haste and impossible timelines. One local male worker spoke of the "secondary injury" incurred by the "profoundly disempowering" recovery process. This mirrors Tierney's (2012, 252) evaluation of the post-Katrina experiences of people affected – "they were made dependent and then given nothing to depend on". This led to significant anger within the communities, perhaps partly the result of the dawning realization that things would never be the same.

> It was inhumane . . . citizens cannot be heard – that's the real frustration. The absolute aftermath when you become a child to professionals, and bureaucrats [become the] parents – that is the disaster, that is inhumane. (local male community worker)

> I felt we were in prison and the guards were walking around with their hands behind their backs watching. (women's focus group)

> The communities so wanted to be healed, they so wanted to be back to being what they were. (male government employee)

Given the trauma suffered, time lines for decisions about rebuilding and decisions about the future (and hence access to government funding and for what purpose) were unrealistic. While governments and other agencies had consulted widely, given information regularly and sought advice constantly, community members were moving at a different pace and were contending with significant grief and post-fire trauma. Participants in our research argued that this put community people and those who had come to help, out of step. Worryingly as the months had passed, the energy required to keep the community going took its toll on women. One local community worker noted – "women got burnt out". This suggests that voluntary work in traumatic circumstances takes a significant toll and consideration must be given to local people through additional funding and infrastructure to undertake time-consuming community activities such as the provision of meals and community space.

As rebuilding continued, informants noted that men had come to dominate the decision-making space and were shaping the public understanding of the fire.

> A lot of "angry men" with "very loud voices" . . . they weren't necessarily speaking on behalf of the community but they were the most heard. (local female community worker)

In summarizing the medium term stages, it appears that the cohesion noted in the immediate aftermath of the fires had ruptured, the hours of volunteer community work had affected women's energies, relationship tensions were boiling over, disunity was evident in families as decisions were being made about staying or going, and livelihood strategies, bureaucratic processes, the slowness of the restoration phase, the countless forms that had to be filled in and the unpredictability of the future were having a significant impact on community members. As a result our informants note that this phase was marked by grief at the loss of people and place, ongoing trauma, neglect of relationships, broken marriages, escalating violence against women, a rise in mental health issues, rising drug and alcohol use, fear as another summer approached and the reality of lost community. People spoke of losing the routines and patterns of life.

> We didn't have mums to run the tuckshop; we didn't have someone to be the kick coach. (local male community worker)

Women in the focus group noted that this was the period when they questioned, "what is home?" They articulated a need at this point to assess what their "place" was, what it meant to them, where it might be and how they might construct "home". They noted this was the period when they needed to move from survival to well being but that the structures essential to enabling this were missing. The number of issues that concerned them expanded as they moved beyond survival – for example water quality and environmental health became causes for concern.

They noted that one of the most critical issues and needs during this phase was childcare. Many were living in sheds or caravans, often without electricity and the children needed a safe place away from the tensions at "home".

> Some children started kinder (school) with no home. (female NGO case manager)

Long term

In paraphrasing the words of our informants, the lessons of this post-disaster experience are many. Gender played a significant part in how people reacted and responded to the fires and the ongoing trauma of the post-disaster experience. Family dysfunction that may have occurred on the day of the fire was rarely resolved effectively. Mental health issues, violence and drug and alcohol issues persisted. Men and women reacted to the fires as individuals – and even five years after the event, because of the trauma they had been through, it was evident many were still viewing their lives from their own individual perspective rather than as part of a family/community.

Many thousands of people were displaced and homeless for long periods. The four temporary villages established as a short-term measure were finally wound up two and a half years after the fires despite the initial expectation of six months. This brought additional problems of further displacement. People noted the loss of control over their lives and the power exerted by government officials. This was particularly evident around decisions relating to who would receive compensation and under what conditions. These decisions were critically disempowering, had divided communities and had led to heightened suspicion and anger towards government employees and case managers. This loss of control over decisions that were so critical to the future also exacerbated mental health issues.

> People need control over their destiny. Local control is essential. We have to foster resilience and lower vulnerability. The ability of people to recover is directly related to their ability to access and use assets. (male government employee)

Those appointed to community committees reported stress and burnout as a result of being asked to do too much too quickly, making decisions hurriedly with minimal consultation and causing further community divisions. Participants also note that the bureaucratic processes required people to decide about rebuilding quickly and this led to regret from many. In hindsight community members note that they were too traumatized to make lasting decisions about rebuilding or moving in the short timeframe given. Considering the position of governments wishing to ensure the safety and security of people, this haste appeared reasonable. However the too urgent haste was recognized by government employees tasked with providing the assistance.

> It's a marathon not a race but it feels like a race at the start. (female government employee)

> We thought it would take six months – that was a mistake – it took years. (female government employee)

Gender continues to be a factor in the long term. There is a widespread acknowledgement of the value of women's community work and their wisdom in terrifying times. Of particular significance is that women are now calling for their own training in fire prevention and the use of equipment such as water pumps. They are no longer prepared to leave this to men. Further they are adopting their own empowering language – referring to themselves as "survivors" not victims and to their lives as "the new normal".

There have been long-term health impacts noted by several informants. These include a rise in suicides, heart attacks, stress and general unwellness, ongoing trauma, hyper-vigilance, and anxiety. Men note they feel "dazed", "shell-shocked", "more emotional", and unable to "watch any disaster on TV". Men noted that it is easier to be with people who have been through the fires – "sometimes we don't have to say anything". Women report ongoing trauma, hyper-vigilance, panic attacks and anxiety, particularly when it gets hot and they worry there are no secure, safe places if there is another fire. Women note the incidence of domestic violence, mental health issues, drug and alcohol issues and relationship breakdowns and worry that community groups are imploding because communities appear fractured and divided, particularly around the issue of who received government funding and who did not. This has made it difficult for community members to regroup with a common purpose. Five years after the event they note "they are not doing well" and have "lost their dignity". Several noted that if they had their time over again they would not rebuild. They felt those who had left the area had managed their post-trauma more successfully.

Recovery is a myth

This research reveals that recovery after a traumatic event such as the Black Saturday fires is not possible. Recovery implies returning to how things were before the disaster. This is unrealistic as nothing can ever be the way it was and this understanding must infuse policies. Community members recognize this and note that they prefer the term "new normal" to describe the future – one that is different. Survivors understand this and articulate strong feelings about the use of the term.

> Recovery is rubbish . . . recovery is not neutral – it's very very loaded with meaning . . . maybe it's something like aftermath . . . it's the same with resilience or preparedness – they're nonsensical words. (female community worker)

Another local male community worker noted:

> I don't believe in recovery – the word is a dog whistle for social workers and welfare workers and churches to come in . . . no one goes back to where they were – everybody is transformed for better or worse. So recovery is an inadequate term. It's a renewal process not a recovery.

Another male government employee noted that this new phase, whether it be renewal or the new normal, is:

> about politics and people and community and religion and gender and power.

Discussion

For some time researchers across the developing world have been aware that gender is a significant factor in vulnerability during disasters. Yet there is a critical lack of understanding about gendered impacts in a developed world context. This chapter provides insights into the experiences of women and men who are survivors of the Black Saturday bushfires in Australia. Our research indicates that there were differences on the day of the fires in the way women and men reacted and that gender differences shaped vulnerability and resilience in the months and years that followed. Gender is revealed as critical in all stages of the disaster from the experience itself, to the immediate post-disaster period and to the medium and long-term periods.

Following the fires, Australia instigated a major post-disaster response that, in its scope and breadth, was one of the most extensive recovery efforts ever undertaken in this country. The Australian community also responded giving such a wealth of donations of goods and clothing that they had to be housed in warehouses. As well, a fund established to assist those affected raised over $400 million. There is no doubt that the government and the general community were extremely moved by the impacts of the fires and wanted to assist the survivors in tangible ways. It is important to acknowledge these efforts and to note that respondents to this research did not wish to deny the significance of the response.

Nonetheless, because there is likely to be a rise in the numbers and frequency of climate-related catastrophic events such as this, it is critical that we learn from the experiences of the survivors to ensure that future efforts are well targeted. Following the fires, the Victorian state government held a Bushfires Royal Commission in order to understand the issues and challenges on the day and in the post-disaster period. Critical to our developing understanding of the impacts of disasters is that community members refer to the post-disaster efforts as the "secondary injury". This requires a deeper understanding of processes that were too cumbersome, potentially uncoordinated between agencies and which failed to empower locals to act on their own behalf. They also suggest that the processes were designed to restore communities, with little time for communities to revision their "new normal" and their new sense of place.

In this context it is critical to acknowledge the gendered factors that shape vulnerability regardless of context. The rise in violence against women, and in drug and alcohol abuse and mental health issues suggests the need for interventions early to assist and empower women and men to have greater control over their lives. Gendered experiences have a major impact on individual and community resilience and resourcing with gender sensitivity may reduce these mal-adaptations. The way communities are resourced in the post-disaster period can provide powerful assistance to individuals and communities suffering significant post-trauma.

At all stages of post-disaster recovery gender analysis is necessary – there are significant gendered factors revealed here that require decision-makers to be aware of the need for gender-sensitivity and to adopt a gender mainstreaming approach to policies and practices. This includes assessing who are appointed to community advisory committees; assessing how resourcing is distributed, and on whose advice and for what purpose; undertaking ongoing consultation and adopting empowering strategies; allowing the communities to help themselves; slowing processes where necessary; and being sensitive to individual differences within families. Critically the need for gender mainstreaming of all policies, practices and actions is essential to moving forward with resilience. In this regard it is pleasing to see that the Municipal Association of Victoria (MVA 2011) has instituted a gender and emergency management strategy, the objective of which is to:

> Improve emergency management and reduce the negative consequences of gender-blind practices in local government by integrating gender into emergency management planning, decision-making, policy development and service delivery.

The stated outcomes are that:

> a. gender differences are considered and reflected in emergency management policy analysis and advocacy;
>
> b. Municipal Emergency Management Planning Committees and local government incorporate best practice knowledge regarding gender differences into emergency management planning, decision-making, policy development and service delivery.

These statements indicate a willingness to adopt gender mainstreaming and to act with gender sensitivity. In adopting this framework, governments at all levels should examine their own views of recovery and be open to working towards a new normal that encompasses gender equality and empowerment of all community members.

Conclusion

While developing countries adopt gender-sensitive practice in disaster areas, developed countries such as Australia have been less aware of the need for gender mainstreaming in post-disaster policies. This chapter has focused on the experiences of Australians following the Black Saturday bushfires in 2009. Participants in our research recognize the significant gendered implications in the responses to the fires and the actions taken in the months and years that followed. This work reveals that there are critical gendered differences in the way women and men responded to the fires and in the immediate aftermath. These actions shaped the ongoing resilience within communities and reveal the critical need for gender sensitivity in policy and practice responses.

I argue that gender mainstreaming should be mandatory practice in this field and that gender sensitivity be incorporated into actions and strategies designed to assist communities in post-disaster sites. Significantly governments and other institutions engaged in rebuilding work should be open to seeing the post-disaster space as one where gender inequalities, or the gender order, can be challenged through the way resources are distributed, by appointing a diverse range of community members to decision-making bodies and by ongoing attention to issues such as violence against women and mental health factors. Further services must be designed to assist women and men negotiating new relationships in the intimate gender regimes in which they operate. By facilitating gender sensitive responses, a new normal is indeed possible – and one that incorporates and nurtures both women and men.

References

Alston, M 2015a, "Social work, climate change and global cooperation", *International Social Work*, vol. 58, no. 3, pp. 355–363.

Alston, M 2015b, *Women and climate change in Bangladesh*, Routledge, London.

Alston, M, Hazeleger, T & Hargraves, D 2014, "Bushfires in Australia", *Presentation to the International Social Work Conference*, Melbourne.

Bishop, J 2014, "The new aid paradigm", Speech to the National Press Club, (http://foreignminister.gov.au/speeches/Pages/2014/jb_sp_140618.aspx).

Bushfires CRC 2010, "Evaluation of the 'stay or go' policy", (http://www.bushfirecrc.com/projects/c6/evaluation-stay-or-go-policy).

Connell, RW 1987, *Gender and power*, Allen & Unwin, Sydney.

Connell, RW 1995, *Masculinities*, Polity Press, Cambridge.

Dankelman, I & Jansen W 2010, "Gender, environment and climate change: Understanding the linkages", in I Dankelman (ed.), *Gender and climate change*, pp. 21–54, Earthscan, London.

Enarson, E 2012, *Women confronting natural disaster: From vulnerability to resilience*, Lynne Reinner Publishers, Boulder, CO.

Firefoxes 2014, "5 years on – Firefoxes Australia continues to grow", (http://firefoxes.org.au/).

Garnaut, R 2011, "Garnaut climate change review – update", *Australia in the global response to climate change*, (http://www.garnautreview.org.au/update-2011/about-review.html).

Hazeleger, T 2013, "Gender and disaster recovery: Strategic issues and actions in Australia", *Australian Journal of Emergency Management*, vol. 28, no. 2, pp. 40–46.

Lane, R & McNaught, R 2009, "Building gendered approaches to adaptation in the Pacific", *Gender and Development*, vol. 17, no. 1, March, pp. 67–80.

Lynch, K. 2011, "Spike in domestic violence after Christchurch earthquake", *National*, (http://www.stuff.co.nz/national/christchurch-earthquake/4745720/Spike-in-domestic-violence-after-Christchurch-earthquake).

MVA (Municipal Association of Victoria) 2011, "Gender and emergency management strategy", (http://www.mav.asn.au/policy-services/emergency-management/Pages/gender-emergency-management.aspx).

Neumayer, E & Plümper T 2007, "The gendered nature of natural disasters: The impact of catastrophic events on the gender gap in life expectancy, 1981–2002", LSE Research Online, (http://eprints.lse.ac.uk/3040/1/Gendered_nature_of_natural_disasters_(LSERO).pdf).

Owen, C 2013, "Gendered communication and public safety: Women, men and incident management", *Australian Journal of Emergency Management*, vol. 28, no. 2, pp. 3–10.

Parkinson, D 2011, "The way he tells it: Relationships after Black Saturday", *Women's Health*, Goulburn, North East, (http://www.whealth.com.au/documents/publications/whp-TheWayHeTellsIt.pdf).

Parkinson, D & Zara C 2011, "Beating the flames: Women escaping and surviving Black Saturday", *Women's Health*, Goulburn North East, (http://www.whealth.com.au/documents/environmentaljustice/BeatingTheFlames-Book.pdf).

Tierney, K 2012, "Critical disjunctures: Disaster research, social inequality, gender and hurricane Katrina", in D Emmanuel & E Enarson (eds), *The women of Katrina: How gender, race and class matter in an American disaster*, pp. 245–528, Vanderbilt University Press, Nashville, TN.

True, J & Mintrom M 2001, "Transnational networks and policy diffusion: The case of gender mainstreaming", *International Studies Quarterly*, vol. 45, no. 1, pp. 27–57.

Tyler, M & Fairbrother P 2013, "Gender, masculinity and bushfire: Australia in an international context", *Australian Journal of Emergency Management*, vol. 28, no. 2, pp. 20–25.

VBRC (2010) "The lessons learnt", Victorian Bushfires Royal Commission 2009, (http://www.royalcommission.vic.gov.au/Finaldocuments/volume-1/HR/VBRC_Vol1_Chapter21_HR.pdf).

Whittenbury, K 2011, "Climate change, women's health, wellbeing and experiences of gender-based violence in Australia", in M Alston & K Whittenbury (eds), *Research, action and policy: Addressing the gendered impacts of climate change*, pp. 207–222, Springer, London.

10

CLIMATE CHANGE, TRADITIONAL ROLES AND WORK

Interactions in the Inuit Nunangat

Michael Kim

Introduction

The Canadian context often sees climate change framed as the result of over-consuming and unsustainable lifestyles being complemented by high levels of resource extraction in "dirty" industries. Less attention is paid to the impact of climate change in vulnerable communities and less still to the different experiences therein. Reusing a term from a 2014 UN climate summit, this chapter is about the "front lines" of climate change in Canada. Specifically, it looks at the implications of climate change for the Canadian Arctic (henceforth "the Arctic") and the Inuit communities that reside there.

Existing research in this realm has been conducted at the micro-level and through the scope of qualitative case studies and anthropological research. This chapter begins with a summary of this research and identifies common themes that exist across Inuit communities. Themes of poverty, insecurity and diverging gendered experiences emerge as common stories, as does the fact that climate change can exacerbate each of them. The chapter takes a critical next step – analyzing macro-level indicators to corroborate the stories that have emerged so far. Publicly available survey data suggests that the themes highlighted anecdotally are endemic to the entire region.

It is possible that increased participation in the wage labour market can offer a livelihood that would help to offset the impacts of climate change. On the other hand, this chapter cautions that traditional divisions of labour may condition Inuit vulnerability and adaptive capacity to climate change along gendered lines – including one's "employability". Compounding the problem is that, in spite of indications of diverging gender experiences, existing observations remain blurred by a shortage of gender-based analysis.

Climate change in the Arctic: Symptoms

The Artic is simultaneously referred to as an *environment of risk* and an *environment at risk* (Nuttall 2005, 19). The first condition refers to climate variability and changes in the physical environment and distribution of wildlife all influencing the predictability and reliability of traditional land and resource use. As an *environment at risk* the Arctic is susceptible to pollution, climate change and industrial development – factors that will exacerbate the aforementioned riskiness of the Arctic. These two conditions are not independent of one another and a vicious cycle emerges wherein the susceptibility of the Arctic to climate change increases the region's unpredictability and, subsequently, the vulnerability of the populations that live there. Though situated in a "rich" country, the region shares many characteristics (lack of economic resources and industrialization, high reliance on subsistence foods) with developing countries. Its economy is also a mixture of subsistence and wage work.

The Arctic is also the region most threatened by climate change and increasing temperatures (Pachauri & Meyer 2014, 8). This is especially problematic for a region that has evolved to be dependent on freezing temperatures and has led to claims that, in Canada, the Inuit will experience climate change first and most severely (Nuttall 2005, 23). Climate change has and will effect change in weather and wildlife patterns, the physical environment (coastal erosion, melting permafrost, glacial/sea ice), and routes and modes of transportation. Observed changes include warmer temperatures (both in the air and the ocean) leading to increased precipitation (mostly in the form of rain), retreating sea ice and changing wind patterns (Furgal & Prowse 2008).

Arctic populations, both human and other, are particularly vulnerable to climate change because there are fewer substitutes for species that decline because of habitat loss than in warmer regions. The displacement of one or few species can have drastic implications for others that depend on them for sustenance – an effect that trickles up the food chain. A variety of factors contribute to this displacement and to diminishing health and population levels for local wildlife (Hassol 2004, 58–77).

Arctic populations are also adversely affected by the thawing of the ground caused by milder temperatures. This thawing tends to disrupt patterns of migration and routes of transportation throughout the region. Unique here is that, "unlike most parts of the world, arctic land is generally more accessible in winter, when the tundra is frozen and ice roads and bridges are available" (Hassol 2004, 86). These warmer temperatures imply a narrower window of accessibility for industrial, commercial and Indigenous transportation. This limited accessibility is compounded by a thawing permafrost and the increased prevalence of extreme weather events (floods, mudslides, avalanche) – phenomena that threaten existing infrastructure and decrease the reliability of arctic transportation (ibid., 86–91).

Exacerbating the situation is the fact that these disruptions are unpredictable in nature, leading to increased risk and uncertainty. Effects like this are detrimental to Indigenous populations who have established an equilibrium with the land.

This equilibrium is changing rapidly, culminating in the population increasingly transitioning to non-traditional substitutes.

The data

Existing research into climate change's impact on the Inuit relies on interviews and community-level case studies (see, for example: Beaumier & Ford 2010; Dowsley et al. 2010; Ford 2009). These case studies provide insight into local knowledge and experiences. Emphasizing the local also serves as a reminder that the research that has been conducted may not be generalizable across the entire region. With this said, the results of these qualitative investigations are bolstered by similar findings across the Nunangat, as well as in other polar locations such as Alaska, Greenland, Sweden and Russia. At a minimum, their findings help to raise awareness and identify potential challenges, threats and opportunities that exist in the region.

The lack of attention that has been given to gendered issues in the Arctic necessitates the drawing of data from a variety of sources including the Aboriginal Peoples Survey (APS), the Canadian Census, Labour Force Surveys and National Household Surveys (NHS). The majority of the quantitative data in this chapter was generated from the 2006 and 2012 APS – a postcensal survey of social and economic indicators about Aboriginal peoples in Canada aged 6 and over.[1] Regrettably, the elimination of the long-form census after 2006 means that while the 2006 APS used Census information to determine eligibility, the 2012 APS was reliant on NHS data for this.[2] While it is less than ideal to be comparing statistics across sources, issues stemming from methodology are tempered by the fact that the same agency – Statistics Canada in this case – was responsible for conducting the relevant data collection with methodologies being relatively consistent across the data sets.

The Inuit

Background

As of 2011 there were nearly 60,000 Inuit living in Canada. The population is quite young with a median age of 22 (Wallace 2014, 6). Roughly three quarters of this population lives in one of the four regions known collectively as the Inuit Nunangat and more than three-fifths of this population resides in the territory of Nunavut. The communities in the Nunangat vary in size from 150 to 5,000 people and span an area that covers over one-third of Canada's land mass (Statistics Canada 2015). A lack of development in the region combines with an inhospitable climate, making it accessible only by air and ice roads in winter and by boat in the summer.

Inuit food systems

Inuit food systems are reliant on both subsistence and store-bought foods. This allows for some substitution across food sources when there are stresses to either of

TABLE 10.1 Inuit food systems

	Country food	*Store food*
Production/ processing	Household hunting, fishing and meat preparation	Industrial food systems outside of community
Distribution	Household/community sharing guided by kinship rules	Cash transfer from individual to store
Consumption	In household, often in groups	In household, often individually

Source: Ford (2009, 85).

the individual food systems. The specific characteristics of these two sub-systems are summarized in Table 10.1.

Country foods

The heightened importance of meat and lack of agricultural opportunity that characterize the region has resulted in a specialized form of food procurement centred around hunting animals such as seal, caribou, whale and narwhal (complemented by harvested berries and fish) (Ford 2009, 84). Collectively these traditional foods are known as country food and can be understood to be the food that is harvested from the land, rivers, lakes and seas of the region (Power 2008).

Reliance on country food has led to the emergence of sharing networks throughout Inuit communities. These networks, typically managed by female elders, play a vital role in enhancing food security. As hunting is especially prone to volatility (both anticipated and not), in times of need country foods have been shared across wider (extended family and community) networks.[3] These practices have performed the dual role of insuring against idiosyncratic risk and strengthening social and community bonds (Ford, Smit & Wandel 2006; Mulrennan 2014).

Country foods remain an important part of the Inuit diet – supplying up to 40 per cent of daily consumption (Ford 2009). Harvesting also remains a popular activity as over 70 per cent of Inuit adults harvest country foods for subsistence purposes (Furgal & Seguin 2006). In addition to being a relatively affordable option, country foods have been identified as being of high nutritional and sociocultural value compared to available substitutes (Nunavut Food Security Coalition 2014).

Store foods

Store-bought food is typically flown in from the south. Over the past half-century, store foods have been increasing in importance in the Nunangat – prompting claims of a "nutritional transition", particularly in younger generations. While becoming increasingly prevalent in the Inuit diet, store foods are characterized as being less nutritionally and culturally fulfilling than country food – especially given a growing preference for nutritionally deficient, high sugar/fat products (Chan et al. 2006). This transition has been highlighted against a backdrop of rising levels of diabetes and obesity in the region (Ford 2009).

Owing to long and volatile supply chains, store foods are exceptionally expensive in the Arctic.[4] Because of this, household food insecurity remains high – even after large subsidies are applied.[5] The region's remoteness means there is frequent shortage, with products arriving near or past their best before dates (Ford et al. 2009). Thus, even when they can be afforded, store foods suffer from a quantitative and qualitative volatility in supply. This was the case in July 2015, when store shelves were left bare because of airport construction and adverse weather conditions (Stevenson 2015).

Store food, by definition, requires purchasing power and a corresponding source of income. As Inuit communities become increasingly reliant on store foods, so too have food systems become more centred around the individual/household rather than the community. This has eroded the aforementioned sharing networks and affected their substitution for exchange-oriented systems (including for money) and altered conceptions of how appropriate it is to ask for help.

Despite a growing reliance on store foods in the Inuit diet, they are not considered an equal trade-off for country food (Ford et al. 2009). Interviews have highlighted the preference in Inuit communities for country food and the social and cultural benefits that stem from its procurement and consumption (Lambden, Receveur & Kuhnlein 2007). These are important considerations in the context of food security.

Interactions: Food (in)security

The presence of these two food systems allows for the substitution of one for the other during times of shortage. During times of environmental (country food) stress, Inuit households substitute store food for country food. The inverse is true during times of economic (store food) stress (Ford 2009). When functioning properly, these dual food systems act as a safeguard of food security. Unfortunately, these systems are both prone to insecurity (climate-induced and other) and, rather than a seamless move to store-bought alternatives when country foods are scarce, an unexpected decline in country food supply can necessitate the skipping of meals, thinner spreading of available food and liquidation of physical assets when possible. This implies a rising sense of insecurity as country food systems are stressed by changing environmental conditions (Beaumier & Ford 2010).

The 2012 APS includes a measure of food security that is derived from a series of six questions pertaining to a household's access to food. Table 10.2 summarizes this food insecurity for Inuit men and women. The APS codes respondents as having either high/marginal, low or very low food security and 52.9 per cent of all respondents (54.1 per cent of women and 51.6 per cent of men) were classified as food insecure. It is also worth pointing out that women are overrepresented at the level of very low food security.[6]

TABLE 10.2 Food security in the Inuit Nunangat

	Men	*Women*	*Total*
High	48.4%	45.9%	47.1%
Low	30.9%	29.8%	30.3%
Very Low	20.8%	24.3%	22.6%

Source: APS (2012).

While some of this difference likely owes to socioeconomic inequalities, Beaumier and Ford suggest that gendered norms exacerbate the gap, pointing to the role that women play in rationing food for the household and caring for others. Country foods were found to be particularly stressed as the female experience is characterized by a tendency and willingness to put the rest of their family first – including ensuring that male family-members have enough energy to hunt. These coping mechanisms are not unique to the Arctic, as more generally it has been found that women tend to eat last and least in impoverished households (Arora-Jonsson 2011, 745).

Gender and work in the Nunangat

Inuit gender roles have been relatively well-defined in the Arctic. While men have traditionally been responsible for the hunt, women have been responsible for the remainder of work and in performing tasks that are complementary to the hunt (Shannon 2006). While this speaks in very general terms, it is a consistent theme in the literature. Changes in the twentieth century (including the forcing of populations into fixed settlements) have greatly impacted this complementarity. This results in "inbalances in the status of Inuit men and women that have affected the relations between them" (Morgan 2008). These influences have also led to social problems such as substance abuse and domestic violence that, while not the focus here, are quite salient (Rasmussen 2009).

These problems are symptomatic of the link between gender and traditional work in the Arctic. For better or worse, hunting has been the most revered work in Inuit society and a high level of respect, influence and political authority is donned on the most successful hunters (Mancini Billson & Mancini 2007, 35–38). This has led to a situation where masculinity and hunting are associated with one another. Stemming from this is the perception that hunting is men's "real work" and a way of affirming masculinity. Work that is not associated with the hunt is not held to as high a level of respect and relatively little prestige is attached to things like wage labour and formal education. These conditions appear to have resulted in a masculine reluctance to transition away from traditional work – even as it decreases in viability (Kafarowski 2009; Rasmussen 2009).

On the other hand, the traditional work of women has often been characterized as complementary to the hunt. Responsibilities associated with the female domain include caring for the sick and elderly, gathering fish and berries, processing animals for food or clothing, work associated with childrearing and the household, and the management of food-supplies and sharing networks (Dowsley et al. 2010; Mancini Billson & Mancini, 29-50). Important here is a more transferable skillset and a conditioned disposition that might be understood as being more adaptable to circumstances. The prestige associated with hunting is notably absent and, thus, the resistance to change associated with Inuit men has not been as evident with women.

In spite of its shortcomings, this division has been responsible for increasing household security, especially vis-à-vis country food procurement. However, the dichotomy of male specialization and female diversity is suggested to have led to varying abilities to cope with stresses to traditional work.

> To use an analogy, males seem to be socialized into path dependency and have difficulty accepting other paths and changes, while females tend to be socialized into situations in which adjustment and change are required, leaving them prepared to move between job categories and job options. (Rasmussen 2009, 529)

Empirical evidence

With a variety of contemporary influences having decreased the returns that Inuit households are getting from country food systems, it is unsurprising that there has been a shift to wage work. Less expected perhaps is that there are unique gender trends in employment in the Arctic that diverge from comparable statistics in the rest of Canada. To this point, research has highlighted a unique interaction between gender and wage labour in the Arctic where the gendered division of traditional work is postulated to have led to divergent attitudes and adaptability vis-à-vis non-traditional forms of work.

Figure 10.1 compares labour force statistics for Inuit, Aboriginal, and non-Aboriginal populations in Canada.[7] In all categories Inuit statistics lag behind the other two groups. Further, and in spite of relatively similar labour force participation rates, Inuit unemployment rates are significantly higher than the other two populations.[8]

There are noticable differences in the participation of men and women in the labour market. Table 10.3 shows that Inuit women are more engaged in wage labour than either of the other two female populations. Further, Inuit women are the only female group with a higher employment rate than their male counterparts (absolute value notwithstanding).

The other two statistics tell a similar story. While men have higher unemployment rates across the board, the five percentage point difference between Inuit men and women is five times greater than the corresponding figure for the other two population groups. Finally, while labour force participation is higher for men in all three groups the Inuit gap is markedly lower than the other two.

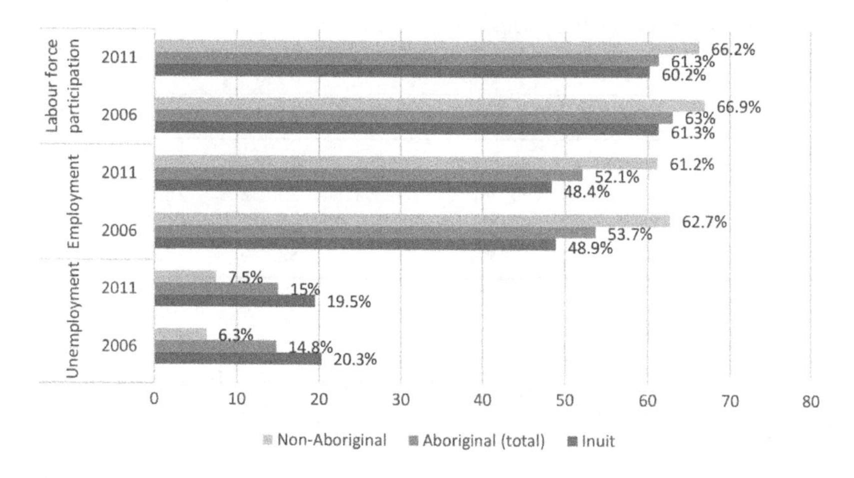

FIGURE 10.1 Labour force statistics (2006 and 2011)

Source: The National Aboriginal Economic Development Board (2015).

Thus, while the region is host to exceptionally high rates of unemployment and low rates of labour force participation, the difference in employment levels and participation rates is largely experienced by gender. While female participation rates in the Nunangat are nearly identical to the Canadian average, Inuit males are more than 10 per cent lower (relative to the Canadian rate). As such, the difference in labour force participation between men and women at the national level is less pronounced in the Nunangat (10.2 to 2.6 per cent respectively).

A similar story emerges here as labour force statistics for Inuit men lag behind the other two groups of men at a far more pronounced rate than Inuit women do other women. Much of this could be related to the significant role that government/administrative work plays in the Arctic – accounting for nearly one-third of Inuit female employment (compared to 8.5 per cent for Inuit men, and 10.7 per cent of all Canadians) in 2011.[9] These differences are drastic and suggest that the non-public sector labour market is especially problematic.

It is also worth noting that these trends have increased in recent history. Table 10.4 compares Inuit labour force statistics from 2006 and 2011 and shows employment statistics that are more encouraging for women than they are for

TABLE 10.3 Labour statistics: Inuit, other Aboriginal, Canada (2011)

	Inuit		*Other Aboriginal*		*Canada*	
	Men	*Women*	*Men*	*Women*	*Men*	*Women*
Employment	48.9%	50.2%	63.9%	57.0%	65.7%	57.8%
Unemployment	23.4%	18.4%	11.7%	10.7%	8.0%	7.0%
Labour force participation	63.9%	61.6%	72.4%	63.9%	71.4%	62.2%

Source: APS (2012); Statistics Canada (2016a).

TABLE 10.4 Inuit labour force participation (2006 and 2011)

	2006			*2011*		
	Male	*Female*	*Total*	*Male*	*Female*	*Total*
Employment	51.6%	49.6%	50.6%	48.9%	50.2%	49.6%
Unemployment	19.8%	17.7%	18.8%	23.4%	18.4%	20.8%
Participation rate	64.4%	60.2%	62.3%	63.9%	61.6%	62.7%
Full-time	75.6%	75.3%	75.5%	82.7%	77.0%	79.6%

Source: APS (2006, 2011).

men. With this said, increasing unemployment is obviously not encouraging, especially considering the high unemployment rates that already plague Inuit communities.[10]

For those who were employed, there was a slight shift away from part-time and toward full-time employment. This was the case for both men and women, though the male increase of seven percentage points was roughly four times greater than the female shift. Thus, while overall employment changed little, there was a slight increase in hours worked. This evokes some optimism as Statistics Canada has highlighted that roughly half of Aboriginal part-time workers work the hours that they do because they are unable to find full-time employment (Bougie, Kelly-Scott & Aarriagada 2013, 48).

An argument could be made that the relatively low employment and labour force participation rates of men is negatively impacted by time spent gathering country food. Specifically, an individual engaged in full-time hunting may be less inclined to work or look for work which would deflate each of these statistics. With this said, the high unemployment rate for Inuit men (23.4 per cent in 2012) indicates that many men are looking for work but unable to find it.

The APS also reports on barriers to employment and shows that the labour market was actually more difficult to access in 2011 than in 2006. Specifically, results for each obstacle to employment increased by 8.6 per cent and 9.4 per cent for

TABLE 10.5 Challenges faced by Inuit jobseekers (2006 and 2011)

	Men		*Women*	
	2006	*2011*	*2006*	*2011*
Not knowing where to look	30.9%	42.3%	30.8%	40.5%
Not knowing what type of job wanted	27.7%	39.8%	30.1%	39.9%
No required experience	41.8%	52.4%	39.1%	50.9%
Not enough education	51.4%	55.1%	50.9%	56.1%
No transportation	22.5%	34.0%	17.9%	30.3%
Shortage of jobs	77.9%	85.1%	74.1%	76.9%
Other	11.6%	12.0%	13.9%	10.6%

Source: APS (2006, 2012).

TABLE 10.6 Inuit education levels in the Nunangat (2011)

	No degree/diploma	*High School*	*Post-secondary*		
			Trades/ apprenticeships or other non-university	*University below Bachelor degree*	*University degree*
Men	61.8%	12.4%	23.1%	1.1%	1.6%
Women	60.7%	14.0%	20.0%	2.0%	3.4%

Source: APS (2012).

women and men respectively on average (omitting "other" from the calculation). That over half of respondents reported "shortage of jobs" and "not enough education" as challenges is significant.

Finally, similar challenges were reported on the labour demand side. A document by the Conference Board of Canada (Howard, Edge & Watt 2012) reported on the challenges that the private sector identified with regards to hiring Aboriginal workers. Topping the list was low skill level (70.2 per cent of respondents) and lack of work experience (59.6 per cent). No other category garnered more than 50 per cent response rates, with mobility issues coming in a distant third at 39.1 per cent. Amongst Aboriginal groups, businesses identified Inuit workers as being the most challenging in terms of employability.

To summarize, the Nunangat is plagued by a lack of labour demand and a lack of skills, education and experience in its supply. These challenges are undoubtedly related to one another and education levels in the Nunangat are especially troublesome. Table 10.6 reflects education levels in the Nunangat, by gender.

More than three-fifths of both populations have not completed their high school education. For those who have, women are more likely to have completed a higher level, which has led some to highlight a (relative) feminization of human capital in the Arctic (Larsen & Fondahl 2014). With this said, educational attainment in the region lags far behind Canadian averages and the absence of a university in the region is contributing to a skills drain out of the Nunangat and a corresponding gender imbalance for those who stay behind (Rodon et al. 2014).[11]

Climate change, gender, and work

The most oft-cited impact of climate change for the Inuit is in the traditional domain and relating to an enduring reliance on foods harvested from the environment. Climate change is threatening this food system and traditional roles lead to different experiences for men and women. There is a tendency in the literature to focus on the male experience (Shannon 2006), but it becomes clear that these changes have implications for females as well, primarily because of existing gendered divisions of work and adaptive strategies.

Impact

A significant amount of attention has been paid to how climate change impacts the procurement of country food (see, for example Pearce et al. 2010). One reason for this is that the causal pathways are relatively direct and easy to measure (changes in the physical climate on the distribution of wildlife). Additionally, studies of this nature are compatible with a broader research agenda (and funding) as the observations of Inuit hunters have been used to corroborate technical findings related to climate change in the Arctic (Laidler et al. 2009).

Analysis of the impact of climate change on country food procurement typically highlights three themes. First, as equipment like GPS locators and snowmobiles are adopted to mitigate the impact of changing weather conditions, the monetary costs assocated with the hunt have increased. In this context it has been reported that waged work has been taken up in order to support the escalating costs of the hunt (Rasmussen 2009).

Second, Inuit traditional knowledge is celebrated for its role in facilitating the hunt and mitigating the risks associated with it. This knowledge relates to land skills, access routes, the prediction of weather conditions and the interpretation of the physical environment. Climate change is undermining this knowledge as environmental conditions shift in the ways discussed above. As such, hunters have reported that many of the traditional indicators that they rely upon are growing less reliable in the wake of a changing climate (Laidler et al. 2009). This growing uncertainty leads to the hunting season being delayed each year and subject to increasing time spent waiting for conditions to turn for the better.

Third, the returns from hunting are decreasing. Changing weather patterns and warmer temperatures are stressing food systems throughout the Arctic and resulting in lower returns to time spent hunting. The overall effect of this is to discourage individuals from hunting. For some this results in less time being spent hunting, while for others it can lead to a complete withdrawal from the activity (ibid.).

Overall, for those who have the means and willingness to subject themselves to such risks, climate change implies a diminished return to hunting. Because of this, less country food is being harvested by fewer hunters, increasing food insecurity (Ford 2009). Simply to maintain the already low level of food security in the region, other sources of food are required to compensate. Thus, the primary impact of climate change is seen as being situated in the traditionally male domain since it makes hunting more costly, more risky and less rewarding.

The impact of climate change on women is less pronounced in the literature. Pearce et al. (2009) suggest that some of this might stem from the tendency for Inuit men to speak on their family's behalf, the predominance of men in research teams, and the higher proportion of men in hunter and elder groups. An alternative explanation is that a lack of attention and funding is given to domains that are traditionally female (Desbiens 2010).

Research in the female domain generally revolves around the role that women play in supporting the hunt, and the secondary effects that stressed country food systems can have on female work (these include diminishing fur quality and food stocks). Attention has also been paid to the impact that climate change is having on health levels – another traditionally female domain. Accordingly, it is important that female perspectives are given more attention in research (Furgal & Seguin 2006).

Going forward

Increasing monetary costs (for hunting equipment and store food) imply a growing need to engage in wage labour. While a division between men as hunters and women as primary wage earners has been observed in the Arctic, the decreasing viability of country food systems suggests that this division will be stressed. Adaptive strategies in the Arctic would be best served by taking these dynamics into account.

These issues are critical given the high prevalence of food insecurity that suggests that Inuit food systems are currently not meeting the populations' needs. As country food systems become increasingly stressed, improving (or even maintaining current levels of) food security will necessitate a further shift to store food consumption. To facilitate this, the impediments to employment need to be addressed and the effect of this shift needs to be better understood. Questions emerge in both of these domains.

From the data above, a lack of engagement with formal education and skills development is undermining the adaptive capacity of the Inuit. Understanding what is causing this and finding community-driven means by which to improve these conditions is essential to breaking down barriers to employment. It is possible that the aforementioned feminization of human capital speaks more to a lack of engagement by Inuit men in education and the wage labour force than it does engagement by women. Alternatively, that Inuit respondents pointed at job shortages as the biggest challenge to finding employment suggests that problems are not only supply-driven. These factors are not exogenous of one another, and a deeper understanding of the supply and demand of labour in the Nunangat is essential if the vicious cycle is to be escaped. This is especially true in the private sector vis-à-vis the region's overreliance on work in the public sector.

It is also critical to consider the role of Inuit women in the household. While relatively high engagement in the wage labour market is suggested to be a means to redefine their role in society (Röhr 2007), this cannot be looked at in isolation. To this point, Table 10.7 reflects hours of upaid work carried out in the household. Evident here is a significantly higher burden of work at home, with Inuit women dedicating an average of 17 hours per week more to housework and childcare than men do.

The danger is that women are filling two significant roles in supporting the household and that additional responsibilities in the wage labour market will be unlikely to correspond with less unpaid work. With this in mind, it is important to consider that increasing one's workload without a corresponding increase in compensation risks exacerbating existing inequalities (Wong 2009).

TABLE 10.7 Hours of unpaid domestic work in the Inuit Nunangat

	Hours of unpaid childcare		*Hours of unpaid housework**	
	Men	*Women*	*Men*	*Women*
No. hours	45.2%	32.4%	17.4%	8.3%
< 5	12.0%	8.1%	26.0%	15.8%
5 to 14	11.1%	10.6%	27.9%	27.0%
15 to 29	8.9%	9.5%	15.8%	22.1%
30 to 59	7.6%	11.2%	7.7%	15.1%
> 60	15.1%	28.3%	5.6%	11.6%
N	15,520	16,295	15,525	16,290
Weighted average	16	25	13	21

Source: Aboriginal Affairs and Northern Development Canada (2012).

* Household work includes cleaning, cooking, household maintenance, yard work.

While there are normative questions that can be raised with regards to replacing traditional activities with waged work, the rampant and increasing food insecurity in the region leaves few viable alternatives. One way to enhance food security is to ensure good sources of income through increased participation in waged labour (Damman, Eide & Kuhnlein 2008; Dietz 2009). To date, this has not been happening and the data suggests that the situation may be worsening.

There is an obvious role for policy in this context and it has been noted that the Canadian government has a moral and legal obligation to the Inuit (amongst other groups) to ensure a right to food and to commit resources to support regional development more broadly (Ford et al. 2010). With this said, current conditions suggest that these programmes have been inadequate to date.

More concerning is the fact that the policies that do exist suffer from a lack of gender-attentiveness. At the federal level, a collection of publicly funded programmes exists for the purpose of spurring economic development in the region. These range from the Northern Aboriginal Basic Education Program (NABEP), which focuses on improving labour market participation, to the Strategic Investments in Northern Economic Development (SINED), which focuses on strengthening the driver sectors of the economy in the Canadian Territories. Unfortunately, CanNor – the federal agency mandated to help develop Canada's North – does not track gender in the application process nor are they able to provide any gender-related statistics.[12] This is reflective of a lack of gender focus in the region and an acknowledged *opportunity for further study* according to the National Aboriginal Economic Development Board (2015).

This is not to neglect the role that government subsidy has and will continue to have in bolstering country and store food systems in the region. Income subsidy represents a considerable portion (18.5% in 2010) of household income on top of the subsidies that go into store foods and there is little reason to believe that these will not be needed in the future (Ford et al. 2010). Here again, things seem to be trending in the wrong direction. Specifically, the Nunavut Harvester Support

Program – designed to provide financial assistance to hunters, was suspended in April of 2014 pending a review of the programme that was to be completed by October of the same year. As of late 2015, this review was still ongoing and the programme's offering in 2016 has yet to be determined.[13]

There may be hope on the horizon, however. In 2014, a Nunavut coalition representing the territorial government as well as a collection of Inuit organizations produced the territory's first ever *Food Security Strategy and Action Plan* (Nunavut Food Security Coalition 2014). The document highlighted high levels of food insecurity and acknowledged that both country and store foods have a role to play in the local food system. Strategic areas outlined in the report include agricultural production and the development of life skills in the population. Conspicous in its absence is that, outside of "mak[ing] life skills programming available to broader audiences that include a diversity of ages and genders", little attention is paid to gender in the document. This is ominous and a route through which gender should be introduced into the discussion.

Conclusion

A relative dearth of gendered analysis exists with regards to climate change and work in the Inuit Nunangat as existing research has focused on the traditionally male domain of hunting. This chapter highlights that the wage labour market in the Arctic is inadequate for the purpose of ensuring food security for the Inuit – particularly vis-à-vis stressed country food systems. It also shows that females in the Arctic are relatively more engaged with the wage labour market than their male counterparts. This may corroborate Rasmussen's claim that Inuit women are conditioned to flexibility and men to rigidity, or speak to other dynamics in the region (for example, the nature of the work that presently exists – especially in the public sector). Next steps for research and policy work will be to gain a better understanding of this trend, and to ensure that neither gender need shoulder a disproportionate burden.

Notes

1 This chapter focuses on individuals aged 15 and over.
2 The newly elected Liberal government is reinstating the long-form census in 2016, suggesting that this problem might be unique to 2011–2012.
3 For an examination into the evolution of these sharing networks, see Harder and Wenzel (2012).
4 It costs $360–$450 per week to feed a family in northern regions and $200–$250 to provide the same basket of goods in southern Canada (Aboriginal Affairs and Northern Development Canada 2012).
5 The Government of Canada provided $64.9 million dollars in food subsidy in 2014–2015 through its Nutrition North programme (Government of Canada n.d.).
6 These levels are derived from the calculation of a composite food security indicator from questions asking about food run-out and affordability and if meals were skipped/reduced because of a lack of resources (Statistics Canada 2011).

7 These statistics are based on national populations and the Inuit figures include respondents residing outside the Nunangat. They constitute an upper bound as economic conditions in the Nunangat are consistently worse than in the rest of the country.
8 The higher Inuit unemployment rate has an inflating effect on the Total Aboriginal figure meaning that the Aboriginal Non-Inuit figure would actually be lower than is reflected.
9 Data sources: Statistics Canada 2016b and APS 2012.
10 Inuit unemployment rates the highest of the three Canadian Aboriginal groups and more than twice as high as the non-Aboriginal rate (The National Aboriginal Economic Development Board 2015, 15).
11 With a 106:100 ratio, Canada was found to have the highest male:female ratio of all Arctic regions.
12 Correspondence with CanNor, 9 August 2015.
13 Correspondence with Nunavut Tunngavik Inc., 4 September 2015.

References

Aboriginal Affairs and Northern Development Canada 2012, *Aboriginal women in Canada: A statistical profile from the 2006 Census*, (http://www.aadnc-aandc.gc.ca/DAM/DAM-INTER-HQ/STAGING/texte-text/ai_rs_pubs_ex_abwch_pdf_1333374752380_eng.pdf).

APS 2006, "Aboriginal Peoples Survey, 2006: Inuit health and social conditions, no. 89-637-X", Statistics Canada, Ottawa.

APS 2012, "Aboriginal Peoples Survey, 2012: The education and employment experiences of First Nations People living off reserve, Inuit and Métis – Selected findings from the 2012 Aboriginal Peoples Survey, no. 89-637-X", Statistics Canada, Ottawa.

Arora-Jonsson, S 2011, "Virtue and vulnerability: Discourses on women, gender and climate change", *Global Environmental Change*, vol. 21, no. 2, pp. 744–751.

Beaumier, MC & Ford, JD 2010, "Food insecurity among Inuit women exacerbated by socio-economic stresses and climate change", *Canadian Journal of Public Health*, vol. 101, no. 3, pp. 196–201.

Bougie, E, Kelly-Scott, K & Arriagada, P 2013, "The education and employment experiences of First Nations Peoples living off reserve, Inuit, and Métis: Selected findings from the 2012 Aboriginal Peoples Survey", Statistics Canada, Ottawa.

CanNor 2015, "Correspondence", 9 August.

Chan, HM, Fedluk, K, Hamilton, S, Rostas, L, Caughey, A, Kuhnlein, H, Egeland, G & Loring, E 2006, "Food security in Nunavut, Canada: Barriers and recommendations", *International Journal of Circumpolar Health*, vol. 65, no. 5, pp. 416–31.

Damman, S, Eide, WB & Kuhnlein HV 2008, "Indigenous peoples' nutrition transition in a right to food perspective", *Food Policy*, vol. 33, no. 2, pp. 135–155.

Desbiens, C 2010, "Step lightly, then move forward: Exploring feminist directions for northern research", *The Canadian Geographer*, vol. 54, no. 4, pp. 410–416.

Dietz, T 2009, "Ecospace, humanspace and climate change", in MA Mohamed Salih (ed.), *Climate change and sustainable development: New challenges for poverty reduction*, pp. 47–58, Edward Elgar Publishing, Inc., Northampton, MA.

Dowsley, M, Gearheard, S, Johnson, N & Inksetter, J 2010, "Should we turn the tent? Inuit women and climate change", *Études/Inuit/Studies*, vol. 34, no. 1, pp. 151–165.

Ford, JD 2009, "Vulnerability of Inuit food systems to food insecurity as a consequence of climate change: A case study from Igloolik, Nunavut", *Regional Environmental Change*, vol. 9, no. 2, pp. 83–100.

Ford, JD, Smit, B & Wandel, J 2006, "Vulnerability to climate change in the Arctic: A case study from Arctic Bay, Canada", *Global Environmental Change*, vol. 16, no. 2, pp. 145–60.

Ford, JD, Gough, WA, Laidler, GJ, MacDonald, J, Irngaut, C & Qrunnut, K 2009, "Sea ice, climate change, and community vulnerability in northern Foxe Basin, Canada", *Climate Research*, vol. 38, no. 2, pp. 137–54.

Ford, J, Pearce, T, Duerden, F, Furgal, C & Smit, B 2010, "Climate change policy responses for Canada's Inuit population: The importance of and opportunities for adaptation", *Global Environmental Change*, vol. 20, no. 1, pp. 177–191.

Furgal, C & Prowse, TD 2008, "Northern Canada", in DS Lemmen, FJ Warren, J Lacroix & E Bush (eds), *From impacts to adaptation: Canada in a changing climate 2007*, pp. 57–118, Government of Canada, Ottawa.

Furgal, C & Sequin, J 2006, "Climate change, health, and vulnerability in Canadian northern aboriginal communities", *Environmental Health Perspectives*, vol. 114, no. 12, pp. 1964–1970.

Government of Canada n.d., Nutrition North Canada 2014–2015: Full fiscal year, (http://www.nutritionnorthcanada.gc.ca/eng/1453219591740/1453219765675).

Harder, MT & Wenzel, GW 2012, "Inuit subsistence, social economy and food security in Clyde River, Nunavut", *Arctic*, vol. 65, no. 3, pp. 305-318.

Hassol, S 2004, *Impacts of a warming Arctic: Arctic climate impact assessment*, Cambridge University Press, Cambridge.

Howard, A, Edge, J & Watt, D 2012, *Understanding the value, challenges, and opportunities of engaging Métis, Inuit, and First Nations Workers*, The Conference Board of Canada, Ottawa.

Kafarowski, J 2009, "Gender, culture, and contaminants in the North", *Signs*, vol. 34, no. 3, pp. 494–499.

Laidler, GJ, Ford, JD, Gough, WA, Ikummaq, T, Gagnon, AS, Kowal, S, Qrunnut, K & Irngaut, C 2009, "Travelling and hunting in a changing Arctic: Assessing Inuit vulnerability to sea ice change in Igloolik, Nunavut", *Climate Change*, vol. 94, no. 3–4, pp. 363–397.

Lambden, J, Receveur, O & Kuhnlein, HV 2007, "Traditional food attributes must be included in studies of food security in the Canadian Arctic", *International Journal of Circumpolar Health*, vol. 66, no. 4, pp. 308–19.

Larsen, JN & Fondahl, G (eds) 2014, "Arctic human development report: Regional processes and global linkages", Nordic Council of Ministers, Denmark.

Mancini Billson, J & Mancini, K 2007, *Inuit women: The powerful spirit in a century of change*, Rowman & Littlefield, Lanham, MD.

Morgan, C 2008, "The Arctic: Gender issues", Parliament of Canada, Social Affairs Division, Ottawa.

Mulrennan, ME 2014, "On the edge: A consideration of the adaptive capacity of Indigenous peoples in coastal zones from the Arctic to the tropics", *Geological Society*, London, vol. 388, no. 1, pp. 79–102.

The National Aboriginal Economic Development Board 2015, *The aboriginal economic progress report 2015*, Gatineau, Quebec, (http://www.naedb-cndea.com/reports/NAEDB-progress-report-june-2015.pdf).

Nunavut Food Security Coalition 2014, "Nunavut food security strategy and action plan 2014–2016", Nunavut.

Nunavut Tunngavik Inc. 2015, "Correspondence", 4 September.

Nuttall, M 2005, "An environment at risk: Arctic Indigenous peoples, local livelihoods and climate change", in JB Ørbæk, R Kallenborn, I Tombre, EN Hegseth, S Falk-Petersen & AH Hoel (eds), *Arctic alpine ecosystems and people in a changing environment*, pp. 19–35, Springer-Verlag, Berlin.

Pachauri RK & Meyer LA (eds) 2014, *Climate change 2014: Synthesis report, Contribution of Working Groups I, II and III to the Fifth Assessment Report of the Intergovernmental Panel on Climate Change*, IPCC, Geneva.

Pearce, T, Smit, B, Duerden, F, Ford, JD, Goose, A & Kataoyak, F 2010, "Inuit vulnerability and adaptive capacity to climate change in Ulukhaktok, Northwest Territories, Canada", *Polar Record*, vol. 46, no. 2, pp. 157–177.

Power, EM 2008, "Conceptualizing food security for Aboriginal people in Canada", *Canadian Journal of Public Health*, vol. 99, no. 2, pp. 95–97.

Rasmussen, RO 2009, "Gender and generation: Perspectives on ongoing social and environmental changes in the Arctic", *Signs*, vol. 34, no. 3, pp. 524–532.

Rodon, T, Walton, F, Abele, F, Dalseg, SK, O'Leary, D & Lévesque, F 2014, "Post-secondary education in Inuit Nunangat: Learning from past experiences & listening to students' voices", *Northern Public Affairs*, SI, pp. 70–75.

Röhr, U 2007, Gender, climate change and adaptation: Introduction to the gender dimensions, *Both Ends Briefing Paper Series* 2 August, (http://hqweb.unep.org/roa/amcen/Projects_Programme/climate_change/PreCop15/Proceedings/Gender-and-climate-change/Roehr_Gender_climate.pdf).

Shannon, KA 2006, "Everyone goes fishing: Understanding procurement for men, women and children in an Arctic community", *Études/Inuit/Studies*, vol. 30, no. 1, pp. 9–29.

Statistics Canada 2011, "About the data: National Household Survey (NHS) Aboriginal population profile", (https://www12.statcan.gc.ca/nhs-enm/2011/dp-pd/aprof/help-aide/about-apropos.cfm?Lang=E#a3).

Statistics Canada 2016a, "CANSIM Table 282-0087: Labour Force Survey Estimates", (http://www5.statcan.gc.ca/cansim/a26?id=2820087).

Statistics Canada 2016b, "CANSIM Table 282-0144: Labour Force Survey Estimates", (http://www5.statcan.gc.ca/cansim/a26?id=2820144).

Stevenson, V 2015, "Iqaluit grocery shelves left bare as ice and fog hinder food shipments", *The Globe and Mail*, (http://www.theglobeandmail.com/news/national/iqaluit-grocery-shelves-left-bare-as-ice-and-fog-hinder-food-shipments/article25543820/).

Wallace, S 2014, "Inuit health: Selected findings from the 2012 Aboriginal Peoples Survey", Statistics Canada, Ottawa.

Wong, S 2009, "Climate change and sustainable technology: Re-linking poverty, gender, and governance", *Gender and Development*, vol. 17, no. 1, pp. 95–108.

11

TOWARDS HUMANE JOBS

Recognizing gendered and multispecies intersections and possibilities

Kendra Coulter

Introduction

Gendered and feminist analyses offer valuable perspectives on the entanglements of work, political economy and climate change. There are, however, significant multispecies dimensions, which, so far, have been under-examined. Accordingly, in this paper, I propose an intersectional approach to the labour politics of climate change which takes nonhuman animals and human–animal relations seriously. There is potential to expand our ideas about green labour to include what I call humane jobs: jobs that benefit both people and animals. As a result, my goals are to offer generative, conceptual fodder that can propel scholarship and political action, and to encourage a nuanced, ethical, sustainable and just approach to multispecies labour.

Crucially, a humane jobs agenda would help move the labour force away from worker–animal–environmental harming practices and towards more beneficial work. As such, humane jobs are both reactive and proactive; they are about what should be critiqued and dismantled, and what should be created. Humane jobs are a response to human–animal harm, and an opportunity to envision and foster more positive political economic relations and multispecies experiences. Humane jobs are thus both a crucial extension of the green jobs agenda, and essential to building more just, caring and sustainable societies and economies. By enlisting both a gendered and a multispecies lens, we are challenged to think differently about labour processes, economic patterns and political possibilities. In that spirit, my proposals for humane jobs build from, but also go beyond, existing paradigms in labour studies, feminist political economy and research on nature and culture. There are compelling social, economic, environmental and ethical reasons to not only to theorize, but also pursue the creation of new sustainable and equitable areas of employment.

A combination of intellectual analyses, political and ethical arguments, and quantitative and qualitative evidence informs this project. The data on the negative effects of contemporary industrialized agriculture on workers, animals, surrounding communities and the global climate are clear, yet such processes are under-considered in much green labour thinking, particularly in solutions-oriented scholarship. The effects of the industrial farming of animals on people and the environment are persuasive enough on their own to demonstrate why dominant agricultural patterns warrant rethinking as part of a comprehensive approach to the climate. However, my arguments about humane jobs are also underscored by a commitment to intersectionality that includes nonhuman animals not only as objects worthy of study, but as sentient subjects worthy of ethical consideration and inclusion in initiatives that prioritize equity, justice and sustainability. Such a commitment is a necessary extension of what I call interspecies solidarity, the interpersonal and political expansion of spheres of empathy across species lines (Coulter 2016a).

This approach shares the spirit of Claire Jean Kim's (2015) call for multi-optic vision. She points out "most social justice struggles mobilize around a single-optic frame of vision. The process of political conflict then generates a zero-sum dynamic. . . a *posture of mutual avowal* – an explicit dismissal of and denial of connection with the other form of injustice being raised" (19). Indeed, many labour, environmental and animal advocacy projects rely on a more myopic lens. In contrast, Kim argues that a commitment to multi-optic vision recognizes connections and synergies, and emphasizes the importance of seeing from within various perspectives, including those of nonhumans. Such an approach does not negate differences or tensions, but rather is informed by the need to forge more diverse and inclusive analyses and alliances. This idea also expands on Josephine Donovan's (2007) call for including the "standpoint" of animals in feminist research and ethical deliberations. Such a commitment not only builds more multi-faceted and thorough understanding, but also encourages broader political coalitions.

Feminist approaches to labour-climate politics begin down a better path by highlighting unevenness and gendered inequities which can be reproduced or transcended. Yet moving beyond androcentrism, while remaining shackled to anthropocentrism, is insufficient and thus a broader approach is needed. Anthropogenic causes of climate change affect "the environment". The environment is comprised of ecosystems, and all members thereof, including sentient nature, are shaped by the actions of the human species. Tackling climate change is about people, including women, the working class, Indigenous communities and the poor, but it is also about other species and their well-being. Accordingly, humane jobs are about people, but they are not only about people.

Health care researchers, including veterinarians, have developed an approach to therapeutic and preventative medicine that interweaves human, animal and environmental health (Mackenzie et al. 2013; Woldehanna & Zimicki 2014). Called the One Health Model, this framework recognizes the connectedness of health and well-being at individual, interpersonal and communal levels, within and across species lines. The idea of humane jobs stems, in part, from the prospects of comparable theory and practice for the world of work. It recognizes that "there has never been any purely human space in world history" (Nance 2013, 7), and that human and nonhuman lives are intermingled in biological and social communities, and in the many intersections of nature and culture, including in the crucial nature-labour nexus (Coulter 2014a). Humane jobs have applicability and relevance around the world, and the agricultural patterns of rich countries affect the global climate, which disproportionately harms the global south. In keeping with this volume's focus, here I concentrate on the global north, and address colonial industries and practices.

Industrial agriculture, labour and climate change

Agricultural production in the global north has been greatly shaped by processes of industrialization, consolidation and corporatization. The image of a small or modest sized family-owned farm which dominates the public imagination is increasingly inaccurate. These kinds of farms are being replaced by larger "factory farms" or corporate agribusinesses, where crops are grown or very high numbers of animals are kept. "Intensive (or Industrial) Livestock Operations" and "Concentrated Animal Feeding Operations" are being expanded in order to maximize the number of animals that can be kept and then killed to accelerate the accumulation of profit (Weis 2007, 2013). Because of corporate consolidation and the strategic purchasing of smaller facilities and companies, a handful of agricultural corporations own much of the food production and processing system. In many cases, the same company may own all or many stages of the "food production" process.

In 1961, there were 500,000 farms operating in Canada. In 2011, that number had dropped to 205,000, and it is projected to continue declining (Statistics Canada 2011a, 2011b). Farm operators hire employees from their local communities, and/or migrant workers who labour and then are usually sent back to their home countries. Most are men, but women are actively involved in various kinds of farm work. Migrant women in particular are often assigned the lowest paying tasks, and women in farm-owning families regularly generate the essential off-farm income generation which allows their businesses to survive (Preibisch and Grez 2014; Price 2012). In Canada's most recent census, nearly half of all individual farmers reported that their primary income source was not their farms.

About 85,000 farms Canada grow crops, 37,000 house cattle who will become meat, 12,000 have cows who will produce milk (and then become meat after a few years), and 8,000 grow fruit (Agriculture and Agri-Food Canada 2011). The rest raise or grow chickens, turkeys, pigs, rabbits, mink, foxes, horses, vegetables, grains or plant proteins, or are mixed animal-crop farms. 700 million farmed animals are killed annually in Canada, a country with a population of about 35 million people. This is noteworthy for a number of reasons, including the fact that there are only about 100,000 farms where animals are raised for human consumption which reflects the large size of contemporary farming operations (Canadian Federation of Humane Societies n.d.; Statistics Canada 2011a).

The Department of Agriculture calculates that nine billion animals are killed annually within the United States, but that number does not include horses, rabbits, fish and crustaceans so is much higher in actuality (Humane Society of the United States 2014). The number of animals killed climbs by tens of billions when considering the global situation.

The industrialization of agriculture and slaughterhouses has meant that many of the rural jobs available are highly precarious, low paying, difficult and dangerous (Lipschitz and Gardner 2014; Pini and Leach 2011; Stull and Broadway 2013). The many effects of slaughterhouse labour on the workers therein and on surrounding communities are troubling. These include significant physical and psychological harm to workers directly (Adam, Gibson & Cook 2016; Dillard 2008), and increasing crime rates and pollution in the areas around slaughtering facilities (Fitzgerald 2010; Fitzgerald, Kalof & Dietz 2009; Jacques 2015). Injury and illness rates in slaughterhouses of all kinds are notably higher than manufacturing averages; in plants killing and processing larger animals like cows, horses and pigs, workers are twice as likely as the average worker in manufacturing to get hurt on the job (Stull & Broadway 2013). White, US or Canadian born male workers have increasingly left jobs in these facilities, and now racialized workers, women, immigrants, migrant workers and/or undocumented people heavily populate slaughterhouse workforces (Stull & Broadway 2013; Pachirat 2011).

In Canada, the United States and many other countries of the global north, the majority of animals intended to be consumed or to produce food with their bodies today live indoors in large, uniform, often window-less shed-like facilities. Cows intended to be beef and sheep used for wool and/or meat (lamb and mutton) will often be permitted to graze outdoors for some or all of the time. Most animals on farms, including pigs, chickens, turkeys, geese, rabbits and a majority of cows used for milk production, are always kept inside. Fur "farms" also house animals inside. Dominant Canadian animal-keeping practices today include gestation crates for female pigs that allow them to stand up and lie down, but not move or turn around for a year or more. Such structures are not permitted in barns being built henceforth due to regulatory changes initiated by animal advocates (and are illegal in a number of European countries), but are accepted and widely used in existing facilities. Dave Wager, the communications director for the National Pork Producers Council in the United States said the following: "So our animals can't turn around for the 2.5 years that they are in the stalls producing piglets. I don't know who asked the sow if she wanted to turn around" (Friedrich 2012, n.p.).

Similarly, most chickens are kept stacked in rows of tiny battery cages within which the birds cannot spread their wings. Milk is produced only by cows who have given birth, thus the animals are impregnated repeatedly over a 3–4 year period after which they, too, are slaughtered. After birth, female calves are taken to become milk producers themselves, while the males are usually kept alive and mostly immobile and solitary for a few months before being slaughtered for veal. Calves of both sexes are almost always taken away from their mothers the day of their birth. Most of these highly restrictive animal-keeping practices are newer developments. For many decades, different methods for raising and housing animals were used across Canada – and are currently used in a number of countries. There is some heterogeneity across North American farms, but intensive confinement is widely used and normalized.

The industrialization of agriculture also has noteworthy environmental implications. The housing of hundreds or thousands of animals together in cramped conditions affects how much food and water are needed, how much waste is produced and where it is stored, and how diseases can transfer among animals, as well as to humans (the latter are called zoonoses). Public health researchers, including the World Health Organization (2010), identify industrial agriculture, and how animals are treated therein, as a significant risk to human health due to direct transfer of pathogens and because of the environmental impacts (Akhtar 2012; Cutler, Fooks & Van Der Poel 2010; Heffernan, Salman & York 2013). Pig farms, for example, are notorious in many rural regions for posing a risk to ground water and air quality in the surrounding areas, namely because of the need for large manure lagoons (Johnsen 2003; Ramsey, Soldevila-Lafon & Viladomiu 2013; Wing, Horton & Rose 2013).

Although a somewhat longer history of research exists, the role industrial animal agriculture plays in climate change gained particular attention following the release of the United Nation's Food and Agriculture Association's report "Livestock's long shadow: Environmental issues and options" (Steinfeld et al. 2006). The report concentrated on the industrialization of agriculture and its impacts on water pollution, air quality, land use, biodiversity loss and climate change. Specifically, the authors found that industrial agriculture is a major driver of climate change, one that contributes more greenhouse gas emissions than the transportation sector. In fact, they argue that "the livestock sector has such deep and wide-ranging environmental impacts that it should rank as one of the leading focuses for environmental policy" (Steinfeld et al., xxiv). Subsequent research has assigned varying percentages of the total quantity of greenhouse gases emitted through the agricultural system globally (from 10 per cent to close to 50 per cent), which Herrero et al. (2011) argue stems from what precisely is included in the calculation and which research methods are used. Most researchers approach industrial agriculture as a larger system that involves not only the raising of animals, but deforestation for grazing, the allocation of other land for growing food which will be fed to farmed animals, the corresponding use and moving of fertilizers, the transportation of living animals to slaughterhouses, and waste management. In other words, there are livestock specific effects, but the agricultural causes are also entangled with other significant industries, particularly fossil fuels and forestry.

Notably, while scientists debate the exact percentages of agriculture's contribution to GHG emissions, there is widespread recognition that industrial agriculture is a "major" contributor to climate change, and one which is increasing rapidly (Caro et al. 2014; Gerber et al. 2013; Heffernan, Salman & York 2013; Herrero et al. 2011). However, despite the initial and subsequent UN reports which reaffirm key findings, the broader scientific data pool and researchers' arguments that industrial agriculture warrants great attention, the climate–livestock links are not a prominent issue in many branches of environmental thinking or political action. Certain researchers and advocates call for decreased meat and dairy consumption, some call for changes to the size and operation of animal agriculture, and a small number call for its elimination. In many cases, however, the negative impacts of industrialized animal farming on the climate are ignored (Bristow 2011). Industrial agriculture's effects on the climate do not need to usurp fossil fuel-focused work, but given the evidence, a failure to identify and address the sector is very problematic.

In addition to the persuasive scientific data, there are compelling ethical reasons to recognize the harm which is wrought by industrial animal agriculture. Women who work in such systems are more likely to experience what Jocelyne Porcher (2011) calls "shared suffering", as they empathize with and internalize the physical and psychological harm being done to animals (see also Wilkie 2010). Indeed, animals are the most damaged in such systems, yet among humans, it is women, working class communities and people, racialized workers, and poor people who are particularly negatively affected, because they are disproportionately represented among the workforces of industrialized animal agriculture facilities and slaughterhouses, and/or because they must live near the facilities (Coulter 2016a; Fitzgerald 2010; Nibert 2013; 2014). As climate change worsens and its effects deepen and expand, it is also these very groups, as well as Indigenous peoples, who will be most harmed, particularly in the global south.

At the same time, labour and environmental researchers and advocates must confront the fact that sentient nature – animals – are being unnecessarily subjected to short lives of intense misery. It is difficult to overstate the suffering rendered by the extreme devaluation that occurs behind the terms "industrial animal agriculture" and "factory farming". Male chicks deemed superfluous within the egg industry are killed within hours of their birth, often by being dropped onto a twirling blade called a macerator. Calves intended to be veal, and chickens of both sexes destined for prompt slaughter as meat have among the shortest lives at 2–3 months (egg-laying hens are often killed for meat after 1–2 years). Whether tasked with producing milk, eggs or babies, females must endure particularly restrictive and painful practices as their bodily labours are exploited (Adams 2010; Coulter 2016a; Gruen 1993). All animals intended to become food or fashion in countries like Canada, from the smallest chickens to the largest cows, are sentient, social beings who possess individual personalities, are capable of rich, interactive relationships, and have their own needs and desires, including an interest in performing their own, self-initiated and controlled care work (Coulter 2016b). Yet in most cases, contemporary agricultural practices prevent animals from interacting with their own offspring, performing natural behaviours, and achieving a modicum of dignity, prompting Barbara Noske (1989, 18–20; see also 1997) to see these animals as an example of alienated labour. She argues that animals in industrial agriculture are alienated from the product (their own offspring or parts of their body), productive activity (not being able to turn around, for example), fellow animals and their social nature, surrounding nature and their species life.

The roots of feminist scholarship are in movements for gender equality and intersectional social justice. Feminist research is about identifying the ideas and practices which reproduce hierarchies of oppression and privilege, and contesting, disrupting and resisting violence and subjugation. Nonessentialist animal ecofeminists have long argued that there is no defensible reason for feminist principles to stop at the species boundary, and have identified the cultural and material practices which underscore both gendered and interspecies domination (Adams 2007; 2010; Adams & Gruen 2014; Gaard 1993, 2011). In addition to the empirical evidence about environmental damage and worker and community harm, there are insuppressible ethical and empathetic reasons for tackling the ills of industrial agriculture, and, crucially, for feminists to play a central role in identifying, developing and promoting alternatives. Violence against animals, whether individualized and targeted, or normalized and systemic, is a feminist issue.

The politics of supply and demand, as well as individual consumption choices are not straightforward, but decreased demand for food, clothing and other products derived from animals can, undoubtedly, affect production practices, and thus the corresponding environmental and interspecies patterns to some degree. For example, the increase in vegetarian and vegan diets seen across the global north can play a role in rewarding particular farming practices and food businesses, while harming the bottom line of others. Given the disproportionate environmental impact of the production of "meat", those who care about the planet should adjust their diets accordingly. Feminist environmentalists have double the reason to do so. At the same time, there is a pressing need for social and political solutions that recognize the political, economic and labour potential of ethical and sustainable alternatives to the status quo.

The promise of humane jobs

Ideas of green-collar jobs and good, green jobs challenge the perception that environmental protection and decent work are antithetical (Cohen & Calvert 2010; Forstater 2004; Hess 2012; Räthzel & Uzzell 2012). Yet, as other contributors to this volume note, particular assumptions about gender and both paid and unpaid work inform much solutions-oriented climate change scholarship and policy-making. Green jobs can simply reproduce existing pay gaps and other inequities, or the challenge of climate change can help inspire us to create new ways of organizing our economies to promote respect for nature and other people. The latter is laudable, and I argue that it must also include sentient nature.

The idea of humane jobs thus extends ideas about equitable and sustainable work. Humane jobs involve responsive or proactive alternatives to worker–environmental–animal harming industries and occupations, and, distinctly, place nonhuman animals into the mix of those considered. Succinctly, humane jobs are those that benefit both people and animals. The expansion and creation of humane jobs would help move the labour force away from damaging and unsustainable labour practices, towards more positive, beneficial work. Humane jobs could offer new areas of paid work for those who experience jobs loss as polluting industries are eliminated, and for those who are currently unemployed or underemployed.

In order for a positive multispecies vision of work to be promoted and achieved, we need to envision, assess and propose humane jobs. There are opportunities to learn from workers about how to improve existing positions, as well as to envision new, innovative areas of employment; such a task will require further research. The following complementary and intersecting principles should guide this crucial intellectual and political work.

Humane jobs prioritize both human and animal well-being

Because humane jobs promote multi-optic vision (Kim), analyses and plans must consider both human and nonhuman animals. Labour research on job quality highlights the importance of material and experiential dimensions (Carré et al. 2012; Coulter 2014b). In other words, tangible working conditions such as pay, benefits and hours of work figure, but so do qualitative aspects like feeling respected, having dignity and being proud of what you do. The small body of research on job quality for those who work with/for animals reveals that many, if not most people who work with/for animals are often driven by love, and a sense of a "calling" can play a significant role in people's motivations and experiences (Bunderson & Thompson 2009; Miller 2008; Sanders 2010).

The prospects for linking human and animal well-being ought to be interconnected; one group must suffer in order for the other to thrive. As is true in many human-focused caring and service sectors, in certain multispecies workplaces there is some evidence that animals benefit when people's working conditions are improved (see, for example, Coulter 2016a; McShane & Tarr 2007). Yet simply improving the working conditions of people is no guarantee that animals will benefit, particularly if harmful practices and killing are structured into the workplace or industry. The jobs in some nonviolent industries can be strengthened to better benefit people and/or animals. There is also a need to create new humane jobs. Sectors with potential include animal cruelty investigations/prevention, humane education, conservation, ecotourism and health care, whether for animals, or for people (and animals are involved). In that vein, green care is a particularly intriguing area. Green care is umbrella term for health care programmes that incorporate positive interactions with nature, such as care farming and animal-assisted therapy (Berget et al. 2012; Sempik, Hine & Wilcox 2010). There is a growing body of research on the positive effects of interspecies green care programmes, particularly for vulnerable socioeconomic groups and people with post-traumatic stress, dementia, autism and depression. A small number of programmes are underway in countries like Canada, but green care is much more developed in western and northern Europe, where care farming in particular is supported by local, national and transnational governments, farmers' groups and health care organizations.

Care farming (or therapeutic farming) is seen as playing an important role in the delivery of preventative and therapeutic health care, child care and job training, and in propelling rural development, economic diversification and income generation. A thoughtful expansion and diversification of green care in Canada could offer meaningful opportunities for creating new humane jobs.

There are also possibilities for agriculture and other kinds of food production. A humane jobs agenda would ensure rural livelihoods and the needs of farm workers, farm operators and farming communities were recognized. There are many kinds of agriculture which do not cause harm to animals and that instead cultivate nutritious proteins, grains, fruits and vegetables. In such a vision, there are clear roles for the public and private sectors. The public sector can help create fertile ground for innovative and ethical practices, including by shifting subsidies away from fur farming, commercial seal hunting and industrial animal agriculture, towards sustainable avenues that will create humane jobs.

Many humane job possibilities only task people with working; animals benefit from people's care and protection, not because they are working. In other cases, humane jobs may involve multispecies labour. Some critical animal studies scholars (e.g. Francione 2008) argue that an end to all interactive human–nonhuman animal relationships is necessary, but my view is that we can foster beneficial multispecies relationships with domesticated animals, provided that they are underscored by interspecies solidarity, and that animals are afforded both protections and positive entitlements (Coulter 2016a; Donaldson & Kymlicka 2011). If approached thoughtfully, this can include some relationships of work. For example, care farms are more positive for animals than industrialized factory farms, but animals cannot be simply seen as tools in green care contexts either. Animals are sentient beings who actively contribute to multispecies workplaces and the provisioning of care; their needs and desires must be taken seriously (Coulter 2016b).

Even among the species that are widely domesticated and where relationships of partnership are both widespread and possible, such as dogs and horses, this does not mean everything we currently ask (or require) these animals to do is ethically defensible. Similarly, simply because many animals of a particular species are willing to do work for us, that does not mean every individual wants to work, or to do so every day. Accordingly, a vision for humane jobs should not include totalizing blueprints, but rather areas of possibility. Individual specifics and diversity need to be continuously considered. In other words, we can recognize that a number of humane jobs with dogs are possible, for example, because this is a species that can participate in mutually-beneficial work relationships, but that does mean that we should condone all types of canine labour, or assume that every dog is suitable for any position. For example, service, therapy or facility dogs need to have particular temperaments, personalities and attitudes. The emotion work demands of these kinds of positions are great. High-energy dogs are better suited for different tasks, and prefer more physically rigorous work, such as herding or detection/scenting (search and rescue, endangered species or poacher detecting, etc.).

In many ways, these principles reaffirm the importance of context. Local and individual specificities matter. They can also change over time. In that vein, I have proposed a continuum of suffering and enjoyment as a way to understand work from animals' perspectives (Coulter 2016a). Suffering prompts recognition of the unpleasant and the horrific, the deeply negative experiences animals may (and all too often) have at work. Enjoyment allows for the fact that work can be a positive experience for animals. The continuum reflects the breadth and diversity of animals' real working lives and experiences. Where animals' work fits on the continuum will be affected, in particular, by (1) the occupation, (2) the work required, (3) the employers or coworkers, (4) the species and (5) the individual animals' own personalities, preferences, feelings and agency (Coulter 2016a, 84).

With this in mind, particular standards, akin to those we employ or propose for people, have relevance for nonhumans and proposals for humane jobs. Animals deserve protections/freedom from harmful practices, as well as positive entitlements/rights to beneficial experiences. Such a commitment can pertain to hours of work, breaks, time off and so forth. There are many basic labour standards which ought to be afforded to animals, and there is a need to develop guiding principles that can govern animals' work-lives. Yet because animals are also distinct species with their own needs, personalities and preferences, we must approach their well-being with an awareness of their particulars. Evidence-based decision-making is important when thinking about humane jobs. Social work researchers in particular have begun to reflect on the moral and practical dimensions of working with animals (Evans & Gray 2011; Hanrahan 2013; Serpell, Coppinger & Fine 2010; Ryan 2011, 2014). These are good starting places, but should also be expanded.

Accordingly, different research should be enlisted to promote accurate understanding. Much animal welfare theory and applied research, for example, emphasizes the five freedoms: freedom from hunger and thirst; freedom from discomfort; freedom from pain, injury and disease; freedom from fear and distress; and the freedom to express normal behaviours and interact with their own kind (Edgar et al. 2013; Van Dijk, Pradhan & Ali 2013). This could certainly be expanded to include freedom from being killed (except as an act of mercy). Similarly, cognitive ethology (the study of animals' minds) can be used to help correctly understand animals' physical, psychological and emotional well-being (Bekoff 2008; Marino & Colvin 2015).

Humane jobs should not reproduce existing inequities

An important feminist critique of dominant approaches to green jobs is that they have been conceptualized in androcentric ways which reinforce particular ideas about work and people's lives. In contrast to most contemporary green jobs plans, the promotion of humane jobs involves expanding and improving occupations and a number of employment sectors that are already feminized (Irvine & Vermilya 2010; Taylor 2010). Along with gender, other factors including race, citizenship and class must be considered when conceptualizing humane jobs. Undoubtedly, an expansion of vegetable, fruit, grain and plant-based protein farming offers a replacement for rural work that requires animal suffering and death. But such work is not an example of humane work if farm workers, including migrant workers, are subjugated and treated unfairly. This exemplifies an underlying commitment to intersectionality and multi-optic vision (Kim 2015). Simply eliminating one form of harm, while condoning or exacerbating another, is not acceptable.

Patricia Armstrong, Hugh Armstrong and Krista Scott-Dixon's (2008) argument for a holistic framework that emphasizes front-line care-providers, as well as the "behind-the-scenes" but still essential ancillary workers is also instructive. Visions of humane jobs can include a cross-section of occupations requiring different education and training, skills and tasks. This means there is potential for diverse kinds of job creation for men and especially women, workers of different classes and abilities, and both rural and urban regions. There are many possibilities which can be gleaned by studying current practices and comparative examples, and by imagining more caring multispecies societies.

Humane jobs are integral to more sustainable and humane societies

Building in particular on feminist political economy, paid work should be situated within a larger personal and political web that recognizes that workers have bodies, families and lives beyond the job, and that they must engage in unpaid work, including social reproductive labour, as well (Bezanson 2006; Luxton 2009; Luxton & Bezanson 2006). Consequently, humane jobs are about formal occupations, but they are also about work-lives. This includes daily life when not working, lives over time and well-being after formal employment has ended. Such factors matter for both people and animals, and connect the world of work to ideas of "the social", and to the realm of public policy. Programmes and initiatives that improve people's work-lives (such as affordable, universal child care) operate in concert with humane jobs, and help create more humane, sustainable and just societies. Moreover, although not all humane jobs will be in the caring sector, care work is, in many ways, at the heart of such a vision. Humane jobs build on feminist political economists call for a movement towards a caregiving society (Glenn 2000), one wherein we recognize the inextricability of human–animal–environmental well-being. Caregiving cannot stop at the species-boundary. "Genuine human and social progress and betterment cannot be based on the suffering of others, period. A just and caring society cannot be created on mass, unmarked animal graveyard" (Coulter 2016a, 162).

Overall, the prospect and promise of humane jobs are about specifics, and about what policies, programmes and initiatives can be envisioned and promoted, in the private and especially the public sector. Practical, achievable plans of action are possible and needed. But humane jobs are also about an expanded vision of social justice, and are about recognizing that all societies are multispecies societies. Our socioeconomic relations intersect with and significantly affect nature, but are socially constructed. Multispecies suffering and environmental degradation can be sustained, or rejected and changed. Humane and sustainable work-lives, work-places, communities and societies can be envisioned, fostered and made.

Acknowledgements

Thanks to Kirsten Francescone and Anelyse Weiler for their research assistance. I am grateful for the financial support of the Social Sciences and Humanities Research Council of Canada.

References

Adam, K, Gibson, L & Cook, M 2016, "Injury prevention in the meat industry: Limited evidence of effectiveness for ergonomic programs in reducing the severity of musculo-skeletal injuries", *Australian Occupational Therapy Journal*, vol. 63, no.1, pp. 59–60.

Adams, CJ 2007, "The war on compassion", in J Donovan and CJ Adams (eds), *The feminist care tradition in animal ethics*, pp. 21–36, Columbia University Press, New York.

Adams, CJ 2010, *The sexual politics of meat: A feminist-vegetarian critical theory*, The Continuum International Publishing Group Inc., New York.

Adams, CJ & Gruen, L (eds) 2014, *Ecofeminism: Feminist intersections with other animals and the earth*, Bloomsbury, New York.

Agriculture and Agri-Food Canada 2011, "Number of farms by industry type", (http://www.agr.gc.ca/poultry/nofrms_eng.htm).

Akhtar, A 2012, *Animals and public health: Why treating animals better is critical to human welfare*, Palgrave Macmillan, New York.

Armstrong, P, Armstrong, H & Scott-Dixon, K 2008, *Critical to care: The invisible women in health services*, University of Toronto Press, Toronto.

Bekoff, M 2008, *The emotional lives of animals: A leading scientist explores animal joy, sorrow, and empathy–and why they matter*, New World Library, Novato.

Berget, B, Lidfors, L, Pálsdóttir, AM, Soini, K & Thodberg, K (eds) 2012, "Green care in the Nordic countries: A research field in progress", Nordic research workshop on Green Care in Trondheim, June, Health UMB, Norwegian University of Life Sciences, Ås, Norway.

Bezanson, K 2006, *Gender, the state, and social reproduction: Household insecurity in neo-liberal times*, University of Toronto Press, Toronto.

Bristow, E 2011, "Global climate change and the industrial animal agriculture link: The construction of risk", *Society & Animals*, vol. 19, no. 3, pp. 205–224.

Bunderson, JS & Thompson, JA 2009, "The call of the wild: Zookeepers, callings, and the double-edged sword of deeply meaningful work", *Administrative Science Quarterly*, vol. 54, no. 1, pp. 32–57.

Canadian Federation of Humane Societies n.d., "Realities of farming in Canada", (http://cfhs.ca/farm/farming_in_canada/).

Caro, D, Davis, SJ, Bastianoni, S & Caldeira, K 2014, "Global and regional trends in greenhouse gas emissions from livestock", *Climatic Change*, vol. 126, no. 1–2, pp. 203–216.

Carré, F, Findlay, P, Tilly, C & Warhurst, C 2012, "Job quality: Scenarios, analysis, and interventions", in C Warhurst, F Carré, P Findlay & C Tilly (eds), *Are bad jobs inevitable? Trends, determinants and responses to job quality in the twenty-first century*, pp. 1–22, Palgrave Macmillan, Houndsmills.

Cohen, MG & John C 2010, "Climate change and labour in the energy sector", in C Lipsig-Mummé (ed.), *The state of research on work, employment and climate change in Canada* Lipsig-Mummé, pp. 48–79, York University, Toronto.

Coulter, K 2014a, "Herds and hierarchies: Class, nature, and the social construction of horses in equestrian culture", *Society & Animals*, vol. *22* no. 2, pp. 135–152.

Coulter, K 2014b, *Revolutionizing retail: Workers, political action, and social change*, Palgrave Macmillan, New York.

Coulter, K 2016a, *Animals, work, and the promise of interspecies solidarity*, Palgrave Macmillan, New York.

Coulter, K 2016b, "Beyond human to humane: A multispecies analysis of care work, its repression, and its potential", *Studies in Social Justice*, vol. 5 no. 2, pp. 199–219.

Cutler, SJ, Fooks, AR & Van Der Poel, WH 2010, "Public health threat of new, reemerging, and neglected zoonoses in the industrialized world", *Emerging Infectious Diseases*, vol. 16, no. 1, pp. 1–5.

Dillard, J 2008, "Slaughterhouse nightmare: Psychological harm suffered by slaughterhouse employees and the possibility of redress through legal reform", *Georgetown Journal on Poverty Law & Policy,* vol. 15, no. 2, pp. 391–408.

Donaldson, S & Kymlicka, W 2011, *Zoopolis: A political theory of animal rights*, Oxford University Press, Oxford.

Donovan, J 2007, "Caring to dialogue: Feminism and the treatment of animals", in J Donovan & C. Adams (eds), *The feminist care tradition in animal ethics*, pp. 360–369, Columbia University Press, New York.

Edgar, JL, Mullan, SM, Pritchard, JC, McFarlane, UJC & Main, DCJ 2013, "Towards a 'good life' for farm animals: Development of a resource tier framework to achieve positive welfare for laying hens", *Animals*, vol. 3, no. 3, pp. 584–605.

Evans, N & Gray, C 2011, "The practice and ethics of animal-assisted therapy with children and young people: Is it enough that we don't eat our co-workers?" *British Journal of Social Work*, pp. 1–18.

Fitzgerald, AJ 2010, "A social history of the slaughterhouse: From inception to contemporary implications", *Human Ecology Review*, vol. 17, no. 1, pp. 58–69.

Fitzgerald, AJ, Kalof, L & Dietz, T 2009, "Slaughterhouses and increased crime rates: An empirical analysis of the spillover from 'the jungle' into the surrounding community", *Organization & Environment*, vol. 22, no. 2, pp. 158–84.

Forstater, M 2004, "Green jobs: Addressing the critical issues surrounding the environment, workplace, and employment", *International Journal of Environment, Workplace and Employment*, vol. 1, no. 1, pp. 53–61.

Francione, GL 2008, *Animals as persons: Essays on the abolition of animal exploitation*, Columbia University Press, New York.

Friedrich, B 2012, "National pork producers council: Anti-science and anti-animal", Common Dreams; Breaking News & Views for the Progressive Community, (http://www.commondreams.org/views/2012/09/08/national-pork-producers-council-anti-science-anti-animal).

Gaard, G (ed.) 1993, *Ecofeminism: Women, animals, nature*, Temple University Press, Philadelphia.

Gaard, G 2011, "Ecofeminism revisited: Rejecting essentialism and re-placing species in a material feminist environmentalism", *Feminist Formations*, vol. 23, no. 2, pp. 26–53.

Gerber, PJ, Steinfeld, H, Henderson, B, Mottet, A, Opio, C, Dijkman, J, Falcucci, A & Tempio, G 2013, *Tackling climate change through livestock: A global assessment of emissions and mitigation opportunities*, Food and Agriculture Organization of the United Nations, Rome.

Glenn, EN 2000, "Creating a caring society", *Contemporary Sociology*, vol. 29, no. 1, pp. 84–94.

Gruen, L 1993, "Dismantling oppression: An analysis of the connection between women and animals", in G Gaard (ed.) *Ecofeminism: Women, animals, nature*, pp. 61–90, Temple University Press, Philadelphia.

Hanrahan, C 2013, "Social work and human animal bonds and benefits in health research: A provincial study", *Critical Social Work*, vol. 14, no. 1, n.p.

Heffernan, C, Salman, M & York, L 2013, "Livestock infectious disease and climate change: A review of selected literature", *CAB Reviews*, vol. 7, no. 11, pp. 1–26.

Herrero, M. Gerber, P, Vellinga, T, Garnett, T, Leip, A, Opio, C, Westhoek, HJ, Thornton, PK, Olesen, J, Hutchings, N, Montgomery, H, Soussana, J-F, Steinfeld, H & McAllister, TA, 2011, "Livestock and greenhouse gas emissions: The importance of getting the numbers right", *Animal Feed Science and Technology*, vol. 166, pp. 779–782.

Hess, DJ 2012, *Good green jobs in a global economy*, MIT Press, Cambridge.

Humane Society of the United States 2014, "Farm animal statistics: Slaughter totals", (http://www.humanesociety.org/news/resources/research/stats_slaughter_totals.html).

Irvine, L & Vermilya, JR 2010, "Gender work in a feminized profession: The case of veterinary medicine", *Gender & Society*, vol. 24, no. 1, pp. 56–82.

Jacques, JR 2015, "The slaughterhouse, Social disorganization, and violent crime in rural communities", *Society & Animals*, vol. 23, no. 6, pp. 594–612.

Johnsen, C 2003, *Raising a stink: The struggle over factory hog farms in Nebraska*, University of Nebraska Press, Lincoln, NE.

Kim, CJ 2015, *Dangerous crossings: Race, species, and nature in a multicultural age*, Cambridge University Press, Cambridge.

Lipschitz, F & Gardner, S 2014, "Not in my city: Rural America as urban dumping ground", *Architecture MPS*, vol. 6, no. 2, pp. 1–19.

Luxton, M 2009, *More than a labour of love: Three generations of women's work in the home*, Women's Press-Canadian Scholars' Press Inc., Toronto.

Luxton, M & Bezanson, K (eds) 2006, *Social reproduction: Feminist political economy challenges neo-liberalism*, McGill-Queen's University Press, Montréal and Kingston.

Mackenzie, JS, Jeggo, M, Daszak, PS & Richt, JA 2013, *One health: The human–animal–environment interfaces in emerging infectious diseases*, Springer, Berlin.

McShane, C & Tarr, J 2007, *The horse in the city: Living machines in the nineteenth century*, The Johns Hopkins University Press, Baltimore, MD.

Marino, L & Colvin, CM 2015, "Thinking pigs: A comparative review of cognition, emotion, and personality in *Sus Domesticus*", *International Journal of Comparative Psychology*, vol. 28, no. 1, pp. 1–22.

Miller, J 2008, "'We can't join a union, that would harm the horses': Worker resistance in the UK horseracing industry", *Centre for Employment Studies Research Review*, April, pp. 1–4.

Nance, S 2013, *Entertaining elephants: Animal agency and the business of the American circus*, The Johns Hopkins University Press, Baltimore, MD.

Nibert, DA 2013, *Animal oppression and human violence: Domesecration, capitalism, and global conflict*, Columbia University Press, New York.

Nibert, DA 2014, "Animals, immigrants, and profits: Slaughterhouses and the political economy of oppression", in J Sorenson (ed), *Critical animal studies: Thinking the unthinkable*, pp. 3–17, Canadian Scholars' Press, Inc., Toronto.

Noske, B 1989, *Humans and other animals: Beyond the boundaries of anthropology*, Pluto Press, London.

Noske, B 1997, *Beyond boundaries: Humans and animals*, Black Rose Books, Montréal.

Pachirat, T 2011, *Every twelve seconds: Industrialized slaughter and the politics of sight*, Yale University Press, New Haven, CT.

Pini, B & Leach B 2011, "Transformations of class and gender in the globalized countryside: An introduction", in B Pini & B Leach (eds), *Reshaping gender and class in rural spaces*, pp. 1–23, Ashgate, Surrey.

Porcher, J 2011, "The relationship between workers and animals in the pork industry: A shared suffering", *Journal of Agricultural and Environmental Ethics*, vol. 24, no. 1, pp. 3–17.

Preibisch, KL & Encalada Grez, E 2014, "The other side of el Otro Lado: Mexican migrant women and labor flexibility in Canadian agriculture", *Signs*, vol. 40, no. 1, pp. 289–316.

Price, L 2012, "The emergence of rural support organisations in the UK and Canada: Providing support for patrilineal family farming", *Sociologia ruralis*, vol. 52, no. 3, pp. 353–376.

Ramsey, D, Soldevila-Lafon, V & Viladomiu, L 2013, "Environmental regulations in the hog farming sector: A comparison of Catalonia, Spain and Manitoba, Canada", *Land Use Policy*, vol. 32, pp. 239–249.

Räthzel, N & Uzzell, D (eds) 2012, *Trade unions in the green economy: Working for the environment*, Routledge, New York.

Ryan, T 2011, *Animals and social work: A moral introduction*, Palgrave Macmillan, New York.

Ryan, T (ed.) 2014, *Animals in social work: Why and how they matter*, Palgrave Macmillan, New York.

Sanders, CR 2010, "Working out back: The veterinary technician and 'dirty work'", *Journal of Contemporary Ethnography*, vol. 39, no. 3, pp. 243–272.

Sempik, J, Hine, RE & Wilcox, D 2010, *Green care: A conceptual framework*, Loughborough University, Loughborough.

Serpell, JA, Coppinger, R & Fine, AH 2010, "Welfare considerations in therapy and assistance animals", in AH Fine (ed.), *Handbook on animal-assisted therapy: Theoretical foundations and guidelines for practice*, pp. 481–502, Academic Press, London.

Statistics Canada 2011a, "Canada's farm population: Agriculture-population linkage data for the 2006 Census", (http://www.statcan.gc.ca/ca-ra2006/agpop/article-eng.htm).

Statistics Canada 2011b, "Neighourhood income and demographics, taxfilers and dependents with income by total income, sex and age group", CANSIM, Table 111-008 (http://www5.statcan.gc.ca/cansim/a26?lang=eng&id=1110008) & "Family characteristics, summary", CANSIM, Table 111-009, (http://www5.statcan.gc.ca/cansim/a26?lang=eng&id=1110009).

Steinfeld, H, Gerber, P, Wassenaar, T, Castel, V, Rosales, M and de Haan, C 2006, *Livestock's long shadow: Environmental issues and options*, Food and Agriculture Organization of the United Nations, Rome.

Stull, DD & Broadway, MJ 2013, *Slaughterhouse blues: The meat and poultry industry in North America: 2nd edition*, Wadsworth Cengage Learning, Belmont.

Taylor, N 2010, "Animal shelter emotion management: A case of in situ hegemonic resistance?" *Sociology*, vol. 44, no. 1, pp. 85–101.

Van Dijk, L, Pradhan, SK, Ali, M & Ranjan, R 2013, "Sustainable animal welfare: community-led action for improving care and livelihoods", in H Ashley, N Kenton & A Milligan (eds), *Participatory learning and action: Tools for supporting sustainable natural resource management and livelihoods*, pp. 37–50, International Institute for Environment and Development, London.

Weis, T 2007, *The global food economy: The battle for the future of farming*, Fernwood Press and Zed Books, Halifax/London.

Weis, T 2013, *The ecological hoofprint: The global burden of industrial livestock*, Zed Books, London.

Wilkie, RM 2010, *Livestock/deadstock: Working with farm animals from birth to slaughter*, Temple University Press, Philadelphia.

Wing, S, Horton, RA & Rose, KM 2013, "Air pollution from industrial swine operations and blood pressure of neighboring residents", *Environmental Health Perspectives*, vol. 121, no. 1, pp. 92–96.

Woldehanna, S & Zimicki, S 2015, "An expanded one health model: Integrating social science and one health to inform study of the human-animal interface", *Social Science & Medicine*, vol. 129, March, pp. 87–95.

World Health Organization 2010, "The FAO-OIE-WHO collaboration: Sharing responsibilities and coordinating global activities to address health risks at the animal-human-Ecosystems interfaces", A Tripartite Concept Note, (http://www.who.int/foodsafety/zoonoses/final_concept_note_Hanoi.pdf?ua=1).

PART IV

Rural and resource communities

12

"MAYBE TOMORROW WILL BE BETTER"

Gender and farm work in a changing climate

Amber J. Fletcher

Introduction

The gendered dimensions of climate change are highly visible in Canadian prairie agriculture. From a gender perspective, farming in Canada is clearly a masculinized industry. Despite women's participation in a number of farm tasks, their contributions continue to be marginalized and devalued, and the dominant image of a Canadian farmer is still an image of a man. These gendered roles and ideologies shape farmers' attitudes toward climate change as well as their experience of a climate event.

This chapter draws upon the findings of two qualitative studies with farmers in the Canadian prairies. The first study (Fletcher 2013), which I conducted in 2011, involved in-depth semi-structured interviews with 30 Saskatchewan farm women. The study inquired about farm women's views on climate change, environmental views and adaptation to climate extremes. The second study was part of a broader project entitled "Vulnerability and Adaptation to Climate Extremes in the Americas" (VACEA), which was conducted between 2011 and 2016. The VACEA project involved a team of natural and social scientists examining climate vulnerability and adaptation in five countries: Argentina, Brazil, Canada, Chile and Colombia. This chapter draws upon one component of the VACEA study, in which community vulnerability assessments (CVAs) were carried out in four rural communities in the Canadian prairie region. The CVAs involved semi-structured interviews with approximately 100 male and female rural residents living in the southern prairies of Alberta and Saskatchewan. For the purpose of this chapter, the VACEA data were re-coded to include only agricultural producers (farmers and ranchers) and to exclude other rural residents.

In both projects, interview data were transcribed verbatim and coded using NVivo software. In this chapter, interview excerpts from the farm women study are labeled with "FW" and VACEA project interviews with "VC". Participant numbers are used instead of names to protect participants' confidentiality. Taken together, the findings from the two projects show that farmers are deeply divided in their views about climate change and that such divisions cut across gender lines. However, the findings showed notable gendered trends in two key areas: environmental awareness, which relates to mitigation of future climate change, and coping or adaptation to the effects of climate change that is already occurring.

Climate change in the Canadian prairies

The Canadian prairie region has one of the harshest and most variable climates in the world. As a semi-arid area, it is one of Canada's driest regions (Sauchyn 2010; Sauchyn & Kulshreshtha 2008). A significant swath of the region lies within an area called Palliser's Triangle, an expanse of approximately 200,000 square kilometers known for its history of severe and protracted droughts (Marchildon, Pittman & Sauchyn 2009). Western Canada has experienced more than 40 severe droughts in the past 200 years (Environment Canada 2004). It has taken a great deal of human adaptation to convert this drought-prone region into one of Canada's most significant agricultural areas, but the region now contains 81 per cent of Canadian farmland and is therefore crucial to the Canadian food production (Statistics Canada 2011a).

The region and its people will be further tested by future climate change. Climatological data indicate that this already dry area will experience dramatic effects as anthropogenic climate change proceeds. As the global warming trend influences natural climate cycles, increased frequency and severity of climate extremes is expected (IPCC 2015). For the prairies, the overall trend will be hotter average temperatures and lower precipitation, which increase drought risk (Henderson & Sauchyn 2008). Drought can have drastic effects for farmers, including lack of water for cattle and other livestock, crop failures, grasshopper infestation and soil erosion. During the major drought of 2001–2, for example, Saskatchewan farmers experienced crop production losses of $925 million in 2001 and $1.49 billion in 2002, to the point that the province reported overall negative net farm income in 2002 (Wheaton et al. 2008). Furthermore, NASA climatologists recently identified the Canadian prairies as a "hot spot" for biome shifts associated with climate change (Bergengren, Waliser & Yung 2011). Shifting climate conditions will create stress for existing plant and animal life, rendering crops and other vegetation more susceptible to diseases, insects and wildfires (ibid.).

Although drier conditions are expected overall, warmer temperatures can produce more extreme precipitation and flash flooding (Groisman et al. 2005; Henderson & Sauchyn 2008). Such rapid fluctuation between extremes proves challenging for farmers and ranchers, testing their adaptive capacity. Existing research has documented prairie farmers' strategies of adapting to climate extremes (McMartin & Hernani Merino 2014; Pittman et al. 2011; Wandel, Pittman & Prado 2010; Warren & Diaz 2012). These strategies range from immediate coping responses, such as hauling water or hay from distant locations during a drought, to longer-term adaptation strategies like irrigation, stockpiling hay, and minimizing tillage to reduce soil erosion. At the core of these adaptive strategies is work: from operating irrigation infrastructure to learning about new drought-resistant crop varieties, farmers rely on their own labour capacity to cope with, and prepare for, climate extremes. However, little attention has been paid to the gendered dimensions of climate adaptation in this dry region.

Gendering climate change

Existing literature has documented the importance of gender in climate extremes and climate change policy around the world (e.g., Alston & Whittenbury 2013; Cannon 2002; Dankelman 2010; Denton 2002; Enarson & Chakrabarti 2009; Enarson & Morrow 1997; Reyes 2002; Seager 2006). A significant amount of empirical research has focused on low-income countries of the Global South, where global inequality and poverty may exacerbate the consequences of climate disaster (MacGregor 2010; Seager 2006). Although climate events may be experienced differently in the wealthy countries of the Global North, agricultural producers are amongst the most exposed to climate change disasters (Milne 2005).

Recent reviews of the gender and climate change field have called for more contextualized and intersectional analyses (Arora-Jonsson 2011; Moosa & Tuana 2014). Such analyses can challenge the tendency to homogenize "women" as a uniformly vulnerable group. The research discussed here examines climate change belief and experience at the intersection of gender and rurality, with particular focus on the agricultural context of family farming. Researchers have documented the gendered effects of drought for rural people and farmers in the Australian context (Alston 2009, 2006; Boetto & McKinnon 2013; Logan & Ranzijn 2008; Stehlik, Lawrence & Gray 2000), but little gender and climate change research has been done in the context of Canadian agriculture. As a key agricultural zone highly exposed to past and future climate extremes, and marked by historically engrained gender roles, research in the Canadian prairie region provides insights into the gendered intersection of work and climate change.

Gender and farm work

Throughout the Canadian prairies, farming is commonly viewed as a masculine profession. For many farm families, work is divided by a rigid gendered division of labour. Tasks frequently performed by men – such as driving farm equipment, managing livestock, planning crops and making major purchases – are the tasks most recognizable as "farming" and therefore continue to be performed mostly by men, who are most often recognized as "farmers". Although farm women's participation in these tasks has increased over time (Martz 2006), they still tend to perform them at a lower rate (and for fewer overall hours) than do farm men (Fletcher 2013; Kubik 2004; Martz 2006).

Farm women, however, perform a wide variety of other farm-related tasks such as accounting and marketing, driving for tractor parts or transporting workers, and preparing large meals for farm workers (Fletcher 2013; Kubik 2004; Martz 2006). Despite the importance of these tasks and their centrality to the family farm system, farm women's contributions are less likely to be recognized as "farm work". Researchers have documented the invisibility and marginalization of farm women's contributions to agriculture throughout the Global North (e.g. Alston 1998, 1995; Ireland 1983; Keller 2014; Kubik 2005; Sachs 1983; Shortall 1999). This invisibility is the product of numerous intersecting factors, including engrained gender ideologies (farming as a masculine profession) and gendered divisions of labour that have positioned women as responsible for caregiving and social reproduction. In addition, because farm women tend to perform tasks that involve less direct contact with the main farm commodities (e.g. grain or cattle), their contributions are less recognized as farm work.

Recent studies in the Canadian prairies have shown that such marginalization continues to exist (Faye 2006; Fletcher 2013; Kubik 2004, 2005). In the two studies discussed in this chapter, farm women consistently described their own roles using marginalizing terms such as "helper" and "supporter". For example, when describing her roles on the farm one woman stated that, "He [husband] does the *important* things and I take care of the details" (VC-T01, emphasis added). Her husband acknowledged that, "Farming is a man's world. I'm not saying it should be. I'm saying it simply is" (VC-T01).

Despite women's historical and contemporary involvement in farming, a variant of breadwinner masculinity persists. As one farm man in the VACEA study stated, "I personally still believe that [men] should be the breadwinners, especially of a nuclear family" (VC-T12). The deeply embedded masculinization of agriculture means that farm women are often detached from control over key agricultural decisions on their operations. This detachment has meant that many farm women seek control and fulfillment through off-farm employment, which has the added benefit of providing a stable additional income during climate crises like floods or droughts (Fletcher 2013).

Political-economic changes occurring in Canadian agriculture have affected farm women's roles. Increased competition on international markets, combined with rising input costs on everything from seed to transport (ibid.), has caused a drop in the total number of farms and a dramatic increase in the size of those remaining (Statistics Canada 2011b). As farms grow larger in an attempt to stay competitive, many are unable to afford hired help and increasingly rely on farm women's flexible labour, to the point that several farm women identified themselves as "hired men". Other tasks done by farm women, such as providing meals, have become more complex: a participant in the farm women study reported that she regularly drives 40 minutes each way to deliver meals to the field on her family's large operation.

Prairie farmers, the environment, and climate change

Climate change is a sensitive and often inflammatory topic in prairie agricultural communities. This situation is due, in part, to public discourse that has emphasized agriculture's contribution to anthropogenic global warming through the production of greenhouse gases like nitrous oxide, carbon dioxide and methane. Canada's National Greenhouse Gas Inventory reported that agriculture contributed 10 per cent of greenhouse gas emissions in 2013 and, like other industries, agriculture's contribution has risen over the past two decades (Environment Canada 2013). This public attention causes a strong reaction amongst many farmers, particularly those for whom environmental responsibility is a source of pride. Throughout the VACEA project ranchers described themselves as "the original environmentalists" and emphasized their contributions to grassland preservation and soil quality. Some had also participated in the Environmental Farm Plan (EFP) programme provided by the federal government and were working to improve environmental practices on their farms.

Despite these contributions, the current industrialized system forces many producers to engage in environmentally unsustainable practices for farm survival. In a productivist paradigm that emphasizes industrialization and economies of scale, new practices have emerged that prioritize economics over environment. Large farms use small airplanes to spray fields with agricultural chemicals. As grain production exceeds storage capacity, crop producers have started using grain bags – each bag consisting of 300 to 700 pounds of plastic – to store grain (Government of Saskatchewan n.d.). Although grain bag recycling facilities are gradually being developed in some communities, many farmers lack access to these facilities and bags are often burned after use, releasing harmful toxins and emissions into the air (ibid.). Even practices generally accepted as environmentally friendly are fraught with unseen consequences; for example, the widespread use of zero-till, direct seeding and chem-fallow have reduced soil erosion and fuel consumption but may require more use of chemical herbicides for weed control.

These competing discourses – of farmers as major emitters versus "original environmentalists" – and tensions between environmental preservation and the financial bottom line are linked to dramatic divergence in views about climate change in rural areas. In both studies, many participants had perceived some form of localized change, such as milder winters, changing growing seasons and more extreme events like flooding. However, opinion was widely divided about the cause of climatic changes. Some participants held a firm belief in climatological science and anthropogenic climate change, such as one Saskatchewan farm woman who said, "I'm concerned for the planet, for one thing. I'm quite convinced that it's what we're doing that's causing this" (FW-04). Another stated: "They're saying the ice is melting for the polar bears, so there has to be some evidence up there. It's definitely going to change the way of farming" (FW-29). Others denied the possibility of anthropogenic change: "No, I think it's all blown out of proportion"

(FW-14) and "It's not getting any warmer. I think our earth cycles and I think we're in what they're perceiving as a global warming cycle" (FW-7).

The distinction between natural climate cycles (i.e. oscillations) and anthropogenic climate change further complicates the discussion of climate change. In many cases, producers believe that climate change is occurring – a belief often grounded in their own experiences and perceptions of changes over time – but there is far less belief in anthropogenic change as a contributor to increased climatic variability. One farm man in the province of Alberta expressed a common view: "I don't believe in global warming, but it seems like there's no question we're seeing climate change. We're seeing more extreme weather events" (VC-T13). A farm woman also highlighted this distinction: "I definitely think that the earth is warming. I have no problem . . . I can see that. But we're at the end of an ice age, so is it human caused or human sped-up or really not human caused, or not? I don't think we know" (FW-23). This creates a difficult situation in terms of climate change mitigation and environmental practices. If anthropogenic climate change is not recognized as a pressing problem, economic imperatives will continue to supersede environmental ones.

Uncertainty is a major underlying trend in many discussions of climate change with prairie agricultural producers. One farm man said, "As far as this global warming thing, I don't know what to say on that. I am torn on that one" (VC-S07). Climate change is an abstract concept; it is unlikely to influence farmers' daily decisions and working lives in the same way as more tangible considerations like finances. Furthermore, although farmers do notice shifting climatic trends over the longer term, these shifts are associated with significant causal uncertainty and are generally seen as uncontrollable and outside the realm of human activity. Even extreme climate events, such as flooding and drought, are seen either as chance events or the product of natural climatic cycles, which are bestowed a certain legitimacy that anthropogenic change is not.

Existing research has identified some gender differences in views on global climate change. McCright (2010) drew a distinction between climate change knowledge and environmental behaviour. The two studies discussed here support the importance of this distinction for gender analyses. Although the studies did not reveal notable gender differences in views about climate change and its causes at the macro level of global change and global warming, clear gender differences emerged when participants discussed their own environmental practices and experiences of climate extremes at the meso (farm/household) and micro (individual) levels.

At the macro level of climate change perceptions, both farm women and men expressed a wide variety of views across the spectrum. This may be explained by the fact that farm women and men receive similar sources of information about climate change. Several participants in both studies had attended public information sessions advocating climate change denial. Despite the lack of gendered trends in climate change views, significant gendered trends could be seen in two areas: environmental practices and adaptation practices. These trends can be attributed to the highly gender-differentiated work roles in prairie agriculture. In the following sections, I discuss each trend in turn.

"Women just get it": Gender and environmental awareness

Participants in the farm women study were asked about potential gender differences in environmental awareness and concern about the environment. Half of the participants believed that farm women are more concerned with, and aware of, the environment than farm men. While some provided essentialist explanations for women's heightened awareness, such as "women are more intuitive" (FW-01), the majority attributed it to social gender roles. The other half felt that environmental awareness and concern is not gendered and varies by family, with some farm families highly engaged in environmental practices and others uninterested. In particular, the decision to shift toward organic farming – a major transition for a farm operation – was often seen as a family decision based on shared environmental concern. Amongst all of the farm women study participants, however, none felt that farm *men* were more environmentally aware than farm women.

Prairie farm women continue to hold disproportionate responsibility for childcare, household work and food provision (Fletcher 2013). These gendered responsibilities mean that farm women were consistently more likely to connect environmental farm practices to nutrition, safety and health. One farm woman explained that, "Kids are learning more and more about the environment and things like that at school and they're bringing that information home" (FW-06). A 2006 study of farm women in Canada showed that, out of a list of farm tasks, application of chemicals was least commonly performed by farm women (Martz 2006). Participants in the farm women study were similarly reluctant to participate in spraying, a reluctance driven mostly by health concerns. One farm woman explained that, "Women . . . have more time in our brain, I think, to think about, 'oh boy, I hope that spray isn't, you know, going to cause somebody breast cancer down the road'" (FW-14).

Indeed, several farm women discussed how their detachment from farming decisions and control afforded them a critical perspective on farm practices – or "more time in our brain", as the farm woman cited above put it. In contrast, the "main farmers" (often men) were perceived to be more concerned with productivity and the financial bottom line than with environmental or health concerns. As MacGregor (2010) and Arora-Jonsson (2011) have pointed out, there is a risk that the discourse of women's heightened environmental awareness may result in increased responsibility and pressure for women to "green" their homes and farms, thus deepening their responsibilities in this area. In the particular context of Canadian agriculture, gendered divisions of labour may in fact produce gender-differentiated environmental awareness; however, rather than implying women's inherent "virtuousness" in relation to the environment (see Arora-Jonsson for a critique of the virtuousness discourse), these findings reveal the ongoing significance of work in structuring and shaping environmental views.

Conversations with farm men in the VACEA project revealed further complexity. While many farm men were indeed focused on financial sustainability, others engaged critically and reflexively with the environment/economic tension. Discussing chem-fallow practices, one farm man said: "Well, I don't like putting all this chemical down because I still don't think that's good for us in the long run, but yet you're almost being forced to do that, in a way, because when fuel costs are so high, you can't till your land . . . and it's costing you more" (VC-RL05).

It can also be difficult for farm men to reject dominant ideologies of masculinity. As those most often positioned as the "main farmers" on their operation, farm men are subject to dominant ideologies and expectations of "good farming" – an ideal which, in the current industrialized paradigm, has come to mean highly productive fields treated with large amounts of agro-chemicals to become free of weeds (Phillips 1998). One farm man described his own rejection of this dominant ideal; his words illustrate the connection between agricultural chemical use and the productivist focus on the bottom line: "I know some neighbours . . . they know every doggone new chemical on the market and they've got to have their crop just perfect and the fertilizer in there. As far as I'm concerned I'm not going to live and breathe making as many dollars as I possibly can . . . there is more to it than that" (VC-S01).

Gendering coping and adaptation

Coping and adaptation to climate change is another gendered domain. As anthropogenic climate change exacerbates natural climate variability, many regions around the world will experience more frequent and severe climate extremes. The Canadian prairie region can expect an overall drying trend interspersed with flash flooding and extreme precipitation events. Examination of how farm families have coped and adapted to past droughts and floods can provide important insights for future climate preparation.

In both studies, stress was a major immediate issue associated with drought, flooding or other farm crises. This stress was often perceived to hit men hardest (for more discussion of this topic, see Fletcher & Knuttila 2016). Dominant ideologies of masculinity and expectations of productivism as "good farming" result in high levels of stress for farm men when a climate disaster threatens the farm. One farm woman described the stress associated with farm failures, saying, "I think some of the male gender sort of take it as a personal affront on their ability [to farm]" (FW-25).

When a climate event occurs, farm women's role as "supporters" is further entrenched. A conversation about farm stress with a Saskatchewan couple illustrates how these gendered roles operate:

> Farm man: Harvest time is when I get antsy . . . then I get anxious.
> Farm woman: And we just sort of let him do his thing, and we just sort of . . .
> Farm man: Leave me alone.
> Farm woman: Feed him and leave him alone. (VC-RL05)

Several women described their attempts to be supportive while staying unobtrusive. As one woman put it, "We [women] try and keep everybody on the level and you know, don't try to irritate them. Try and keep it a peaceful atmosphere, like 'maybe tomorrow will be better'" (FW-03). A common sentiment amongst farm women was that their relative disconnection from farming helped to buffer them from the worst of the stress; in this way, disconnection was construed as a site of power. However, some women downplayed their own feelings and concerns to focus on the needs of their partners and families. Farm women's emphasis on their supportive role may limit their ability to seek support for themselves.

Furthermore, although farm women's marginalization can sometimes act as a buffer from stress in the short term, it may ultimately limit their agency in longer-term adaptation strategies. As peripheral farmers, farm women are less involved in "the plan" for coping and longer-term adaptation on the farm: "It's still the same role where . . . he has the plan and . . . you're just there in a support role" (FW-13). Farm women were much less likely than men to be involved in longer-term climate preparedness decisions like selecting crop varieties or planning rotations.

Policy implications

Climate change is a controversial issue in the Canadian prairies, but one that is little understood. Given the overall uncertainty and dominant discourse emphasizing natural cycles over anthropogenic change, there is a need for climate change scholars and climatologists to more effectively communicate the relationship between natural cycles and anthropogenic change. It is important for farmers to be aware of how anthropogenic change will affect agriculture in the future. Indeed, many farmers responded positively to the VACEA project, which involved natural scientists and climatologists in addition to the social scientists who interviewed them. Many participants inquired about how to obtain climatological scenarios to assist with their future farm planning, indicating that many farmers are interested in climatological data.

There is a need to shift the broader discourse on climate change mitigation away from blaming farmers and toward a greater recognition of farmers' ability to mitigate future climate change through environmentally responsible farming practices. Such a shift will require complementary policies and programmes that facilitate environmental farm practices while making such practices financially viable. In an increasingly corporatized, competitive system where government policies have expanded market competition and decreased farm subsidies and supports, farmers cannot be expected to prioritize environmental preservation over their immediate survival. The stronger environmental awareness amongst farm women, who are often disconnected from the farm due to gendered roles and ideologies, indicates that environmental practices can be encouraged if farmers are able to focus less on the financial bottom line. This suggests the need for a broader paradigm shift in agriculture towards a system that better supports agricultural producers in both financial and environmental respects. It is crucial that all programmes are designed to ensure participation and leadership by farm women, but women should not be positioned as solely responsible for "green" farm initiatives. Change can begin with a stronger recognition of women's contributions and expertise in agriculture as a whole.

The gendered dimensions of coping and adaptation indicate a strong need for mental health supports in rural communities. These support services must be attentive to gendered dimensions of climate stress while ideally challenging embedded power differentials between farm men and women. Further, extension services and programmes providing information about adaptive strategies must recognize women's role in agriculture and enhance their capacity to participate in climate preparedness planning.

Conclusion: At the crossroads of gender, work and climate

In discussions of climate change mitigation and adaptation, the intersection of gender and work demands our attention. Agriculture in the Canadian prairies provides a useful site for inquiry into this intersection: it is a highly masculinized sector with rigid gendered divisions of labour and, at the same time, agricultural producers are highly exposed to climate change. Furthermore, the sector's contribution to greenhouse gas emissions means that climate change is a particularly controversial issue for agricultural producers, resulting in widely variable beliefs about climate change.

The findings presented in this chapter show how ongoing gendered divisions of farm work produce identifiable gendered trends in two areas: environmental awareness and climate coping/adaptation. On the family farm, economics and farm survival constitute an imminent and pressing concern that often overrides environmental concerns. However, as marginalized farm "helpers", farm women's relative disconnection from farm control may allow them a more critical environmental perspective compared to men, who are often positioned as "main" or "central" farmers responsible for the farm's financial success. Rather than suggesting women's inherent "virtuousness" in relation to the environment, these findings show how gender-differentiated work roles can produce gender-differentiated views on the environment.

Yet, the findings also show that environmental awareness amongst farm women has not yet shifted upward to the macro level of climate change belief. In the two studies discussed here, views on climate change (e.g. belief or disbelief in anthropogenic climate change) do not appear gender-differentiated but are rather highly variable. Climate change is a broad, macro-level issue that is often viewed as outside farmers' agency and disconnected from their daily work; further, climate change is the subject of much uncertainty for farmers. As such, the interconnection between climate change belief and work roles is less apparent and therefore – at least for now – belief in macro-level climate change is less gendered.

This further supports the assertion that gendered work roles strongly determine environmental and adaptive practices. In areas where there is a *tangible* connection between work roles and environment, such as environmental farm practices and adaptation to flooding and drought, gendered dimensions become particularly clear. This indicates the ongoing importance of considering gender and work in discussions about climate change. Recently, concerns have been raised about discourses that universalize or essentialize women within the climate change literature (i.e. women as either helpless victims of climate change or environmentally virtuous 'heroes') (Arora-Jonsson 2011). The analysis presented here suggests that focusing on gendered work roles is a crucial way forward in the discussion. Gendered work roles do not suggest an essential or universal experience shared by all women; however, a work-focused approach allows us to acknowledge trends. Where work is gendered, environmental experiences and attitudes may also be gendered, but these trends may not necessarily transfer upward to macro-level – and often more abstracted – topics like climate change belief. Environmental awareness and climate adaptation are key areas for policy intervention to support mitigation and adaptive efforts over the long term, and these interventions must therefore be designed through a gender lens.

Acknowledgments

The studies discussed in this chapter were supported by grants from the Social Sciences and Humanities Research Council of Canada (SSHRC), which funded the study on farm women in Saskatchewan, and the International Research Initiative on Adaptation to Climate Change – a joint initiative between the International Development Research Centre (IDRC), the Natural Sciences and Engineering Research Council of Canada (NSERC), and SSHRC – which funded the Vulnerability and Adaptation to Climate Extremes in the Americas (VACEA) project.

The author gratefully acknowledges the work of the VACEA research team led by Dr. David Sauchyn at the University of Regina and, in particular, the work of the community vulnerability assessment (CVA) research team: Dr. Harry (Polo) Diaz, Bruno Hernani, Erin Knuttila and Jessica Vanstone.

References

Alston, M 1995, *Women on the land: The hidden heart of rural Australia*, NewSouth Publishing, Sydney.

Alston, M 1998, "Farm women and their work: Why is it not recognised?" *Journal of Sociology*, vol. 34, no. 1, pp. 23–34.

Alston, M 2006, "'I'd like to just walk out of here': Australian women's experience of drought", *Sociologia Ruralis*, vol. 46, no. 2, pp. 154–170.

Alston, M 2009, "Drought policy in Australia: Gender mainstreaming or gender blindness?" *Gender, Place & Culture*, vol. 16, no. 2, pp. 139–154.

Alston, M & Whittenbury, K (eds) 2013, *Research, action and policy: Addressing the gendered impacts of climate change*, Springer, Dordrecht, (http://www.springer.com/environment/global+change+-+climate+change/book/978-94-007-5517-8.)

Arora-Jonsson, S 2011, "Virtue and vulnerability: Discourses on women, gender and climate change", *Global Environmental Change*, vol. 21, no. 2, pp. 744–751.

Bergengren, JC, Waliser, DE & Yung, YL 2011, "Ecological sensitivity: A biospheric view of climate change", *Climatic Change*, vol. 107, no. 3-4, pp. 433–57.

Boetto, H & McKinnon, J 2013, "Gender and climate change in rural Australia: A review of differences", *Critical Social Work*, vol. 14, no. 1, pp. 15–31.

Cannon, T 2002, "Gender and climate hazards in Bangladesh", *Gender & Development*, vol. 10, no. 2, pp. 45–50.

Dankelman, I 2010, *Gender and climate change: An introduction*, Earthscan, London.

Denton, F 2002, "Climate change vulnerability, impacts, and adaptation: Why does gender matter?" *Gender & Development*, vol. 10, no. 2, pp. 10–20.

Enarson, E & Dhar Chakrabarti, PG 2009, *Women, gender and disaster: Global issues and initiatives*, Sage Publications, New Delhi.

Enarson, E & Morrow, BH 1997, "A gendered perspective: The voices of women," in WG Peacock, BH Morrow & H Gladwin (eds), *Hurricane Andrew: Race, Gender and the Sociology of Disaster*, pp. 116–140, Routledge, London.

Environment Canada 2004, "Threats to water availability in Canada", NWRI Scientific Assessment Report Series No. 3 and ACSD Science Assessment Series No. 1, National Water Research Institute, Meteorological Service of Canada, Burlington, ON.

Environment Canada 2013, "National Inventory Report 1990–2013: Greenhouse gas sources and sinks in Canada", Government of Canada, Ottawa, ON, (https://www.ec.gc.ca/ges-ghg/default.asp?lang=En&n=5B59470C-1).

Faye, L 2006 "Redefining 'farmer': Agrarian feminist theory and the work of Saskatchewan farm women", Memorial University of Newfoundland, St. John's, NL.

Fletcher, AJ 2013, "The view from here: Agricultural policy, climate change, and the future of farm women in Saskatchewan", University of Regina, Regina, SK.

Fletcher, AJ & Knuttila, E 2016, "Gendering change: Canadian farm women respond to drought", in H Diaz, M Hurlbert & J Warren (eds), *Vulnerability and adaptation to drought: The Canadian Prairies and South America*, pp. 159–177, University of Calgary Press, Calgary, AB.

Government of Saskatchewan n.d., "Fact sheet: Don't burn grain bags", (http://environment.gov.sk.ca/adx/aspx/adxGetMedia.aspx?DocID=c672faa3-3f48-4ae8-a197-0ebf31768b0c&MediaID=ef5aeefe-e25f-4acd-a9a5-f42146189ea8&Filename=Grain+Bag+Burning+Fact+Sheet.pdf&l=English).

Groisman, PY, Knight, RW, Easterling, DR, Karl, TR, Hegerl, GC & Razuvaev, VN 2005, "Trends in intense precipitation in the climate record", *Journal of Climate*, vol. 18, no. 9, pp. 1326–1350.

Henderson, N & Sauchyn, D 2008, "Climate change impacts on Canada's prairie provinces: A summary of our state of knowledge", no. 08–01, Prairie Adaptation Research Collaborative (PARC), Regina, SK, (http://www.parc.ca/pdf/research_publications/summary_docs/SD2008-01.pdf).

IPCC 2015, *Climate change 2014: Synthesis report; Contribution of working groups 1, II, and III to the Fifth Assessment Report of the Intergovernmental Panel on Climate Change*, RK Pachauri & LA Meyer (eds), Intergovernmental Panel on Climate Change, Geneva.

Ireland, G 1983, *The farmer takes a wife*, Concerned Farm Women, Chesley, ON.

Keller, JC 2014 "'I wanna have my own damn dairy farm! Women farmers, legibility, and femininities in rural Wisconsin, US", *Journal of Rural Social Sciences*, vol. 29, no. 1, pp. 75–102.

Kubik, W 2004, "The changing roles of farm women and the consequences for their health, well being, and quality of life", University of Regina, Regina, SK.

Kubik, W 2005 "Farm women: The hidden subsidy in our food", *Canadian Woman Studies*, vol. 24, no. 4, pp. 85–90.

Logan, C & Ranzijn, R 2008, "The bush is dying: A qualitative study of South Australian farm women living in the midst of prolonged drought", *Journal of Rural Community Psychology*, vol. E12, no. 2, (https://marshall.edu/jrcp/VE12%20N2/jrcp%2012%20 2%20Logan%20and%20Ranzijn.pdf).

McCright, AM 2010, "The effects of gender on climate change knowledge and concern in the American public", *Population and Environment*, vol. 32, no. 1, pp. 66–87.

MacGregor, S 2010, "A stranger silence still: The need for feminist social research on climate change", *The Sociological Review*, vol. 57, October, pp. 124–140.

McMartin, DW & Merino, BHH 2014, "Analysing the links between agriculture and climate change: Can 'best management practices' be responsive to climate extremes?" *International Journal of Agricultural Resources, Governance and Ecology*, vol. 10, no. 1, pp. 50–62.

Marchildon, GP, Pittman, J & Sauchyn, DJ 2009, "The dry belt and changing aridity in the palliser triangle, 1895–2000", in GP Marchildon (ed.), *A dry oasis: Institutional adaptation to climate on the Canadian Plains*, pp. 31–44, Canadian Plains Research Center Press, Regina, SK.

Martz, DJF 2006, "Canadian farm women and their families: Restructuring, work and decision making", University of Saskatchewan, Saskatoon, SK.

Milne, W 2005, "Changing climate, uncertain future: Considering rural women in climate change policies and strategies", *Canadian Woman Studies*, vol. 24, no. 4, pp. 49–54.

Moosa, CS & Tuana, N 2014, "Mapping a research agenda concerning gender and climate change: A review of the literature", *Hypatia*, vol. 29, no. 3, pp. 677–694.

Phillips, EJ 1998, "The social and cultural construction of farming practice: 'Good' farming in two New South Wales communities", Charles Sturt University, Australia.

Pittman, J, Wittrock, V, Kulshreshtha, S & Wheaton, E 2011 "Vulnerability to climate change in rural Saskatchewan: Case study of the rural municipality of Rudy No. 284", *Journal of Rural Studies*, vol. 27, no. 1, pp. 83–94.

Reyes, RR 2002, "Gendering responses to El Niño in rural Peru", *Gender & Development*, vol. 10, no. 2, pp. 60–69.

Sachs, CE 1983, *The invisible farmers: Women in agricultural production*, Rowman & Allanheld, Totowa, NJ.

Sauchyn, D 2010, "Prairie climate trends and variability", in PD Harry, DJ Sauchyn & S Kulshreshtha (eds), *The new normal: The Canadian Prairies in a changing climate*, pp. 32–40, CPRC Press, Regina, SK.

Sauchyn, D & Kulshreshtha, S 2008, "Prairies", in DS Lemmen, FJ Warren, J Lacroix & E Bush (eds), *From impacts to adaptation: Canada in a changing climate 2007*, pp. 276–328, Government of Canada, Ottawa, ON, (https://www.nrcan.gc.ca/environment/impacts-adaptation/assessments/10031).

Seager, J 2006, "Noticing gender (or not) in disasters", *Geoforum*, vol. 37, no. 1, pp. 2–3.

Shortall, S 1999, *Women and farming: Property and power*, Palgrave Macmillan, Basingstoke.

Statistics Canada 2011a, "Census of agriculture, number and area of farms and farmland area by tenure, Canada and provinces", Table 004-0001, Government of Canada, Ottawa, ON, (http://www5.statcan.gc.ca/cansim/a47).

Statistics Canada 2011b, "2011 Census of agriculture", Government of Canada, Ottawa, ON, (http://www.statcan.gc.ca/daily-quotidien/120510/dq120510a-eng.htm).

Stehlik, D, Lawrence, G & Gray, I 2000, "Gender and drought: Experiences of Australian women in the drought of the 1990s", *Disasters*, vol. 24, no.1, pp. 38–53.

Wandel, J, Pittman, J & Prado, S 2010, "Rural vulnerability to climate change in the South Saskatchewan River Basin", in HP Diaz, DJ Sauchyn & Kulshreshtha, S (eds), *The New normal: The Canadian Prairies in a changing climate*, pp. 245–258, University of Regina Press, Regina, SK.

Warren, JW & Diaz, HP 2012, *Defying Palliser: Stories of resilience from the driest region of the Canadian Prairies*, Canadian Plains Research Center Press, Regina, SK.

Wheaton, E, Kulshreshtha, S, Wittrock, V & Koshida, G 2008, "Dry times: Hard lessons from the Canadian drought of 2001 and 2002", *The Canadian Geographer/Le Géographe Canadien*, vol. 52, no. 2, pp. 241–262.

13

UNDERSTANDING THE GENDERED LABOURS OF ADAPTATION TO CLIMATE CHANGE IN FOREST-BASED COMMUNITIES THROUGH DIFFERENT MODELS OF ANALYSIS

Maureen G. Reed

Introduction

What do we know about gender and labour in Canadian forest-based communities that can inform a research agenda about climate change adaptation? Surprisingly little. This chapter suggests that part of the reason for this gap is that in both academic and popular media, both climate change and forestry have been framed primarily as scientific and technical concerns and secondarily as economic and political ones. While understanding the science, economics and politics of climate change and forestry is necessary, such framing is insufficient for identifying community-based impacts and benefits of climate change or local capacities and challenges associated with adaptation. Despite their small size, forest-based communities may have heterogeneous populations, frequently characterized by Aboriginal and settler inhabitants. Understanding and addressing climate change as a social phenomenon introduces multiple social dimensions. The gendered labours of climate change adaptation represent some of those dimensions – and ones to which feminist scholars and practitioners are well positioned to contribute.

The chapter begins by considering research gaps and the challenges of applying feminist scholarship to climate change issues, particularly in the context of Canadian forestry. Next, it describes the context by explaining some of the social characteristics of forest-based communities. This context helps explain challenges they face for adaptation. Research undertaken on gender and adaptation in other rural contexts in post-industrial countries including the effects of drought, flooding and fire is then described, followed by three broad avenues of research. Conclusions focus on very real challenges and requirements for generating useful, applied research on gender and climate change adaptation founded on principles of feminist scholarship.

Research gaps

There has been very little research connecting gender, climate change adaptation and Canadian forestry; indeed an extensive literature review conducted in 2013 revealed only two articles (Davidson, Williamson & Parkins 2003; Klenk, Reed & Mendis-Millard 2012). By contrast, much of the research on climate change in Canada has focused on the north where the effects of climate change are considered to be the most rapid and deeply felt (e.g. Pearce et al. 2009). Yet the northern work been conducted at a relatively high level of abstraction and has been criticized for focusing on technical interventions while "excluding from its frame of reference the broader colonial and political-economic context within which northern Indigenous peoples struggle to respond to climate change" (Cameron 2012, 110). Political scientists have also criticized vulnerability assessments for neglecting the complex interactions between government policy and local-level implementation. For example, in a critique of forest-sector vulnerability assessments for climate change adaptation, Wellstead, Rayner and Howlett (2014) explained:

> adaptation planning involves governments in a complex process of interaction with a variety of stakeholder organizations and interests affected by such plans . . . . these studies ignore . . . the role of "micro" level factors . . . or on-the-ground implementation resources which affect both a government's ability to plan and societal actors' abilities to take part in those plans and planning efforts. (112–113)

Interestingly, research that has explored local planning efforts has found that climate change has not been a high priority for Canadian forest managers (Johnston & Hesseln 2012) or forest-based communities (Davidson, Williamson & Parkins 2003). Similarly, a recent study of rural community planning processes in Saskatchewan suggested that local politicians and even administrators of rural communities do not consider climate change to be of high priority and/or lack the tools to address the effects (Zamchevska 2014).

Notwithstanding this apparent indifference, over the past decade, government scientists and practitioners have conducted applied research on adaptive capacity to support understanding and action by forest-based communities (e.g. Johnston & Williamson 2007). At the local level, new guidebooks have been developed to assess the vulnerability of forest-based communities and help them identify strategies for adaptation (e.g. Pearce & Calihoo 2011). In 2011, the Canadian Model Forest Network (CMFN) sponsored the development of a guidebook and a resource book to help model forest communities assess their vulnerability and identify strategies for adjustment (Pearce & Callihoo 2011). This has been followed by publications supported by the Canadian Council of Forest Ministers (e.g. Gray 2012; Williamson, Campagna & Ogden 2012); however, none of these initiatives has addressed social dimensions of community or attempted a gender-based analysis of adaptive capacity.

Despite a growing literature on gender and climate change in developing countries, research in postindustrial settings has been slow to emerge (Alston & Whittenbury 2013; Eriksen 2014). At least two key challenges arise when applying feminist scholarship to climate change and forestry management in post-industrial settings. The first challenge is making a link between contemporary feminist scholarship and specific policy and programme needs in forestry. A lack of reliable evidence has given rise to tired stereotypes about men and women in forestry contexts (e.g. Arora-Jonsson 2011). Few feminist researchers have focused on forestry topics in Canada, and yet feminist scholars have long argued that gender is one relation through which access to and distribution of natural resources, wealth, work, decision-making and political power are differentiated and should be viewed (e.g. Eriksen 2014; Reed 2003). Recently, feminist scholars have considered whether the emphasis on gender theory has resulted in a lack of "practical strategies in dealing with daily issues" (discussion edited by Hawkins & Ojeda 2011, 247). Identifying almost no uptake of gender analysis in climate change issues, Joni Seager asked provocatively, "are we doing *useful* gender analysis?" (as cited in ibid., emphasis added). A second, and related, challenge is making this subject relevant to people who are responsible for the formulation and implementation of policies, programmes and planning in forest-based communities. As neither climate change nor gender issues figure prominently in the priority list for action by local practitioners, feminist scholars need to generate usable knowledge that is accessible to practitioners who can then apply it to planning and programme design.

Context

Like other resource extractive industries in Canada, during the twentieth century, forestry brought enormous wealth to private companies, public purses, and some groups of workers in resource extraction and processing. The average salary for individuals involved in forest harvesting and processing at one time was several times above the Canadian average (Marchak 1983). Today, forestry jobs are not as abundant or well paid, however, they are still considered "good jobs". Statistics Canada reported that the average income in 2014 from forestry and logging jobs was about equal to all manufacturing jobs, while the average income in pulp and paper positions was about 11 per cent greater than for all manufacturing (Statistics Canada 2015). Drawing from Census data, Natural Resources Canada reported that in 2013, 216,500 direct jobs and 350,000 indirect jobs nationally were attributable to forestry, while the forest sector formed a major driver in 171 census subdivisions (Natural Resources Canada 2015a, 2015b). In 2015, the agency's Annual Report also reported about 9,000 members of the forestry workforce was composed of Aboriginal people. None of the employment data provided was disaggregated by gender (Natural Resources Canada 2015c).

Notwithstanding these contributions, since the late 1970s, the industry has faced economic, environmental and political stress. Overharvesting, globalization of markets, technological displacement, the introduction of flexible work systems, declining union strength and decreasing job security have all led to reduced employment and job security in the industry (Clapp 1998; Marchak 1983). Changing tastes and social values about forests also gave rise to significant political protests about logging practices, resulting in changing regulation and more job losses (Hayter 2000). These protests did not cause adjustments in the industry; they only added to the mix of economic factors and foreshadowed other environmental effects that had appeared on the horizon in the 1970s. For example, the shift from old growth to new growth forests will inevitably lead to a reduced volume of timber available for harvest – an outcome known as the fall down effect (Hayter 2000). In Canada, climate change has begun to increase the frequency and severity of drought, fires, pests and diseases (Hogg & Benier 2005) and alter growing seasons and the availability of harvestable wood (Johnston & Williamson 2007).

Despite these stresses, there has been strong resistance on the part of communities, companies and provincial governments responsible for managing forestry or supporting the industry to fundamentally alter the structures upon which the economies of forest-based communities are based (Clapp 1998; Davidson, Williamson & Parkins 2003). Forest-based communities are particularly challenged because their workforce is highly specialized (Davidson, Williamson & Parkins 2003), local leaders often underestimate the level of risk (Clapp 1998), and communities frequently lack the tools to make the necessary transitions (Zamchevska 2014). For these reasons, "The combined effects of higher potential impacts and lower adaptive capacity mean that forest-based communities tend to be more vulnerable to climate change than other type of communities" (Johnston et al. 2010, 19).

However, a deeper social analysis of forest-based communities reveals that structural inequalities within forest-based communities have always existed. Even in good times, the benefits of resource booms in forestry (as in other resource sectors) have not been evenly distributed. Forest-based communities typically exhibit a bimodal income distribution with high-income earners – typically white men – in the professional and extractive sectors and low-income earners – typically white women – in the service and support sectors – including those that support the primary industry (Martz et al. 2006; Reed 2003, 2008). Women and Aboriginal people have less consistent access than non-Aboriginal people to high-income forestry employment (Mills 2006; Parkins et al. 2006). Aboriginal peoples have less access to employment overall and, where they are employed, are more likely to be hired in part-time and temporary positions (Mills 2006). Aboriginal women have documented experiences of racism from non-Aboriginal women and men; hence, they are more likely share common experiences with their male counterparts (MacPhail & Bowles 2015; Mills 2006). Beyond paid employment, women and men have differential access to community and social services (Martz et al. 2006; Munoz & Boyd-Noël 2008/9) and to service on community committees that give advice to government and industry on forest use and management (Reed & Varghese 2007). These differences suggest that the impacts of climate change will vary and the labours associated with adaptation will not be evenly distributed.

Further disparities exist across communities. Patriquin, Parkins and Stedman (2007) revealed that communities in Canada's boreal region display significantly higher poverty and unemployment rates, and lower educational attainment levels than other resource-based towns. For example, high populations of Aboriginal people had the effect of reducing median family income in boreal communities. In a related study, forest sector employment was found to play a relatively minor role in the enhancement of Aboriginal socioeconomic status across Canada, and the contribution of forestry to the income of Aboriginal families was found to be insignificant (Parkins et al. 2006). These findings compelled Parkins and others to conclude that the ties of Aboriginal people to the forested landscape and increased access to forest lands through land entitlements or the signing of modern treaties did not (yet at least) correspond with secure and high-income forestry employment. Hence, there is not likely to be a uniform assessment of adaptive communities across or within forest-based communities.

An open-ended research agenda

Although the forestry sector does not provide much guidance, parallel research related to climate change in western rural and agricultural communities offers some insights. This literature includes gender-based analysis of long-term and acute "socio-ecological" disasters (e.g. drought, flooding, fire) in countries such as the United States, Canada and Australia (e.g. Alston 2011, David & Enarson 2012; Eriksen 2014; Fletcher 2013; Seager 2006). Researchers have found capabilities for immediate response and long-term adaptation to be highly differentiated by gender, including: perception of risk and security (e.g. Davidson et al.); planning for an event (Enarson 2013); and during and following a disaster (see Alston 2011; David & Enarson 2012; Seager 2006). Effects of acute disasters such as flooding and fires, and slow-moving ecosystem changes such as drought and pest infestations have all been shown to vary by gender (see other chapters). With this understanding, one might be able to apply gender-based analyses and feminist analytical perspectives into research that has focused on adaptive capacity to climate change. Below, three such opportunities are considered.

The capitals framework for adaptive capacity

The term "adaptive capacity" has emerged as a concept for exploring the responses to risks, impacts and opportunities associated with these changes. Adaptive capacity refers to the ability to access, mobilize and deploy assets and endowments in ways that facilitate adaptation to change without degrading those resources (Norris et al. 2008). This includes both the ability to deal with, accommodate or recover from change (Smit & Wandel 2006); and the creative, innovative process of learning to use the opportunities that arise with change to enhance the quality of, and access to, resources to improve livelihoods and other aspects of community well-being (Norris et al. 2008).

Researchers of adaptive capacity have articulated a set of capitals and institutions to characterize the capacity of communities to adapt to climate change. Assets (or capitals) typically include financial, ecological, built/technological, human, social, cultural and political capitals (e.g. Wall & Marzall 2006), while institutions include formal rules and procedures such as policies, programmes and property rights as well as informal relations including power relations, systems of knowledge, cultural norms, values and worldviews. Each capital is then assessed through indicators for which data can be obtained and analysed.

Determining indicators and then acquiring appropriate data are challenging. As these capitals are interrelated and vary within and across household units, their individual and collective contributions to adaptive capacity are not well understood. Additionally, capitals can interact in positive or negative ways, suggesting that a primary focus on economic capital may overlook other elements that are important in a community's adaptive capacity. Gender, family structure and traditions, including systems of inheritance appear to be key to understanding these variations (Iyer, Kitson & Toh 2005). Additionally, selecting defining characteristics of each capital requires a judicious determination of efficiency, accuracy and cultural sensitivity, particularly as Canadian forest regions are co-inhabited by Aboriginal and non-Aboriginal populations. Research methods that include Aboriginal and non-Aboriginal peoples in selecting and interpreting indicators may be required to ensure the findings are valid for these communities. Such nuance will be important to determine how communities of study understand gender, what constitutes appropriate adaptation and interpretations of how culture and nature are connected.

A study by Klenk, Reed and Mendis-Millard (2012) explored how adaptive capacity varies by gender *within* forest-based communities by using Census data to conduct a "rapid assessment" of adaptive capacity in five Canadian Model Forest regions. Using a capitals framework, they found that, despite having higher levels of educational achievement in these regions, women had substantially lower economic capital assets Additionally, they found that women also had less access to social capital, because of higher rates of lone-mother families and higher rates of divorce among women. They also found that Aboriginal populations exhibited different trends between genders. Hence, they recommended that employment/training and social service programmes might be targeted to serve different situations of men and women across regions.

But they also pointed to challenges of selecting appropriate indicators and of interpreting the results. Often indicators that can be measured are simply proxies for more complex concepts and do not capture the richness of meaning intended by the concept. For example, one of the indicators of cultural capital – language diversity – does not necessarily mean that communities can access traditional/local knowledge and skills that might be important components of culture. With reference to Aboriginal communities, they noted: "without knowing how Aboriginal people are utilizing capital assets to adapt to changes in the forest sector, we cannot assume that declines in regional access to economic and human capitals will impact their adaptive capacity in the same way that the literature suggests it will impact non-Aboriginal communities" (Klenk, Reed & Mendis-Millard 2012, 96). Further, they suggested that following up preliminary or rapid assessment with in-depth community studies that directly involve local people might better reveal these relationships and outcomes.

Modelling exercises build on capitals frameworks by articulating relationships among driving factors of capacity. Many of these factors are identified as the "capitals" or "assets" described above. For example, Moser and Ekstrom offered a systematic diagnostic framework by which to identify barriers to undertaking intentional planned adaptation. Their focus on detecting barriers was "to design strategies to circumvent, remove, or lower the barriers" (2010, 22030). Similarly, Cutter et al.'s (2008) model for understanding resilience at the community level identified six categories of community resilience indicators that are remarkably similar to the capitals described earlier: ecological, social, economic, institutional, infrastructure and community competence. Despite these efforts to conceptualize and assess resilience, these metaphorical and theoretical models have not progressed to the operational stages where they effectively measure or monitor resilience or planned adaptation at the local level. Hence, next steps could include using participatory approaches to refine indicators suggested by the capitals frameworks, disaggregate the data by gender and other social categories, and work with communities to test the results in a real-world application.

Undertaking inclusive local planning

Studies just described can potentially be undertaken without direct input from community members, although they may be improved through more inclusive methodologies. However, planning for social change will require that communities be engaged in planning processes. This is challenging because climate change competes with other, often more immediate, challenges and gender concerns are not typically raised within local government structures. An exception is perhaps when communities face crises.

Between 1999 and 2011, the Mountain Pine Beetle (*Dendroctonus ponderosae*), devastated on British Columbia's (BC)'s pine-dominated forest ecosystems and forest-dependent communities responsible for an estimated loss of 53 per cent or 710 million cubic meters of commercially valuable pine (BC Ministry of Forests, Lands and Natural Resource Operations 2012, 3). Although naturally occurring, the insect had come to thrive as a consequence of longstanding management practices and a changing climate that had eliminated killing winter frosts (BC Ministry of Forests, Lands and Natural Resource Operations). The province's then Ministry of Forests and Range estimated that by 2015, 76 per cent of the pine volume in the interior of the province could be dead (BC Ministry of Forests and Range 2008, 2). The chief forester raised the annual allowable cut for affected regions to remove viable timber before it lost value and became standing fuel for wildfire. Sawmills and communities of the BC interior enjoyed a small economic boom, with the knowledge that they would soon experience loss of access to timber and associated losses of income to residents and communities. During the mid-late 2000s, timber prices fell, the Canadian dollar strengthened, and, in 2008, the US housing market plummeted resulting in layoffs, shift losses and mill shutdowns at an unprecedented level. The impacts were swift and severe. Residents around the region reported increasing demands on social services including food banks and soup kitchens. The percentage of residents collecting employment insurance in Quesnel, BC, rose by 22 per cent between December 2008 and December 2009 (Statistics Canada 2010) while residents of Williams Lake reported increasing violence and behavioural issues in schools as well as new housing arrangements aimed at reducing costs (Davis & Reed 2013). In 2008, the University of Northern BC and the Canadian Women's Health Network hosted a forum entitled, "Women's Perspectives on Mountain Pine Beetle" in Prince George to explore "social, economic, and health-related impacts through the eyes of women from beetle-affected communities in northern BC" and "identify action steps to address the impacts of mountain pine beetle on women, families, and communities" (Munoz & Boyd-Noël, 1). The opening session, "Untangling Sex and Gender" pointed out that the shift of governments from gender-specific supports to "'gender neutral' policy formulation and programming at the provincial (e.g. cuts to Women's Centres in 2004) and federal levels (e.g. cuts to Status of Women Canada in 2007)" (Orcherton, Boyd-Nöel & Merrick 2008, 25) ignored the constraints that institutionalized norms and behaviours had placed on women and men, and served to widen power differentials between men and women.

Participants expressed concerns over individuals and families required to relocate when jobs were lost; the loss of health services through regionalization; cuts to post-secondary education, social and legal services; reductions in inter-community transportation services; and losses in face-to-face government services. They also noted that job losses often hit where women were employed and/or placed heavy burdens on women in their places of employment (many of the services named above). They recommended greater focus on services for youth, particularly opportunities to retain youth in the communities and address youth sexual health. They also suggested a focus on women's employment and engagement in community development, political involvement and health provision. Finally, they called for replanting of trees using First Nations and ecological principles. It is not clear, however, whether any of these recommendations entered into mainstream planning. For example, the Climate Action Planning process in BC, established since the forum was held, has not made a single reference to gender or to differential impacts on women and men.

Practitioner strategies have sought to provide tools to municipalities and communities to help them plan for change. These efforts have typically placed emphasis on preparing for changes in physical infrastructure and local environments (e.g. West Coast Environmental Law 2012). Specific to forestry, *Pathways to Climate Change Resilience* (Pearce & Callihoo 2011) is a practical guidebook designed to help Canadian rural and forest-based communities plan for climate change.

Six steps are articulated along the "pathway" to preparing for climate change: (1) getting started getting prepared; (2) learning about climate change; (3) charting and scanning impacts and opportunities; (4) deciding priorities; (5) planning and taking action; and (6) watch(ing), learn(ing) and refin(ing). While the guidebook targets rural and forest-based communities, there is little explicit attention paid to social characteristics that shape communities, such as gender, age or ethnicity. In anticipation of a revision of a planned revision of the guidebook, researchers and practitioners from Sweden and Canada (including the guidebook's lead author) engaged in a workshop in 2014 to brainstorm how gender might be more effectively discussed and addressed in the guidebook. Only two of the 20 participants would have called themselves feminist scholars.[1]

Importantly, workshop participants recommended that the terminology "gender" be broadened to "diversity and inclusion" both to improve understanding and to generate more widespread support locally. This suggestion was in keeping with feminist scholarship that now focuses on "intersectionality" – mutually reinforcing and intersecting elements (e.g. gender, class, race, age) that contribute to one's identity. Addressing diversity and inclusion in planning processes may also broaden the climate change agenda from its current focus on infrastructure planning to social conditions such as effects on aging populations, socio-psychological effects, and social and community service needs. Participants suggested that paying attention to "who" is included through all phases of planning for climate change adaptation will be important for determining "what" is included. Hence, attention at the earliest phases should be given to widening the list of interested or affected participants and stakeholders who might contribute to local planning processes.

Getting prepared, according to the guidebook, involves gathering data about climate, community demographics and so on, and identifying the strategy/ies for moving forward. It is important to pay attention to the duration and stages of climate change effects. Slow-moving (e.g. drought) and acute (e.g. fire) effects will affect social groups differently. Each stage of an event – preparing and planning, during and follow-up – will place different kinds of demands on different kinds of workers within a community. Hence, it is necessary to think through how each event might unfold and who might need to be prepared, what kinds of support are needed, and whether follow-up evaluation, support or de-briefing can be anticipated.

The step, "getting prepared", is not simply technical. It means making decisions about what individuals, organizations and partners will be involved in all stages of the process. Potential participants from social service, community health, Indigenous, educational and even religious organizations and agencies are part of the social networks within a community that are part of, or that serve people at greatest risk from climate change and who have typically been excluded. Identifying such groups in the earliest planning phases may reframe the issue of climate change adaptation from a technical issue focused on infrastructure to a social issue focused on the social challenges (and opportunities) of adaptation. Workshop participants also reinforced that greater inclusion in "getting prepared" will require multiple methods to ensure a broad range of citizens participate. For example, youth and elder representatives will likely respond to different cues and these differences will influence who and what information gets brought to the table. Efforts to reach these groups can broaden understanding of social dimensions and point to information on demographic characteristics of communities and social and community services – the social infrastructure – that will be required to address adaptation needs.

The *Pathways* guidebook suggests that existing community planning structures should be used in adaptation planning rather than add "climate change" as another layer into already-stretched local planning resources and processes. However, given our understanding of social inclusion and exclusion in forest-based communities, local participants of a planning process might first analyse whether and how women and other groups are represented in these structures. Having broad representation at the outset will have ripple effects in subsequent stages such as identifying community impacts of climate change, assessing opportunities from climate change, prioritizing impacts for action planning, identifying potential actions, and monitoring and assessing the actions. Inclusion of previously excluded groups can help ensure that social needs are not overlooked.

The guidebook encourages communities to identify potential community impacts of climate change; however, workshop participants noted it is assumed that these impacts will be uniform throughout the community. This is likely not the case. Thus communities should be supported in identifying ways that women or other minority groups may be impacted by climate change and may contribute to adaptive capacity. Potential impacts of climate change are identified: property loss/damage, crop loss/damage, job disruption, travel/commuting or supply disruption, health impacts (e.g. from heat, flooding) and/or evacuation. At the very least, Census (and other) data should be disaggregated to determine if some groups are more likely than others to experience potential harm because of their geographic proximity to risk or their capacity to mitigate or adapt to it. Visualization tools can also aid understanding. For example, maps generated through participatory geographic information systems can help participants identify *where* impacts are likely to occur and inform residents about *who* might be affected.

Assessing assets and challenges for adaptive capacity must account for differences by gender and ethnicity. As previously described, economic resources and human and social capital are not distributed equally among people in forestry communities, and these differences must be taken into account. Such an accounting can help communities capitalize on previously hidden strengths while supporting those social groups that may be disadvantaged. Additionally, understanding the temporal dimensions of adaptation also points to a range of needs and resources for both women and men. The record forest fires in western Canada in the summer of 2015 will provide opportunities to learn how these tasks and the associated capitals to address the fires were gendered.

Workshop participants noted that the processes by which communities might use the guidebook are as important as the content of their work. The guidebook anticipates variations in how it will be used and its creator has already engaged in modified planning exercises suited to individual communities. Several modes of inquiry such as community-based research and participatory action research involve researchers and communities working together to design questions, learn and search for practical knowledge that can support collective and transformative action. Feminist scholars have long employed and contributed to the development of such methodologies, reflecting on their successes and failures, and emphasizing the importance of egalitarian research methodologies that work towards transformative change (e.g. Moss 2002). Appreciative inquiry (which is typically coupled with participatory and/or community-based research) focuses on working from the strengths of a community rather than its deficits or gaps, as the latter is more likely to lead to defensiveness and even conflict rather than shared understanding (Nyaupane & Poudel 2012). In the context of community-focused adaptation, working from a positive framing may remove the stigma associated with specifying individual limitations and, instead, encourage collective action dedicated to positive social change.

Understanding gender mainstreaming

The issues documented above raise a broader point – that Canadian forestry policy and social response to climate change might be improved through gender mainstreaming. Gender mainstreaming "seeks to institutionalize equality by embedding gender sensitive practices and norms in the structures, processes, and environment of public policy" (Daly 2005, 435). Gender-disaggregated statistics, gender impact assessment and gender budgeting are some tools that can be developed to aid in improving public policy.

In contrast to Canada, Sweden has a particularly broad and robust practice of gender mainstreaming where "responsibility for gender equality is extended to most, if not all, actors involved in public policy" (Daly 2005, 438). Its application has affected Sweden's forestry policy in a number of ways. All decisions must be analysed from a gender perspective; all statistics that are gathered about forestry are disaggregated by gender to enable gender analysis. In 2011, Sweden launched a national strategy for gender equality in the forest sector that includes several actions aimed at increasing women's participation in forestry (Lidestav & Berg Lejon 2013). While still subject to criticism, the policy provides a benchmark against which action (or inaction) can be judged.

But gender mainstreaming has not been a priority in Canada despite the long-standing and durable structural conditions in Canadian forestry that continue to restrict women's participation in forestry occupations and management decision-making (Munoz & Boyd-Noël 2008/9; Reed 2008). Indeed, an informal review of government documentation from British Columbia, where climate change adaptation in forestry has been strongest in Canada, revealed not a single mention of "gender", "women" or "men". Hence, learning how gender mainstreaming has been accomplished elsewhere, along with its challenges and limitations, may provide lessons for making change in Canada.

Conclusion

The phrase "climate change" is a catchall for a multi-faceted, interactive and complex set of biophysical and social changes that will vary across space, over time and according to cultural and local context. We are just beginning to understand the biophysical changes. This understanding is necessary in order to build out the social dimensions of climate change. But fundamentally, to address the human dimensions of climate change adaptation, we must move beyond the environmental, scientific and even some of the political dimensions to consider climate change as a fundamental social issue. It would be easy to classify such efforts as "women's work" or at the very least, the work of social scientists. Indeed, the Mountain Pine Beetle forum in Prince George described earlier was attended by 49 women and three men. Yet, with respect to climate change and forestry, the subject matter is highly interdisciplinary with expertise required of natural and social scientists, as well as humanities scholars. It also requires that academics work directly with community and policy practitioners to ensure that "theory" and "practice" inform each other.

As there has been virtually no research in Canada or other post-industrial countries that explicitly links gender and related social characteristics of forest-based communities to adaptive capacity for climate change; the research canvas is large and blank. It should, however, be drawn with care. If the effects of climate change are urgent and insistent, then more time spent by feminist scholars on theorizing gender identity or intersectionality will not be of immediate assistance. We know enough to act. We owe communities with whom we work action-oriented research with local people that will produce results they can use. This is not to discount theory; indeed, feminist theorizing has helped better understand the multiple social dimensions of difference upon which we can build action research. However, the academic valuation of theoretical over applied research remains a tension within feminist scholarship as well. We must acknowledge this tension and re-value applied research that is dedicated to transformational change.

Useful research will also need to distinguish between rapid onset (e.g. fire) and slow-moving (e.g. pest infestations) disturbances because each will affect different groups of workers and residents. Beyond understanding the when, the how and possibly the why of such disturbances, it will be important to understand the social contexts and meanings associated with disturbances and with changes necessary for adaptation. Understanding how assets are interpreted and used locally will require research methods that engage local people directly in identifying, assessing and implementing options that are tailored to their local contexts.

Making local managers and decision-makers aware and interested in conducting such planning, particularly when faced with multiple competing and seemingly more immediate concerns, is a significant challenge. In some rural communities, these concerns may be couched in the context of physical infrastructure versus social planning. But these two should not be viewed as competitive objectives, but rather, complementary. Ironically, local governments have more levers for the provision of social services locally than they do to regulate environmental management. In Aboriginal and settler communities, however, these concerns are frequently fundamental to their basic well-being, political autonomy, identity and cultural cohesion. Concerns for climate change and gender equality are only a few of the fundamental issues they face such as ensuring clean water, providing safe and accessible housing, securing land and governance "rights" and "responsibilities, providing good and healthy food – things that many other Canadian communities take for granted. Feminist scholars and local residents can work together to understand how these issues are connected and to find strategies to mutually support the advancement of multiple goals.

Feminist scholars can help to advance an agenda for climate research through improved conceptualizations of key terminology such as vulnerability and adaptation, demonstrations of action-based methodologies that support transformative change and research targeted to changing decision-making and policy processes and outcomes from local to national levels. Feminist analyses and methodologies can help sensitize researchers and practitioners to multiple social inequalities and systemic marginalization within forest-based communities that will affect their adaptive capacities. Feminist researchers can also create opportunities for more inclusive analyses, policies and practices. This is not an isolated or merely academic agenda. The expertise of social and natural scientists, practitioners and residents in the research process must be brought together to generate usable research that builds awareness within forest-based communities; identifies requirements for equitable and effective adaptation; provides options for planning processes in government agencies at local, provincial, and federal levels; and contributes to a better understanding of climate change adaptation in all its social dimensions.

Acknowledgements

The Social Sciences and Humanities Research Council funded this research. I thank Cindy Pearce whose work with Model Forest communities on climate change adaptation and whose frank discussions stimulated this work. I also thank those people who participated in the gender forum for stimulating new ways to think about gender, forestry and climate change adaptation. I am particularly grateful to Marjorie Griffin Cohen for this opportunity to share these ideas and to learn from other scholars doing related research on gender, climate change, work and society in rich countries.

Note

1 Further reading about this topic and the workshop can be found online at: International Union of Forest Research Organizations (2014) and Canadian Science Publishing (2014).

References

Alston, M 2011, "Gender and climate change in Australia", *Journal of Sociology*, vol. 47, no. 1, pp. 53–70.

Alston, M & Whittenbury, K (eds) 2013, *Research, action and policy: Addressing the gendered impacts of climate change*, Springer, Dordrecht.

Arora-Jonsson, S 2011, "Virtue and vulnerability: Discourses on women, gender and climate change", *Global Environmental Change*, vol. 21, issue 2, pp. 744–751.

BC Ministry of Forests and Range 2008, *Sustaining communities 2008: 2006–2011 Mountain Pine Beetle action plan progress report*, BC Ministry of Forests and Range, Victoria, BC.

BC Ministry of Forests, Lands and Natural Resource Operations 2012, *A history of the battle against the Mountain Pine Beetle: 2000 to 2012*, BC Ministry of Forests, Lands and Natural Resource Operations, Victoria, BC, (https://www.for.gov.bc.ca/hfp/mountain_pine_beetle/)

Cameron, E 2012, "Securing Indigenous politics: A critique of the vulnerability and adaptation approach to the human dimensions of climate change in the Canadian Arctic", *Global Environmental Change*, vol. 22, pp. 103–114.

Canadian Science Publishing 2014, "Gender, climate change adaptation and Forestry: Are they connected?" (http://www.cdnsciencepub.com/blog/gender-climate-change-adaptation-and-forestry-are-they-connected.aspx).

Clapp, A 1998, "The resource cycle in forestry and fishing", *Canadian Geography*, vol. 42, pp. 129–144.

Cutter, SL, Barnes, L, Berry, M, Burton, C, Evans, E, Tate, E & Webb, J 2008, "A place-based model for understanding community resilience to natural disasters", *Global Environmental Change*, vol. 18, no. 4, pp. 598–606.

Daly, M 2005, "Gender mainstreaming in theory and practice", *Social Politics*, vol. 20, no. 4, pp. 433–450.

David, E & Enarson, E (eds) 2012, *The women of Katrina: How gender, race, and class matter in an American disaster*, Vanderbilt University Press, Nashville, TN.

Davidson, DJ, Williamson, TB, & Parkins, JR 2003, "Understanding climate change risk and vulnerability in northern forest-based communities", *Canadian Journal of Forest Research*, vol. 33, no. 11, pp. 2252–2261.

Davis, EJ & Reed, MG 2013, "Governing transformation and resilience: The role of identity in renegotiating roles for forest-based communities of British Columbia's interior", in J Parkins & MG Reed (eds), *Social transformation in rural Canada: Community, cultures, and collective action*, pp. 249–268, UBC Press, Vancouver.

Enarson, E 2013, "Two solitudes, many bridges, big tent: Women's leadership in climate and disaster risk reduction", in M Alston & K Whittenbury (eds), *Research, action and policy: Addressing the gendered impacts of climate change*, pp. 63–74, Springer, New York.

Eriksen, C. 2014, *Gender and wildfire: Landscapes of uncertainty*, Routledge, New York.

Fletcher, A 2013, "From 'free' trade to farm women: Gender and the neoliberal environment", in M Alston & K Whittenbury (eds), *Research, action and policy: Addressing the gendered impacts of climate change*, pp. 109–22, Springer, Dordrecht.

Gray, PA 2012, *Adapting sustainable forest management to climate change: A systematic approach for exploring organizational readiness*, Canadian Council of Forest Ministers, Ottawa.

Hawkins R & Ojeda D (eds) 2011, "Gender and environment: Critical tradition and new challenges", with K Asher, B Baptiste, L Harris, S Mollett, A Nightengale, D Rocheleau, J Seager & F Sultana (eds), *Environment and Planning D: Society and Space*, vol. 29, pp. 237–253.

Hayter, R 2000, *Flexible crossroads: The restructuring of British Columbia's forest economy*, UBC Press, Vancouver, BC.
Hogg, EH & Bernier, PY 2005, "Climate change impacts on drought-prone forests in western Canada", *The Forestry Chronicle*, vol. 81, pp. 675–682.
International Union of Forest Research Organizations 2014, IUFRO – The Global Nework for Forest Science, (http://www.iufro.org/news/article/2014/07/08/scientific-summary-122-in-iufro-news-vol-43-double-issue-67-mid-july-2014/).
Iyer, S, Kitson, M & Toh, B 2005, "Social capital, economic growth and regional development", *Regional Studies*, vol. 39, no. 8, pp. 1015–140.
Johnston, M & Hesseln H 2012, "Climate change adaptive capacity of the Canadian forest sector", *Forest Policy and Economics*, vol. 24, pp. 29–34.
Johnston, M & Williamson T 2007, "A framework for assessing the vulnerability of the Canadian forest sector to climate change", *The Forestry Chronicle*, vol. 83, no. 3, pp. 358–361.
Johnston, M, Williamson, T, Munson, A, Ogden, A, Moroni, M, Parsons, R, Price, D, & Stadt, J 2010, *Climate change and forest management in Canada: Impacts, adaptive capacity and adaptation options*, Sustainable Forest Management Network, Edmonton, (http://cfs.nrcan.gc.ca/pubwarehouse/pdfs/31584.pdf).
Klenk, N, Reed, MG & Mendis-Millard, S 2012, "Adaptive capacity in Canadian model forest communities: A social and regional analysis", in MS Beaulieu & RN Harpelle (eds), *Pulp friction: Communities and the forest industry in a global perspective*, pp. 85–99, Laurier University Press, Waterloo.
Lidestav, G & Berg Lejon, S 2013, "Harvesting and silvicultural activities in Swedish family forestry: Behavior changes from a gender perspective", *Scandinavian Journal of Forest Research*, vol. 28, no. 2, pp. 136–142.
MacPhail, F & Bowles, P 2015, "Mining and energy in northern British Columbia: Employment, community, and inclusion", in P Bowles & G Wilson (eds), *Resource communities in a globalizing region: Development, agency and contestation in northern British Columbia*, pp. 220–225, UBC Press, Vancouver.
Marchak, P 1983, *Green gold: The forest industry in British Columbia*, UBC Press,Vancouver.
Martz, D, Reed, MG, Brueckner, I & Mills, S 2006, *Hidden actors, muted voices: The employment of rural women in Canadian forestry and agri-food industries*, Policy Research Fund, Ottawa.
Mills, SE 2006, "Segregation of women and Aboriginal people within Canada's forest sector by industry and occupation", *The Canadian Journal of Native Studies*, vol. 26, no. 1, pp. 147–171.
Moser, SC & Ekstrom, JA 2010, "A framework to diagnose barriers to climate change adaptation", *Proceedings of the National Academy of Sciences*, vol. 107, pp. 22026–22031.
Moss, P (ed.) 2002, *Feminist geography in practice: Research and methods*, Blackwell, Oxford.
Munoz, D & Boyd-Noël, S 2008/9, "'Boom, bust and beyond': Women's perspectives on the Mountain Pine Beetle", *Canadian Women's Health Network*, (http://www.cwhn.ca/en/node/39369).
Natural Resources Canada 2015a, "Communities", (http://www.nrcan.gc.ca/forests/canada/aboriginal/17448).
Natural Resources Canada 2015b, "Overview of Canada's forest industry", (http://www.nrcan.gc.ca/forests/industry/overview/13311).
Natural Resources Canada 2015c, "The state of Canada's forests, Annual Report, 25th Anniversary Edition", (http://cfs.nrcan.gc.ca/pubwarehouse/pdfs/36553.pdf).
Norris, FH, Stevens, SP, Pfefferbaum, B, Wyche, KF & Pfefferbaum, RL 2008, "Community resilience as a metaphor, theory, set of capacities and strategy for disaster readiness", *American Journal of Community Capacity*, vol. 41, pp. 127–150.

Nyaupane, GP & Poudel, S 2012, "Application of appreciative inquiry in tourism research in rural communities", *Tourism Management*, vol. 33, pp. 978–987.

Orcherton, D, Boyd-Nöel, S & Merrick, J 2008, "Beyond 'one size fits all'/ Gender perspectives and the mountain pine beetle", *LINK: Forum for Research and Extension in Natural Resources*, vol. 10, pp. 24–26.

Parkins, J, Stedman, R, Patriquin, M & Burns, M 2006, "Strong policies, poor outcomes: Longitudinal analysis of forest sector contributions to Aboriginal communities in Canada", *Journal of Aboriginal Economic Development*, vol. 5, no. 1, pp. 61–73.

Patriquin, MN, Parkins, JR & Stedman, RC 2007, "Socio-economic status of boreal communities in Canada", *Forestry*, vol. 80, no. 3, pp. 279–291.

Pearce, C & Callihoo, C 2011, *Pathways to climate change resilience: A guidebook for Canadian forest-based communities*, Canadian Model Forest Network, Ottawa, ON, (http://www.modelforest.net/images/Guiebook_Climate_Change_ENG_2011.pdf).

Pearce, T, Smit, B, Duerden, F, Ford, JD, Goose, A & Kataoyak, F 2009, "Inuit vulnerability and adaptive capacity to climate change in Ulukhaktok, Northwest Territories, Canada", *Polar Record*, vol. 46, pp. 157–177.

Reed, MG 2003, *Taking stands: Gender and the sustainability of rural communities*, UBC Press, Vancouver.

Reed, MG 2008, "Reproducing the gender order in Canadian forestry: The role of statistical representation", *Scandinavian Journal of Forest Research*, vol. 23, no. 1, pp. 78–91.

Reed, MG & Varghese, J 2007, "Gender representation on Canadian forest sector advisory committees", *The Forest Chronicle*, vol. 83, pp. 515–525.

Seager, J 2006, "Noticing gender (or not) in disasters", *Geoforum*, vol. 37, pp. 2–3.

Smit, B & Wandel, J 2006, "Adaptation, adaptive capacity and vulnerability", *Global Environmental Change*, vol. 16, pp. 282–292.

Statistics Canada 2010, "Employment insurance beneficiaries receiving regular benefits by census metropolitan area and census agglomeration (monthly)", (http://www.statcan.gc.ca/tables-tableaux/sum-som/l01/cst01/labor03a-eng.htm).

Statistics Canada 2015, "CANSIM Table 281-0027: Average weekly earnings", (http://www5.statcan.gc.ca/cansim/a26?lang=eng&id=2810027)

Wall, E & Marzall, K 2006, "Adaptive capacity for climate change in Canadian rural communities", *Local Environment*, vol. 11, pp. 373–397.

Wellstead, A, Rayner, J. & Howlett, M 2014, "Beyond the black box: Forest sector vulnerability assessments and adaptation to climate change in North America", *Environmental Science and Policy*, vol. 35, pp. 109–16.

West Coast Environmental Law 2012, *Preparing for climate change: An implementation guide for local governments in British Columbia*, West Coast Environmental Law, Vancouver, BC.

Williamson, TB, Campagna, MA & Ogden, AE 2012, *Adapting sustainable forest management to climate change: A framework for assessing vulnerability and mainstreaming adaptation into decision making*, Canadian Council of Forest Ministers, Ottawa, ON.

Zamchevska, V 2014, *Strengthening sustainability assessment in town planning in rural Saskatchewan*, Master of Environment and Sustainability thesis, University of Saskatchewan, Saskatoon.

14

THE COMPLEX IMPACTS OF INTENSIVE RESOURCE EXTRACTION ON WOMEN, CHILDREN AND ABORIGINAL PEOPLES

Towards contextually-informed approaches to climate change and health

Maya K. Gislason, Chris Buse, Shayna Dolan, Margot W. Parkes, Jemma Tosh and Bob Woollard

Introduction

It is now widely understood that human health and well-being is affected not only by the social and economic contexts and conditions within which people live, but also by ecological systems and services. Yet dramatic social and ecological challenges to people's health and well-being, and their impacts in particular on vulnerable populations, are not always carefully studied. In this chapter, we consider a range of ways that Aboriginal and non-Aboriginal women and children living in northern British Columbia (BC) are impacted by intensive resource extraction (IRE), and how these complex, regional dynamics need to be taken into account when seeking to understand the dynamics of climate change and health. We focus on the interrelated pathways that make women and children vulnerable to short-term social and ecological changes as well as to their long-term consequences, and why these populations may serve as sentinels which illustrate how sustainably communities are being developed (Poland & Dooris 2010). The task of this chapter is to present both evidence and conceptual tools intended to strengthen research on the interrelated factors influencing vulnerability within communities where intensive resource extracting occurs. The impacts on Aboriginal peoples are foregrounded due to the fact that IRE is known to occur on, or close to, the traditional territories of many Aboriginal communities, especially in Canada. The term "Aboriginal peoples" "collectively refers to the original inhabitants of Canada and their descendants, including First Nations, Inuit, and Métis peoples, as defined in Section 35(2) of the Canadian Constitution Act, 1982" (NCCAH 2013, 1).

Throughout this chapter we return to the importance of conducting research at the intersection between public health, environmental impacts and sustainable development, within which both IRE and climate change are interrelated drivers of change and complexity. The geographical focus of this chapter is northern BC – a place significantly impacted by the global demands to explore for, extract and export a range of natural resources which include mining, forestry, fishing and oil and gas exploration and extraction, and which is also raising attention to the overlay and cumulative impacts of climate and resource development (Gillingham et al. 2016; Picketts, Parkes & Déry 2016). In this chapter we draw examples from the oil and gas sector in particular as this focus offers concrete examples of the ways in which extractive industries and their related processes affect real people and places in northern BC.

Anthropogenically driven climate change, peak oil, biodiversity loss, extreme poverty and growing social inequities are all expressions of a larger number of wicked social-ecological problems which affect the health of humans, animals and the ecosystems that support all species (Brown 2010; Hallstrom, Guehlstorf & Parkes 2015; IAEH 2014). Both Aboriginal and non-Aboriginal women and children are being disproportionally affected at the local level by international forces such as globalization, labour exploitation and environmental degradation in the global North and South. Canadian society continues to be confronted with the need to understand and address the implications of our continued reliance on extractive industries as a primary source for our national "wealth" but not necessarily our health and well-being and the links between these practices and climate change.

In this chapter we proceed on the understanding that the development of the oil and gas sector has complex social, environmental and public health impacts. In addressing these and related challenges, the chapter contributes to newly emerging work which is developing integrative approaches to research, policy development and public health practice by making explicit the importance of understanding the interplay between social and ecological determinants of health. Our strategy deploys insights from health research – the area of expertise of the authors – in order to consider how the global demand for natural resources is impacting the health of people and environments in northern BC communities and to begin to imagine what this impact is for Aboriginal and non-Aboriginal women and children living in the region. We argue that placing attention on how social and ecological interplay is essential if the goal is to foster a more prevention oriented response to ameliorating the direct and indirect health impacts of climate change (Hancock, Spady & Soskolne 2015; IAEH 2014; MEA 2005). We suggest that health must be studied within the interactions between the environmental, social, political and economic contexts within which vulnerable populations live, work, play and love (Poland & Dooris 2010). This integrative, contextually informed approach moves research beyond dualistic framings, for example, of the environment versus the economy, and offers opportunities to observe how both social and ecological processes drive anthropogenically driven climate change. It also moves the analysis farther upstream, for example by considering how to create conditions that enable healthy people living in healthy communities.

Background: Unconventional natural gas development in northern British Columbia

Northern BC comprises over 70 per cent of the landmass of the province of British Columbia, less than 10 per cent of the total population, over 50 per cent of the Aboriginal population and is the site of approximately 65 per cent of all the active or proposed projects related to intensive resource extraction and manufacturing. In this region, whether from mining, fishing, forestry or oil and gas, communities are directly and/or indirectly impacted by intensive resource extraction (Foster et al. 2011). People in northern BC live far away from capital cities in the south of the province and outside of the social imagination of most urbanites. Unlike in cities, health and well-being is especially sensitive to ecosystem degradation, especially for Aboriginal peoples (Greenwood et al. 2015; Loppie Reading & Wien 2009; Parkes 2011; Parlee, O'Neil & Lutsel K'e Dene Nation 2007).

Health Canada recognizes that children, pregnant women, Aboriginal peoples and seniors are the populations with the greatest vulnerably to health harms from environmental risks. These populations appear to be suffering the worst impacts of growing health inequities and are the least likely to be employed directly by resource sectors such as the unconventional natural gas sector (Greenwood et al. 2015; Health Canada 2015). Despite its social, ecological and health impacts, BC elected officials continue to place concerted effort into the oil and gas sector, framing it as a good strategy for transforming the BC economy.

Unconventional natural gas (UNG) is the dominant mode of extraction and development in BC. Natural gas development in the province still includes a variety of traditional drilling methods but now predominantly utilizes a combination of hydraulic fracturing (injecting water, sand and chemicals to fracture rock and "stimulate" a well), directional drilling and liquefying natural gas for shipping to domestic and international markets. The promise of this suite of technological innovations – widely referred to as "UNG" development is to secure BC's place as an energy superpower, and export BC's gas to Asian markets where there is a purported demand for a "cleaner" energy alternative to transition away from coal and its resulting air quality issues (Heffernen & Dawson 2010). Ironically, despite being marketed as "clean", UNG extraction, production and transportation all increase greenhouse gas emissions, do not effectively supply the natural gas needs of the region and also produce a range of direct and indirect health impacts not only for humans but also for animals and ecosystems (Kniewasser & Horne 2015).

Due to the scale and technologies in this region, UNG activity is dramatically transforming social and environmental landscapes in northern BC with far-reaching implications for the health and wellbeing of those living in the region. The expansion of the province's UNG industry occurs through a combination of increased drilling, pipeline development, workforce training and approving the construction of processing facilities capable of liquefying natural gas so that it can be transported to overseas markets. As a result of these efforts, northern BC has experienced (until recently) a natural gas "boom". Natural gas contributed $6.4 billion to the BC economy in 2013. More than $63 billion in capital investment has been spent on natural gas development in the province since 2000, and at present, twenty UNG projects are proposed across the province. Twelve are currently approved for export licenses. In a 2012 report the BC government committed to having at least five LNG plants operating by 2020, a venture they estimated could have a $1 trillion impact on the provincial economy (British Columbia Ministry of Energy and Mines 2012). However, these benefits are speculative and debatable (Lee 2016). In fact, Canada's natural gas reserves may not only pale in relation to untapped reserves found in China, one of its primary export targets, but also more recent forecasts suggest that the economic promise of the industry are less robust than anticipated even a year earlier. The Canadian Centre for Policy Alternatives has also shown that if Canada carries out a range of probable expansion scenarios for its oil and gas industry that the nation will be unable to meet the *Paris Agreement* targets of reducing its greenhouse gas emissions to 30 per cent below 2005 levels by 2010 (Hughes 2016).

Far away from the rhetoric of budgets and profits bantered about in capital cities, the lived experience of communities in northern BC offers a complex and nuanced set of insights into what it is like to live within a landscape contoured by the clustering of intensifying industrial activity (Badenhorst et al. 2014; Mitchell-Foster and Gislason 2016, Kinnear et al. 2013). Activity oriented around four shale basins in northeastern BC illustrates the scope of impact. In this region the Montney Play Trend, the Horn River Basin, the Cordova Embayment and the Liard Basin are estimated to contain 219 trillion cubic feet of marketable, unconventional natural gas, making this group one of the largest shale gas reserves in the world (Government of Canada 2016). Commercial shale gas production is taking place in the Montney Play Trend and the Horn River Basin; however, much of this is also Treaty 8 territory – a territory of 840,000 kilometres governed by a treaty between 39 First Nations communities in BC, Northern Alberta, Northwestern Saskatchewan, and Southwest Northwest Territories (Treaty 8 Tribal Association 2015). Within Treaty 8 territory there are an estimated 23,419 operating wells and another 4,000 dry and abandoned wells. This is in addition to multiple other forms of resource development, including a long history of coal and mineral mines, industrial agriculture, forestry and tourism/recreation. Increasingly, Aboriginal communities are issuing urgent calls for regional strategic environmental assessments to look at the cumulative impacts of all existing and proposed intensive resource extraction projects (Loppie Reading & Wien 2009; Macdonald 2016).

Booms in single resource economies are characteristically unstable and lead to eventual economic busts. In June 2015, the UNG industry in northern BC entered a bust cycle. In addition to the decline in oil prices and a reduced demand globally, Canadian UNG is turning out to be costly to produce and difficult to sell competitively. It is an unenviable task to weather boom and bust cycles. People living in communities in northern BC have no control over fluctuations in the global economy and little impact of the activities of multinational corporations. In an economic downturn, temporary workers leave and long-term residents of host communities are left to find ways to live within towns that are being dismantled socially and economically. For all residents of the region, but most significantly for Aboriginal peoples, it also means that the local populations are left to source essential ecosystem services, such as clean air, water, soil, and country foods and wildlife, from landscapes which have been dramatically altered and often contaminated.

Agencies responsible for the public health and well-being of people in northern BC, the Northern Health Authority (NH) and the First Nations Health Authority (FNHA), are reporting increased public health service needs in areas where there are intensive resource extraction initiatives and large temporary work camps, whether or not the sectors are in a boom or bust cycle (Northern Health 2012, 2013). As health authorities try to assemble a picture of precise health needs in order to enable their response to the increasing rate and scale of emergent requests, research attention is being placed on understanding the mechanisms through which public health is and will be impacted by intensive resource extraction writ large, by boom and bust cycles, and by social and gender inequity (Buse et al., 2016; Parkes 2016).

Poignantly, northern BC is becoming a region not only where companies oriented to making a profit in global markets have exploited local people and environments but also a place where climactic changes are already impacting local economies, transportation corridors, access to country foods, wildlife health and cultural practices. In the next section, four areas of specific impact of UNG on communities in northern BC are described. Many of the impacts are interrelated and are increasingly acknowledged to be producing cumulative environmental, community and health impacts (CIRC 2016; Gillingham et al. 2016).

Environmental and biophysical impacts

Both the exploration for natural resources and the extraction of these resources (e.g. oil, gas and minerals) significantly impact the environment as a range of intensive techniques are utilized which require heavy machinery, seismology (compressed air, thumper trucks, explosives), and/or the boring of test wells, to name a few methods. The impacts include increasing air pollution, soil contamination, land degradation, noise and light pollution, the contamination of water sources, excessive water use in areas where there is short supply and the precipitation of small earthquakes through hydraulic fracturing (Benusic 2013; Powers et al. 2014). Compromised air, water and food, through environmental pathways of exposure, can also dramatically impact health (Gillingham et al. 2016; Greenwood et al. 2015).

The introduction of contaminants and pollutants into the natural environment has direct and indirect impacts on the health and well-being of northern BC populations. For instance, natural gas extraction is known to be a source of carcinogenic and neurotoxin compounds that leads to illness in the local human population as well as to animal deaths (Allen et al. 2013; Colborn et al. 2011; Finkel & Law 2011). Vibrations, offensive odours, "deafening noise" and light pollution can disrupt sleep patterns (Korfmacher et al. 2013), as can high amounts of endocrine-disrupting chemicals that are found in water in natural resource extraction areas, which can also impact reproductive health and increase rates of cancer.

Psychological impacts

The introduction and operation of intensive resource extraction industries, which precipitates dramatic changes to natural, social and sensory landscapes, impacts on the mental well-being of local populations (Saberi 2013). While some residents may initially feel hopeful due to the promise of increased income, others may experience anxiety and stress due to factors such as housing shortages, noise and light pollution, and community conflict (Brasier et al. 2011), to the degree that it can overwhelm local mental health services (Moss, Coram & Blashki 2013). Substance use, depression and suicide have also been identified as related to intensive resource extraction (Shandro et al. 2014). The psychological impacts of intensive resource extraction are linked to the social impacts, as depression, anxiety and substance use are associated with increases in violence and social inequity. Aboriginal communities are disproportionately impacted (Gone & Trimble 2012) as the loss of connection to land, cultural practices and the compounding factors of colonization can intensify the impacts of intensive resource extraction (King, Smith & Gracey 2009).

Social impacts

The public health implications of boom and bust economies related to natural resource extraction remains a largely understudied phenomenon (Schmidt 2011). Due to the inescapable realities of "boom" towns associated with the resource industry in Canada, it is clear that the public health system is facing a critical moment where new interventions are needed. Social impacts affecting health are triggered by issues such as housing shortages and exacerbated living costs that can push out long-term residents (Brasier et al. 2011; Schmidt 2015) and render workers into insecure accommodation. For example, in a survey of youth aged 15–25 in Fort St John, BC, participants showed that despite a gas "boom", problems related to education, addictions and housing were the key areas affecting them by virtue of living in an oil and gas town (Goldenberg et al. 2010). In addition, increases in crime (Moss, Coram & Blashki 2013) and domestic violence, violence against women and child neglect (Eckford & Wagg 2014) can be linked to the inability of rural and remote resource communities to absorb the social impacts of changeable population sizes, economic trends and a lack of social infrastructure, as well as the influx of predominantly young men who work in the gas industry. Community conflict is a concern highlighted by residents due to perceived increases in inequality between the "haves" and the "have-nots", with a widening divide between those who benefit financially from the introduction of intensive resource extraction and those who experience the exacerbation of multiple social inequities (Brasier et al. 2011).

Research on social and health inequities is providing a powerful framework for advancing insights into what is happening for Aboriginal and non-Aboriginal women and children in communities impacts by IREs in northern BC. The examples described above primarily capture the direct and indirect health impacts associated with environmental contamination yet comparatively little research has addressed impacts to vulnerable populations. Below, we outline some of these vulnerabilities in the context of the social dynamics of communities impacted by IREs and natural gas development.

Vulnerable populations: Considering the intersections of health

In the Canadian context, the communities vulnerable to impacts of IREs provide an appropriate unit at which to analyse the interaction between human activities and natural resource extraction and to appreciate how these activities map onto issues of climate change. In northern BC, Aboriginal peoples and women and children are among the most vulnerable. In this section some of the key vulnerabilities and the impacts of social inequity are presented.

Aboriginal health

Aboriginal health and well-being is affected by all the factors mentioned above but these impacts are compounded through deep cultural and historical connections between Aboriginal peoples and the natural world (FNHA 2013; Greenwood et al. 2015; Larsen 2006; Wilson 2003). In BC, a significant amount of UNG extraction has occurred on Treaty 8 land, although the unceded lands of other First Nations groups will be impacted across the supply chain. While UNG development has brought opportunities for First Nations, the physical extraction and operation of natural gas processing has disproportionately affected their ability to exercise treaty rights (Garvie et al. 2014). The social and geographical displacements and cultural and material losses caused by large-scale resource extraction, have at times led to forced resettlement, loss of territory and loss of subsistence resulting in intergenerational and collective trauma related to colonization and environmental degradation (Niezen 2009). This has occurred to the degree that certain Aboriginal communities allege that these impacts lead to their people not being able to practice their treaty rights (e.g. hunting and fishing) and functions therefore, as a breach of the Federal and Provincial government's fiduciary obligations – as has been argued in some civil claims Aboriginal communities have made against the province of BC (Supreme Court of British Columbia 2015).

Women and children

In "The burden of being 'employable': Underpaid and unpaid work and women's health", Reid and LeDrew (2014) demonstrate that there are "inextricable links among gender, location, unpaid caregiving and cultural expectations", that they intersect in "multiple and overlapping ways", and interact with "role expectations and threats of violence and discrimination to limit women's opportunities for well-paid employment across a range of British Columbian communities" (84). The authors go on to demonstrate that the lack of child care, access to stable and well-paying employment, and the "daily challenge of juggling unpaid and paid work limited women's opportunities financially" and produced chronic worry of living in material scarcity which accrue over the lifecourse (86). There are numerous other examples of how adverse health effects from human-induced environmental changes are being unequally distributed. For example, the gendered health impacts of the "boom town impact model" (McQueen et al. 2012) include: the relationship between natural resource extraction and violence against women and children (Downey, Bonds & Clark 2010); the effects of "rigger culture" that promotes "problematic gender relations" which impact on "sexual dynamics between workers and local women" (Goldenberg et al. 2008, 352); the creation of contexts conducive to sexual violence and related health harms (Tosh & Gislason 2016); and the effect of being an "oil and gas widow" (Reid & LeDrew 2013). Moreover, "gender-related exposures" (e.g. occupation, domestic work), sex-related features (e.g. body size and hormonal changes) and culture result in varied health outcomes (Chakravartty, Wiseman & Cole 2014). The noted increases in domestic abuse and child neglect, in addition to the potential for contaminants and pollutants impacting children during "critical periods of development" (Kassotis et al. 2014) means that women and children's physiological, psychological and cultural/spiritual health is being impacted and that action is urgently needed (see Allison 2016).

The Chief Medical Health Officer (CMHO) for Northern Health observes that current challenges to child and maternal health in northern BC provide a reminder that "childhood adversity is a root cause for many chronic diseases and health behaviours across the lifespan" (Allison 2016, 36). In order to address this, the CMHO proposes a framework that integrates evidence and responses to children's biological, social, cultural and physical environments. This report identifies remoteness, the fluctuations of undiversified economies, transient populations and significant pressures on services contour this health setting as factors impacting children's health. The report concludes with a call for widespread action based on northern BC's high rates of family poverty, infant mortality, poor oral health, hospitalization due to injury and child abuse, neglect, children in protection and low levels of readiness of children to enter school. In this region, pathways between social and health equity and climate change can be contemplated.

These realities are unfolding in an era where the World Health Organization (WHO 2010) framework for national and local health authorities on "Managing the public health impacts of natural resource extraction activities" advocates for integrated, interdisciplinary and mixed methods research into risk analysis, control and upstream prevention. Canadian and international attention to "upstream" determinants of health is creating demands for the health sector to work with new and unusual allies (Hancock, Spady & Soskolne 2015; Hallstrom et al. 2015;

Northern Health 2012; Parkes et al. 2016). This need is especially pronounced in northern BC communities where converging health impacts from socio-economic, environmental and demographic change demand new levels of collaboration and integration to optimize limited social and ecological resources. Accordingly, resource extraction governance is being recognized as an increasingly important "upstream" determinant of health. However, policies have not yet developed ways of accounting for the myriad forms of paid and unpaid labour that women engage in, which limit their financial opportunity and contribute to their living in conditions of economic instability, and increase their vulnerability to exploitation and violence (Reid & LeDrew 2013). Nor has there been sufficient attention paid to the impacts of macroeconomic changes which have continued, until recently, to strip away the Canadian social safety net and transfer responsibilities for child, family and community care to the "already-overburdened shoulders of individual women" (ibid., 91). All of these issues underscore the highly gendered dimension to the practices and impacts of intensive resource extraction and the importance of listening and responding to the experiences of Aboriginal and non-Aboriginal women and children.

Theories that foster integrative, contextually-informed approaches: Ecohealth, critical feminist theory and life-course approaches

Research and intersectoral collaboration have an important role to play in linking the lived realities, the data and the policy processes being developed to respond to the public health impacts of intensive resource extraction on vulnerable populations. One of the challenges is to identify theoretical and methodological frameworks that are well suited to the task of producing evidence based decision-making. The brief overview presented above reinforces the importance of developing work that links the cumulative and iterative interplay of the social and the ecological determinants of health, with particular attention to their impacts on Aboriginal and non-Aboriginal women and children. To advance more focused research that links insights into the social and ecological determinants of health of women and children in communities impacts by IREs, we propose that three theoretical and methodological approaches be used in concert: (1) ecosystem approaches to health, (2) critical feminist theory and (3) life course perspective. We first introduce ecosystem approaches to health (ecohealth) and second extend this approach by drawing upon insights from critical feminist theory and life course perspectives. We have found this triangulation of theoretical approaches enables researchers to consider the subjectivity and agency of women and children as they go about living their daily lives, navigating the impacts of social and health inequities and engaging in community capacity building within communities impacted not only by IREs but also incrementally by climate change.

Ecosystem approaches to health (also known as ecohealth) has been recognized as a milestone in public health research (Webb et al. 2010) as it makes explicit links "between interacting social, economic and ecological processes and their influence on human health" (Charron 2012, 19). Ecohealth theoretical frameworks and methodological tools are ideally suited to the task of conducting community-based research as they reconnect ecosystems with social dynamics by paying attention to the principles of transdisciplinarity, participation, equity and sustainability, and to the challenges of learning and working across sectorial, disciplinary, gender and cultural boundaries (Webb et al. 2010). Ecohealth principles also share significant common ground with Aboriginal perspectives on wellness (FNHA 2013) and can be thought of as a knowledge framework that reconnects with age-old insights into the links between people, places and health (Parkes 2011; Parkes, De Leeuw & Greenwood, 2011).

Critical theoretical engagement with ecohealth principles and practices can apply an equity lens to understanding the interplay between the earth systems and social systems (Mertens et al. 2005; Hallstrom et al. 2015; Webb et al. 2010). For example, critical feminist theory illuminates the organizational and structural dimensions of women's intersecting social identities (e.g. by race/ethnicity, gender, class), mutually constituted oppressions (e.g. racism, sexism) and lived experiences (embodiment, narrative). Critical feminist theory is therefore useful in surfacing power relations that construct masculine dominance in the oil and gas sector (Filteau 2014), feminize poverty and gender division of labour in the UNG industry (McKee 2014). The work of Black feminist theorists, who initially advanced intersectionality theory, which theorizes the convergence of gender and race/ethnicity and extends to consider the interplay between discrimination at the nexus between multiple social categories, serves to further highlight how discrimination experienced at the individual level reflects larger societal systems of oppression and privilege (Bowleg 2012, Crenshaw 1991). Aboriginal communities in northern BC have experienced devastating structural violence through colonialism that has ongoing expressions as gendered and sexualized violence. As but one illustration, in BC "from 1989 to 2006 nine young women went missing or were found murdered along the 724 kilometre length of highway 16 – now commonly referred to as the Highway of Tears. All but one of these victims were Aboriginal women" (CSFS 2016). Correlations have been made between increases in sexual exploitation and areas with high levels of industrial traffic (Shandro et al. 2014).

Given the requirement to account for the gendered nature of social life and to find ways to systemically account for social hierarchy, specific tools are required to capture how disparities manifest across time and space, including in northern BC communities. The life course perspective (LCP) is a tool that provides ways to understand and address health disparities across socioeconomic and racialized or ethnic groups, while making links between particular exposures that occur in childhood and how these early-life experiences shape health across a lifetime. Accordingly, LCP aligns well with the "systems thinking" pillar of ecohealth practice. By systemically considering the role of context (e.g. social, economic, environmental), biophysical factors, time and the impact of family and lived experiences, the LCP helps construct insights into health over generations. Bringing these theoretical perspectives together helps to address dangerously neglected areas in research that relies upon frameworks that over-emphasize environmental health or the social determinants of health, and underemphasize pathways and interactions between the two. To consciously seek out and interrogate these areas is essential to understanding the breadth of, and intersections between, the impacts of intensive resource extraction on communities and their cumulative impacts on human, animal and environmental health.

Conclusion

This chapter has sought to link local and global realities within the context of both intensive resource extraction economies and climate change. This pursuit is spurred by a desire to build a nuanced understanding of the complex ways in which economic, political, social and ecological systems weave together in the lives of both Aboriginal and non-Aboriginal peoples in northern BC, with a focus on women and children. Canadian society continues to be confronted with the need to understand and address the implications of our continued reliance on extractive industries as a primary source for our national "wealth" but not necessarily our health and well-being.

Amidst a growing awareness of the significance of climate change as a driver of health issues and through gathering evidence of its myriad pathways of impact, the tenability of our dominant economic systems is being called into question and their culpability in accelerating anthropogenically driven climate change named. The impacts of large-scale intensive resource extraction projects on the health and well-being of actual people in the landscapes in which they live, work, play and love moves us beyond dualisms such as "the environment versus jobs" or "feeding families versus saving trees". Studying the health impacts of intensive resource extraction is a foundation-building opportunity that requires researchers to tackle system-wide issues if the goals are to bring about primary prevention, to reduce human pressures on the environment *and* to lessen existing risks for vulnerable populations. It is increasingly clear that research at the intersection between public health, environmental change and sustainable development needs to simultaneously consider the importance of healthy people living in healthy communities who are sustained by resilient ecosystems. Collaboration between health, climate change and resource management could play an important role in adding nuance to efforts to link global economic patterns with the lived realities, adding to our understanding of the determinants of health for women, children and Aboriginal peoples living in communities impacted by IRE.

To make informed decisions about our future, we must understand the true costs of IRE in terms of the health of our environment, economies and communities long into the future. This understanding must be robust and our capacity to act and adapt to existing, predicted and unpredictable change must be enhanced. This chapter has called on three theoretical approaches – ecosystem approaches to health (ecohealth), critical feminist theory and the life course perspective (LCP) – which we propose can be used together to develop integrative, contextually-informed knowledge and evidence. We highlight the importance of working across scales – from the individual, the family and the community through to the national and the earth system scale where we see ourselves as component parts of an overarching whole. We suggest that these theories could be activated to help us understand the complex and interacting factors that influence the health of our planet, our place in it and ourselves.

References

Allen, DT, Torres, VM, Thomas, J, Sullivan, DW, Harrison, M, Hendler, A, Herndon, SC, Kolb, CE, Fraser, MP, Hill, AD, Lamb, BK, Miskimins J, Sawyer, RF & Seinfeld, JH 2013, "Measurements of methane emissions at natural gas production sites in the United States", *Proceedings of the National Academy of Sciences*, vol. 110, no. 44, pp. 17768–17773.

Allison, S 2016, "Chief Medical Health Officer's health status report on child health", Northern Health, (https://northernhealth.ca/Portals/0/About/Community_Accountability/documents/Northern-Health-CMHO.pdf).

Badenhorst, CJ, Mulroy, P, Thibault, G & Healy, T 2014, "Reframing the conversation: Understanding socio-economic impact assessments within the cycles of boom and bust – Strategies for developing a comprehensive toolkit for socio-economic impact assessments as part of current environmental impact studies in British Columbia Canada", *International Journal of Translation & Community Medicine (IJTC)*, vol. 2, no. 3, pp. 21–26, (http://www.scidoc.org/IJTCM-2333-8385-02-301.php).

Benusic, M 2013, "Fracking in BC: A public health concern", *BC Medical Journal*, vol. 55, no. 5, pp. 238–239.

Bowleg, L 2012, "The problem with the phrase women and minorities: Intersectionality– An important theoretical framework for public health", *American Journal of Public Health*, vol. 102, no. 7, pp. 1267–1273.

Brasier, KJ, Filteau, MR, McLaughlin, DK, Jacquet, J, Stedman, RC, Kelsey, TW & Goetz, SJ 2011, "Residents' perceptions of community and environmental impacts from development of natural gas in the marcellus shale: A comparison of Pennsylvania and New York cases", *Journal of Rural Social Sciences*, vol. 26, no. 1, pp. 32–61.

British Columbia Ministry of Energy and Mines 2012, "Liquefied Natural Gas: A strategy for BC's newest industry", (http://www.gov.bc.ca/ener/popt/down/liquefied_natural_gas_strategy.pdf).

Brown, VA, Harris, JA & Russell, JY 2010, *Tackling wicked problems: Through the transdisciplinary imagination*, Routledge, New York.

Buse, C, Jackson, J, Nowak, N, Fyfe, T & Halseth, G 2016, "A scoping review on the community impacts of unconventional natural gas for northern BC", Cumulative Impacts Research Consortium, University of Northern British Columbia, Prince George, BC.

Chakravartty, D, Wiseman, C & Cole, D 2014, "Differential environmental exposure among Non-Aboriginal Canadians as a function of sex/gender and race/ethnicity variables: A scoping review", *Canadian Journal of Public Health*, vol. 105, no. 6, pp. 438–444.

Charron, DF 2012, "EcoHealth: Origins and approach", in DF Charron (ed.), *Ecohealth research in practice: Innovative applications of an ecosystem approach to health*, Insight and Innovation in International Development, pp. 1–32, Springer, New York.

CIRC 2016, "Cumulative Impacts Research Consortium", UNBC, Prince George, (http://www.unbc.ca/cumulative-impacts).

Colborn, T, Kwiatkowski, C, Schultz, K & Bachran, M 2011, "Natural gas operations from a public health perspective", *Human and Ecological Risk Assessment: An International Journal*, vol. 17, no. 5, pp. 1039–1056.
Crenshaw, K 1991, "Mapping the margins: Intersectionality, identity politics and violence against women of color", *Stanford Law Review*, vol. 43, no. 6, pp. 1241–1299.
CSFS (Carrier Sekani Family Services) 2016, "Highway of tears, Preventing violence against women", (http://highwayoftears.org/about-us/highway-of-tears).
Downey, L, Bonds, E & Clark, K 2010, "Natural resource extraction, violence and environmental degradation", *Organization & Environment*, November, pp. 296–316.
Eckford, C & Wagg, J 2014, "The Peace Project: Gender-based analysis of violence against women and girls in Fort St. John", Fort St. John Women's Resource Society, Status of Women Canada, (https://thepeaceprojectfsj.files.wordpress.com/2014/03/the_peace_project_gender_based_analysis_amended.pdf).
Filteau, MR 2014, "Who are those guys? Constructing the oilfield's new dominant masculinity", *Men and masculinities*, vol. 17, no. 4, pp. 396–416.
Finkel, ML & Law, A 2011, "The rush to drill for natural gas: A public health cautionary tale", *American Journal of Public Health*, vol. 101, no 5, pp. 784–785.
FNHA (First Nations Health Authority) 2013, "About the FNHA", *First Nations Health Council*, (http://www.fnha.ca/about/fnha-overview).
Foster, LT, Keller, CP, McKee, B & Ostry, A 2011, "The British Columbia demographic context", *The British Columbia atlas of wellness, 2nd edition*, pp. 30–46, Department of Geography, University of Victoria, Western Geographical Press, Victoria, (http://dspace.library.uvic.ca:8080/bitstream/handle/1828/3838/toc.pdf?sequence=2&isAllowed=y).
Garvie, KH & Shaw, K 2014, "Oil and gas consultation and shale gas development in British Columbia", *BC Studies*, vol. 184, pp. 73–102.
Gillingham, MP, Halseth, GR, Johnson, CJ & Parkes, MW 2016, *The integration imperative: Cumulative environmental, community and health impacts of multiple natural resource developments*, Springer International Publishing AG, Switzerland.
Goldenberg, SM, Shoveller, JA, Koehoorn, M & Ostry, AC 2010, "And they call this progress? Consequences for young people of living and working in resource-extraction communities", *Critical Public Health*, vol. 20, no. 2, pp. 157–168.
Goldenberg, SM, Shoveller, JA, Ostry, AC & Koehoorn, M 2008, "Sexually transmitted infection (STI) testing among young oil and gas workers: The need for innovative, place-based approaches to STI control", *Canadian Journal of Public Health/Revue Canadienne de Santé Publique*, vol. 99, no. 4, pp. 350–354.
Gone, JP & Trimble, JE 2012, "American Indian and Alaska Native mental health: Diverse perspectives on enduring disparities", *Annual Review of Clinical Psychology*, vol. 8, no. 1, pp. 131–160.
Government of Canada 2016, "The unconventional gas resources of Mississippian-Devonian shales in the Liard Basin of British Columbia, the Northwest Territories and Yukon", Energy Briefing Note, National Energy Board, (https://www.neb-one.gc.ca/nrg/sttstc/ntrlgs/rprt/ltmtptntlbcnwtkn2016/index-eng.html).
Greenwood, M, de Leeuw, S, Lindsay, NM & Reading, C (eds) 2015, *Determinants of Indigenous peoples' health in Canada: Beyond the social*, Canadian Scholars Press, Toronto.
Hallstrom, LK, Guehlstorf, NP & Parkes, MW 2015, "Convergence and diversity: Integrating encounters with health, ecological and social concerns", in LK Hallstrom, NP Guehlstorf & MW Parkes (eds), *Ecosystems, society and health: Pathways through diversity, convergence and integration*, pp. 3–28, McGill Queens University Press, Montreal.

Hancock, T, Spady, D & Soskolne, C 2015, "Canadian Public Health Association discussion document global change and public health: Addressing the ecological determinants of health", (http://www.cpha.ca/uploads/policy/edh-discussion_e.pdf).

Health Canada 2015, "Environmental and workplace health: Vulnerable populations", (http://www.hc-sc.gc.ca/ewh-semt/contaminants/vulnerable/index-eng.php).

Heffernan, K & Dawson, FM 2010, "An overview of Canada's natural gas resources", Canadian Society for Unconventional Gas Report, (http://www.csug.ca/images/news/2011/Natural_Gas_in_Canada_final.pdf).

Hughes, DJ 2016, "Can Canada expand oil and gas production, build pipelines and keep its climate change commitments?" CCPA, Ottawa, (https://www.policyalternatives.ca/sites/default/files/uploads/publications/National%20Office%2C%20BC%20Office/2016/06/Can_Canada_Expand_Oil_and_Gas_Production.pdf).

IAEH (International Association for Ecology & Health) 2014, "Editorial: EcoHealth2014 call to action on climate change", *EcoHealth*, vol. 11, no. 4, pp. 456–458.

Kassotis, CD, Tillitt, DE, Davis, JW, Hormann, AM & Nagel, SC 2014, "Estrogen and androgen receptor activities of hydraulic fracturing chemicals and surface and ground water in a drilling-dense region", *Endocrinology*, vol. 155, no. 3, pp. 897–907.

King, M, Smith, A & Gracey, M 2009, "Indigenous health part 2: The underlying causes of the health gap", *The Lancet*, vol. 374, no. 9683, pp. 76–85.

Kinnear, S, Kabir, Z, Mann, J & Bricknell, L 2013, "The need to measure and manage the cumulative impacts of resource development on public health: An Australian perspective", in A Rodriguez-Morales (ed.), *Current topics in public health*, pp. 125–144, INTECH Science Technology and Medicine, (http://www.intechopen.com/books/current-topics-in-public-health/the-need-to-measure-and-manage-the-cumulative-impacts-of-resource-development-on-public-health-an-au).

Kniewasser, M & Horne, M 2015, "BC shale scenario tool technical report", The Pembina Foundation and The Pembina Institute, Calgary.

Korfmacher, KS, Jones, WA, Malone, SL & Vinci, LF 2013, "Public health and high volume hydraulic fracturing", *NEW SOLUTIONS: A Journal of Environmental and Occupational Health Policy*, vol. 23, no. 1, pp. 13–31.

Larsen, SC 2006, "The future's past: Politics of time and territory among Dakelh First Nations in British Columbia", *Geografiska Annaler: Series B, Human Geography*, vol. 88, no. 3, pp. 311–321.

Lee, M 2016, "Pipelines vs. Paris: Canada's climate conundrum, Behind the numbers", CCPA, Ottawa, (http://behindthenumbers.ca/2016/04/22/pipelines-vs-paris-canadas-climate-conundrum/).

Loppie Reading, C & Wien, F 2009, *Health inequalities and social determinants of Aboriginal peoples' health*, National Collaborating Centre for Aboriginal Health, Prince George, BC.

Macdonald, E 2016, "Atlas of cumulative landscape disturbance in the traditional territory of Blueberry River First Nations", David Suzuki Foundation, Ecotrust Canada, (http://www.davidsuzuki.org/publications/Blueberry%20Atlas%20report_final.pdf).

McKee, LE 2014, "Women in American energy: De-feminizing poverty in the oil and gas industries", *Journal of International Women's Studies*, vol. 15, no. 1, p. 167.

McQueen, D, Wismar, M, Lin, V, Jones, CM & Davies, M (eds) 2012, *Intersectoral governance for health in all policies: Structures, actions and experiences*, Observatory Studies Series 26, European Observatory on Health Systems and Policies, Copenhagen.

MEA 2005, *Ecosystems and human well-being: Desertification synthesis*, Millennium Ecosystem Assessment, (http://www.millenniumassessment.org/documents/document.355.aspx.pdf).

Mertens, F, Saint-Charles, J, Mergler, D, Passos, CJ & Lucotte, M 2005, "Network approach to analyzing and promoting equity in participatory ecohealth research", *EcoHealth*, vol. 2, no. 2, pp. 113–126.

Mitchell-Foster, K & Gislason, MK 2016, "Vignette 7: Lived reality and local relevance: Complexity and immediacy of experienced cumulative long-term impacts", as cited in "Exploring cumulative effects and impacts through examples", in MP Gillingham, GR Halseth, CJ Johnson, & MW Parkes (eds), *The integration imperative: Cumulative environmental, Community and health impacts of multiple natural resource developments*, Chapter 6, pp. 173–175, Springer International Publishing AG, Switzerland.

Moss, J, Coram, A & Blashki, G 2013, "Is fracking good for your health? An analysis of the impacts of unconventional gas on health and climate", Technical brief no. 28, The Social Justice Initiative, The Australian Institute, (http://www.tai.org.au/sites/defualt/files/TB%2028%20Is%20fracking%20good%20for%20your%20health.pdf).

NCCAH (National Collaborating Centre for Aboriginal Health) 2013, "An overview of Aboriginal health in Canada", UNBC, Prince George, BC.

Niezen, R 2009, "Suicide as a way of belonging: Causes and consequences of cluster suicides in Aboriginal communities", in J Kirmayer & G Valaskakis (eds), *Healing traditions: The mental health of Aboriginal peoples in Canada*, pp. 178–196, University of British Columbia Press, Vancouver, BC.

Northern Health 2012, "Part 1: Understanding the state of industrial camps in northern BC – A background paper", (https://northernhealth.ca/Portals/0/About/NH_Reports/documents/2012%2010%2017_Ind_Camps_Backgrounder_P1V1Comb.pdf).

Northern Health 2013, "Part 2: Understanding resource and community development in northern British Columbia – A background paper", (https://northernhealth.ca/Portals/0/About/PositionPapers/documents/IndustrialCamps_P2_ResouceCommDevel_WEB.pdf).

Parkes, MW 2011, "Ecohealth and Aboriginal health: A review of common ground", National Collaborating Centre for Aboriginal Health, UNBC, Prince George, BC, (http://www.nccah-ccnsa.ca/docs/Ecohealth_Margot%20Parkes%202011%20-%20EN.pdf).

Parkes, MW 2016, "Cumulative determinants of health impacts in rural, remote, and resource-dependent communities", in MP Gillingham, GR Halseth, CJ Johnson & MW Parkes (eds), *The integration imperative: Cumulative environmental, community and health effects of multiple natural resource developments*, pp. 117–149, Springer International Publishing, Switzerland.

Parkes, MW, De Leeuw, SA & Greenwood, M 2011, "Warming up to the embodied context of First Nations health: A critical intervention into and analysis of health and climate change research", *International Public Health Journal*, vol. 2, no. 4, pp. 477–485.

Parlee, B, O'Neil, J & Lutsel K'e Dene First Nation 2007, "'The Dene way of life': Perspectives on health from Canada's North", *Journal of Canadian Studies/Revue D'études Canadiennes*, vol. 41, no. 3, pp. 112–133.

Picketts, I, Parkes, MW & Déry, SJ 2016, "Climate change and resource development impacts in watersheds: Insights from the Nechako River Basin, Canada", *The Canadian Geographer*, doi:10.1111/cag.12327.

Poland, B, & Dooris, M 2010, "A green and healthy future: The settings approach to building health, equity and sustainability", *Critical Public Health*, vol. 20, no. 3, pp. 281–298.

Powers, M, Saberi, P, Pepino, R, Strupp, E, Bugos, E & Cannuscio, C 2014, "Popular epidemiology and 'fracking': Citizens' concerns regarding the economic, environmental, health and social impacts of unconventional natural gas drilling operations", *Journal of Community Health Problems: Through the Trandisciplinary Imagination*, Earthscan, London, (http://link.springer.com/article/10.1007/s10900-014-9968-x#page-2).

Reid, C & Ledrew, RA 2013, "The burden of being 'employable': Underpaid and unpaid work and women's health", *Affilia: Journal of Women & Social Work*, vol. 28, no. 1, pp. 79–93.

Saberi, P 2013, "Navigating medical issues in shale territory", *New Solutions: A Journal of Environmental and Occupational Health Policy*, vol. 23, no. 1, pp. 209–221.

Schmidt, C 2011, "Blind rush? Shale gas boom proceeds amid human health questions", *Environmental Health Perspectives*, vol. 119, no. 8, pp. A349–A353.

Schmidt, G 2015, "Resource development in Canada's North: Impacts on families and communities", *Journal of Comparative Social Work*, vol. 9, no. 2, pp. 2–25.

Shandro, JA, Jokinen, L, Kerr, K, Sam, AM, Scoble, M & Ostry, A 2014, "Ten steps ahead: Community health and safety in the Nak'al Bun/Stuart Lake Region during the construction phase of the Mount Milligan Mine", University of Victoria, Norman B. Keevil Institute of Mining Engineering, Monkey Forest Social Performance Consulting, Fort St James District, Nak'azdli Band Council (http://www.piplinks.org/system/files/Nak'al+Bun-Stuart+Lake+Mount+Milligan+Construction+Phase+Report+December+2014.pdf).

Shandro, JA, Veiga, MM, Shoveller, J, Scoble, M & Koehoorn, M 2011, "Perspectives on community health issues and the mining boom–bust cycle", *Resources Policy*, vol. 36, no. 2, pp. 178–186.

Supreme Court of British Columbia (N. Smith J.) 2015, *Blueberry River First Nations v. British Columbia*, BCSC 1302, (http://www.blg.com/en/newsandpublications/publication_4217).

Tosh, J & Gislason, MK 2016, "Fracking is a feminist issue: An intersectional ecofeminist commentary on natural resource extraction and rape", *Psychology of Women Section Review*, vol. 18, no. 1, pp. 54–59.

Treaty 8 Tribal Association 2015, "Treaty 8 Agreement", (http://treaty8.bc.ca/treaty-8-accord/).

Webb, J, Mergler, D, Parkes, MW, Saint-Charles, J, Spiegel, J, Waltner-Toews, D, Yassi, A & Woollard, RF 2010, "Tools for thoughtful action: The role of ecosystem approaches to health in enhancing public health", *Canadian Journal of Public Health*, vol. 101, no. 6, pp. 439–441.

Wilson, K 2003, "Therapeutic landscapes and First Nations peoples: An exploration of culture, health and place", *Health & Place*, vol. 9, no. 2, pp. 83–93.

World Health Organization (WHO) 2010, "Managing the public health impacts of natural resource extraction activities: A framework for national and local health authorities", Discussion draft, (http://www.gahp.net/new/wp-content/uploads/2013/07/WHO_health-in-EI.pdf).

PART V

Public policy and activism

15

HOW A GENDERED UNDERSTANDING OF CLIMATE CHANGE CAN HELP SHAPE CANADIAN CLIMATE POLICY

Nathalie J. Chalifour

The effects of climate change on the planet and its inhabitants are undeniable. From more frequent and extreme adverse weather events to melting permafrost, the changes brought on by the disrupted climate system will be felt across the globe (Intergovernmental Panel on Climate Change 2014). While these changes will impact everyone, they will be experienced differently on the basis of not only where people live, but also their socio-economic status, age, gender, race and culture. This renders climate change an important social justice and human rights issue (Ugochukwu 2015). While the North–South dimension of climate policy has featured prominently in UNFCCC negotiations from the beginning, the inter- and intra-generational components have garnered less attention. However, human rights, intergenerational equity and gender equality featured prominently in negotiations in Paris at COP 21 and their importance is reflected in the Preamble of the *Paris Agreement*.

Understanding about how gender intersects with climate has grown significantly in recent years (Alston & Whittenbury 2013; GenderCC 2007). For instance, research shows that women and men affect the climate differently (Cohen 2014; Ergas & York 2012), have different attitudes about climate and its solutions (Röhr 2001), are impacted in dissimilar ways (Denton 2002), and participate differently in climate negotiations and policy development (Haigh & Vallely 2010). While the research has often focused on the South, there is a growing body of research about the gender implications of climate change in industrialized countries (Johnsson-Latham 2007). This book is an important contribution to that scholarship.

This chapter focuses on Canada and the importance of taking a gender perspective in developing and implementing Canada's climate policies. Applying a gender perspective to the climate issue underscores the need for Canada to take strong and immediate action to mitigate and adapt to climate change, since women (especially those living in the North and/or in poverty) are, in general, more vulnerable to the effects of climate change. At the same time, policy responses must be designed to ensure those policies do not exacerbate existing inequalities. Canada is required to conduct a gender-based analysis of its climate policies which can help ensure that its responses to climate change (including the processes by which it determines those responses) support gender equality. We also have a constitutional obligation to ensure government policies are not discriminatory.

The analysis in this chapter shows that gender has been neglected in Canada's response to climate change. Gender considerations were completely absent from climate plans and policies across the country, and there is no evidence of any gender-based analysis having been conducted on any climate-related policy. The silver lining is that Prime Minister Trudeau's government, elected in October 2015, has not only promised more action on climate change, but has renewed the government's commitment to gender equality as evidenced in his mandate letter to the Minister for the Status of Women. This means a gender-sensitive approach to climate policy in Canada is still possible. Such an approach would contribute positively not only to Canada's commitments in relation to gender, but also to equality and justice more broadly, and generate co-benefits such as improvements in health and well-being.

The chapter is divided into four sections. The first section highlights Canada's commitments to gender equality at the international and domestic levels, including the requirement to conduct gender-based analysis and the equality guarantee in the *Canadian Charter of Rights and Freedoms*. The second section examines the role of gender from a procedural perspective, and specifically the participation of women in decision-making in areas of public policy relevant to the climate issue. The third section presents the results of my analysis of existing federal climate regulation and selected climate reports, as well as provincial climate policies. The analysis aims to empirically assess whether gender was considered in any of these policies, and to support a discussion of what issues might arise from a gender-based analysis. The fourth and final section offers some concluding thoughts.

Methodologically, the analysis is based on a literature review (including peer-reviewed scholarship, government reports and research from civil society), legal doctrinal analysis and review of some statistical data. I also systematically review the plans and policies identified in Table 15.1 for their assessment of gender considerations.

Background

Canada's commitment to gender equality

Equality and social justice are fundamental Canadian values for which there is a long-standing commitment. Substantive equality is one of the central guarantees enshrined in the *Charter*, creating an obligation on governments to avoid discriminatory laws and policies. This means that climate policies need to be crafted so they are not discriminatory and thereby avoid disadvantaging vulnerable members of Canadian society.

Canada has reiterated its commitment to gender equality at the international level for several decades. For instance, Canada is a signatory to the 1981 *UN Convention on the Elimination of all Forms of Discrimination against Women* (CEDAW), which requires Parties to take steps to eliminate discrimination in their country. Canada further committed to gender equality when it adopted the *UN Beijing Platform for Action* in 1995 and developed the 1995 *Federal Plan for Gender Equality*. These commitments require Canada to ensure that policies are analysed for their impact on women and men, respectively, before decisions are taken. Gender equality is a key goal of the recently adopted Sustainable Development Goals, of which Canada is a signatory.

The Canadian *Charter* includes a guarantee of equality that has been interpreted progressively as protecting substantive equality. Section 15(1) states that "every individual is equal before and under the law and has the right to the equal protection and equal benefit of the law without discrimination", including discrimination based on sex. The concept of substantive equality is different from formal equality, where one looks for the mere presence of similar or differential treatment. Substantive equality requires a contextual analysis of a person or group's experience to determine whether differential treatment might have ensued due to pre-existing disadvantage or negative stereotyping. It requires taking the social, political and economic context into account, and considering whether an ameliorative policy might be required to address a particular person or group's situation.

The *Charter* also includes a broad provision on gender equality in section 28, which states that "[n]otwithstanding anything in this *Charter*, the rights and freedoms referred to in it are guaranteed equally to male and female persons." While the general equality clause in s. 15 already protects against sex discrimination, s. 28 is interesting in that it seeks to ensure the rest of the provisions of the *Charter* are applied equally between women and men. More importantly, the opening language "notwithstanding anything in this *Charter*" suggests that the limitations allowed by s. 1 of the *Charter* (which would permit discrimination when justifiable in a democratic society) do not apply to s. 28, implying it is never acceptable to discriminate based on sex.

Gender-based analysis

One of the core ways the Government of Canada has to implement its commitment to gender equality is gender-based analysis (GBA). The purpose of GBA, referred to originally as gender mainstreaming, is to ensure that decision-makers understand the implications of any planned actions, policies or programmes before decisions are made. Gender analysis ensures that women's and men's concerns and experiences are considered throughout the design, implementation, monitoring and evaluation of programmes and policies in all areas of public policy (Charlseworth 2005; Sjolander 2005). The idea is for gender issues to become central features of decision-making and not be relegated to peripheral specialist institutions.

Gender budgeting is essentially gender-based analysis applied to government decisions that pertain to taxation and spending, notably as they are presented in government budgets and related policy documents. The practice is critical because the different socio-economic roles and responsibilities of women and men mean that the policies reflected in these budgets often have a gendered, unequal impact when implemented (Bakker 2006). Gender analysis of fiscal decisions opens the door to evaluating work beyond the paid sector of the economy, and including the unpaid provision of care undertaken most often by women. A commitment to conducting gender-based analysis necessarily includes the requirement to do a gender budget analysis.

Ensuring the engagement of women in climate policy development

Integrating gender into climate policy requires that decision-makers understand how gender is relevant and how climate policy can be shaped to promote gender equality. This section considers two tools which are critical for this to happen: GBA and the effective engagement of women in climate policy development.

Gender-based analysis

One of the first, critical steps in ensuring that climate policies promote gender equality is to understand where climate and gender intersect. This is why gender-based analysis (GBA) is so important, as it enables decision-makers to better understand how a given policy might impact women and men differently, and whether there is a way to design that policy in a way that promotes equality.

While GBA has been implemented in some federal government departments, many have criticized its implementation as being patchy and inconsistent. The Auditor General of Canada conducted an audit in 2009 of how GBA was being used in several federal government departments and three central agencies and confirmed its uneven implementation (Office of the Auditor General of Canada 2009). It also found little evidence of its influence on decisions. The audit noted that while briefing documents presented to the Minister of Finance for approval of policies or spending initiatives must contain a section on gender impacts, the Auditor was unable to verify whether gender impacts were reported due to confidentiality. Anecdotally, the 2007 budget proposed eliminating an excise tax exemption that was in place for ethanol and bio-diesel renewable fuels (Canada 2007). The Department of Finance concluded that this was a gender-neutral decision since both men and women drive vehicles. However, policies that impact upon the price of fuel are not necessarily gender neutral. For instance, Professor Lahey aptly pointed out to the Sanding Committee on the Status of Women that "[e]ven two minutes of research will disclose that women drive completely differently [from men]" and suggested that everything from the type of cars women drive and whether they are new or used to whether or not women will be able to pay more for their ethanol once the exemption has been repealed, needed to be considered (39th Parliament). This suggests that the requirement for GBA may not have been taken very seriously.

As a result of the audit, Status of Women Canada along with the Privy Council Office and the Treasury Board Secretariat were required to table an action plan to Parliament. The *2010 Government of Canada Departmental Action Plan on GBA* (the Plan) has resulted in the development of a new policy on GBA, known as *Gender-based Analysis Plus* (GBA+). The plus in GBA+ signifies that GBA now also requires consideration of other identify factors such as age, language, geography, education, income and culture. This type of intersectional analysis is helpful and to be encouraged.

The application of GBA at the provincial/territorial level is mixed (Foreign Affairs and International Trade Canada 2013). No province has legislated that GBA must be used, but all are taking some action to assess policy decisions through a lens of GBA. For example, PEI has guidelines for gender and diversity considerations that must be applied in policy development. In New Brunswick, GBA is mandated by cabinet and applies to all departments, policies and programmes. Ontario, Newfoundland and Labrador, Nova Scotia, Manitoba, Nunavut and the Yukon have established Women's Councils or Women's Directorates.

GBA applied to climate

GBA is a process that should be applied consistently to all policies and plans, and climate policies are no exception. Doing this creates an opportunity to bring gender considerations into the development of policies at the outset, ensuring not only that the gender implications of climate-related decisions are understood, but also that gender equality is promoted by these policies. Given the lack of implementation of GBA to date, it is unlikely it will be applied consistently to climate policies. However, civil society could raise the importance of climate justice, including a focus on gender, and point governments to their existing obligations under GBA+. GBA+ could also be integrated into broader sustainability assessment processes.

Prime Minister Trudeau signalled his commitment to gender issues when he gave his new Minister for the Status of Women, Patricia Hajdu, her mandate. In addition to requiring that she ensure that "government policy, legislation and regulations are sensitive to the different impacts that decisions can have on men and women", the Minister was tasked with working with the Privy Council Office to ensure that "a gender-based analysis is applied to proposals before they arrive at Cabinet for decision-making". This provides a clear mandate to apply GBA+ to any new climate policies.

Applying GBA to climate policies would require considering the emerging body of research offering a gendered perspective on GHG emissions (Cohen 2014). For instance, one Swedish study found that "men account for the bulk of energy use, carbon-dioxide emissions, air pollution and climate change – both among the rich and the poor" (Johnsson-Latham 2007), largely due to transportation. In Canada, road transportation accounts for 28 per cent of Canada's GHG emissions. Canadian statistics demonstrate that females account for about 34 per cent of kilometres driven, mainly in passenger vehicles (Cohen 2014). More women use public transportation to get to work than men (eg. 13.3 per cent of women versus 8 per cent of men in 2001, Statistics Canada, 2001). Cohen's research on GHG emissions by gender in Canada found that the overall contribution is uneven (23.5 per cent attributed to women, and 76.5 per cent attributed to men).

It is important not to overgeneralize, since the statistics are blunt and ignore many subtleties. For instance, transportation choices vary widely among women based on income, marital status, parenthood and other factors. Similarly, households are often comprised of male and female couples with one member of the household driving to work to earn income for the household. Evaluating data of this kind is not meant to be an exercise in blaming, but rather an opportunity to understand patterns, choices, impacts and design the most effective and fair policies. The main point is that gender differences in driving habits and use of public transit are relevant considerations in designing equality-promoting climate policies.

Participation and consultation

A common recommendation to improve gender equality is to increase the number of women in decision-making roles, since women are generally under-represented in positions of power and influence, including senior positions in the public and private sector. Research has shown that women continue to be underrepresented at many levels of decision-making relating to climate change. For instance, none of the bodies established under the UNFCCC achieve gender parity, with women representing between 6 and 40 per cent. The gender composition of party delegations to sessions of the governing bodies of the UNFCCC vary between a low of 36 per cent at COP20 to a high of 41 per cent at the Ad Hoc Working Group on the Durban Platform for Enhanced Action 2014 (United Nations Framework on Climate Change 2015).

Women continue to be underrepresented in leadership positions in the private sector, both in developing and developed countries. For instance, less than 16 per cent of board members of Fortune 500 companies are women; more than one-third have no women on their boards at all (Soares et al. 2013). Women account for a small percentage of scientists and engineers (US Bureau of Labor Statistics 2014; Women's Environmental Network 2007). The private sector will inevitably play an important role in mitigating climate change. Ensuring that women's voices within these sectors are heard is important if we are to create progressive, gender-supportive climate policies.

Of course, greater representation of women in decision-making is not a simple numbers game. The issue of representation in decision-making is more complex than initially appears. While some research suggests that greater representation of women leads to more progressive outcomes (Ergas & York 2012), other studies suggest this is not always the case (see Buckingham & Külcür in Chapter 3 of this volume). As such, we need to be careful about simplistic recommendations to balance numbers, and instead investigate power differentials, patriarchal structures and much more. However, since the reality is that women's perspectives are sorely underrepresented in both private and public-sector decision-making related to climate, a starting point would be to ensure better participation of women in these venues. Ensuring that this translates to greater equality would also be important.

Increasing gender representation in the institutional and decision-making processes related to climate change means many things, from involving women in the negotiations of international climate commitments to soliciting the views of women's advocacy groups in the policy development process. There have been some positive examples of gender representation in the Canadian climate policy process in recent years. For example, Inuit women's voices were represented at some international climate change meetings over the last decade. Most recently, the Canadian delegation at the 2015 Paris COP was led by Minister Catherine McKenna. However, in spite of these positive instances, women continue to be under-represented in positions of influence in climate-related areas in Canada. For instance, only one in four members of the Canadian Parliament are women. The dominance of men in decision-making roles within the Canadian private sector, especially the oil and gas sector, is also evident.

Prime Minister Trudeau demonstrated a commitment to gender equality by ensuring that his cabinet is gender balanced. While the former conservative government had only 12 women among its 39 cabinet Ministers, Prime Minister Trudeau appointed 15 women among 31 posts. His mandate letter to the Minister for the Status of Women requires her to make meaningful progress on the wage gap and increasing the number of women in senior decision-making positions and on boards in Canada.

Given this mandate, the Government of Canada has a clear commitment to ensure its climate policies are gender sensitive. While conducting GBA of climate policies is critical for achieving this, it is also essential to ensure women are part of climate-related decision-making at multiple levels of decision-making. One specific opportunity is in the institutional arrangements created to support the pan-Canadian climate policy framework. The government should ensure that women are represented in senior decision-making roles in these institutions. Representation on the four working groups created to develop recommendations under the pan-Canadian framework suggests gender may have been taken into consideration, though gender parity was not achieved: four of the ten co-chairs and 22 of 60 members were women, for gender representation of 40 per cent and 37 per cent, respectively.

Women's groups should be consulted in the development of climate policies, whether these are regulations or carbon pricing. Women's views about how to spend the considerable revenue that would be generated from carbon pricing should be actively solicited and considered. If the government opts to create an emissions trading regime, there is once again an opportunity to ensure that women are represented in the senior positions administering the system. It is important to include a diversity of women's voices, from Aboriginal women, women living in poverty, senior women, working and non-working women, and urban/rural perspectives, among others. This might require directly soliciting and facilitating their participation, especially if the policies are being developed rapidly and behind closed-doors.

Minister McKenna has recognized publicly that women are more vulnerable to climate change, and Foreign Affairs Minister Dion (at the time) recognized in a March 2016 speech that national climate policies must be gender responsive and that GBA is required. These statements are very important and may signal a turn for the federal government, given that the perspective of women's groups had not been solicited or considered in the design of federal climate regulations to date (discussed below). However, the key is now for these statements to be put into action. For instance, although the federal government solicited submissions on a range of specific questions relating to the options contemplated by the four working groups under the pan-Canadian framework, the questions did not mention gender. As of the time of writing, none of the 2,680 submissions referenced on the consultation website offered a gender perspective.

A gender perspective on Canadian climate regulations

Whenever policies are created or modified, we have an opportunity to ensure they are designed in a way that advances not only their main purpose, but also related policy objectives. The opportunity to achieve policy efficiency and co-benefits is recognized. In fact, the federal government requires a *Regulatory Impact Analysis Statement* (RIAS) to be prepared in advance of regulatory changes in order for decision-makers to understand the implications of the regulation, and ensure the best design (Treasury Board of Canada 2014).

All climate policies, whether they are aimed at mitigating GHG emissions or helping adapt to climate change, have gendered components. For instance, mitigation policies such as carbon pricing will increase the costs of energy, food and other carbon-intensive goods and services. Given that women have lower average incomes than men, such policies have to be carefully designed to avoid worsening inequality. Adaptive measures may rely upon women's unpaid labour for their success (Masika 2002), and as such need to be carefully conceived to avoid contributing to inequality. Mitigation and adaptation policies may also have a significant positive impact on women and other groups by reducing their vulnerability to climate disruption (Aguilar 2013). The key is to ensure gender considerations are integrated into all parts of the policy process to ensure Canada's response to climate change is equality-promoting. This section examines whether gender was considered in current Canadian climate policy, and offers some discussion of what a gendered analysis might reveal.

GHG mitigation policies

While there are many ways to reduce emissions, the two most common approaches are through regulations that establish standards (such as vehicle emissions standards or regulations on coal-fired electricity) and carbon pricing. The two main ways to price carbon are through taxation and cap and trade systems, though variations (including fee and dividend and hybrid systems) exist. Governments also spend money in order to encourage the development of clean energy, create incentives for improving energy efficiency and improve public transportation.

Federal climate policy

This section considers Canada's Sixth National Report on Climate Change (Environment and Climate Change Canada 2014), which is the government's report to the UNFCCC on its actions to meet commitments under that Convention, and an extensive government report on sector perspectives on climate change (Warren & Lemmen 2014). Both documents are almost 300 pages in length, yet neither makes any meaningful mention of gender or women, nor the need for gender-based analysis.

The *National Report* notes that the annual average surface temperature in Canada has warmed by 1.7 degrees Celsius, twice the global average, and that warming in the North is even more pronounced at a rate of 2.5 times the global average. It also includes a 25-page vulnerability assessment which signals the existence of several federal departmental climate risk assessments (e.g. Aboriginal Affairs and Northern Development Canada, Natural Resources Canada, and Fisheries and Oceans Canada), as well as a number of provincial initiatives (e.g. in Alberta, Nova Scotia and Yukon). The report also mentions a number of guidebooks aimed at helping municipalities assess and manage climate risks. There is some mention of Inuit and First Nations communities, but only in the context of contaminated marine resources, assessments for adaptation planning and plans for Health Canada to support the development of health adaptation plans and information tools in Northern First Nations and Inuit communities. The *National Report* makes no mention of gender, nor of plans to conduct a gender-based analysis.

The *Sector Perspectives* report includes a detailed chapter on the human health risks of climate change. This chapter surveys risks to air quality, food and water quality, zoonoses and vector-borne diseases (such as Lyme disease), natural hazards and ultraviolet radiation. This chapter draws upon an expansive literature survey to explain many of the ways in which the effects of climate change can impact human health. Vulnerable populations and regions are identified. Canadians living in the North and coastal areas are flagged, as are different vulnerabilities faced by rural versus urban communities. The report identifies a number of other vulnerable groups, including children, the elderly, those with pre-existing health conditions (such as asthma) and socially disadvantaged people, such as those with lower incomes, the homeless or those living alone. Aside from the identification of higher mortality rates among men from non-melanoma skin cancer due to UV radiation and the word gender in a table describing key vulnerability factors, there is no analysis of any potential differences in vulnerability between women and men. Given the level of detail in the chapter, and the government's commitment to gender-based analysis and gender equality, this is a surprising omission.

With respect to federal regulations, I reviewed all of the *Regulatory Impact Assessment Statements* (RIAS) for the following federal sectorial regulations on climate change enacted under the *Canadian Environmental Protection Act 1999*:

1. Passenger automobile and light truck greenhouse gas emission regulations;
2. Heavy-duty vehicle and engine greenhouse gas emission regulations;
3. Renewable fuels regulations;
4. Reduction of carbon dioxide emissions from coal-fired generation of electricity regulations;
5. On-road vehicle and engine emission regulations.

The RIAS process is mandated by the Government under the *Cabinet Directive on Streamlining Regulation* to ensure that policies protect and advance the public interest and lead to policy coherence. The Cabinet Directive also requires the government to undertake thorough consultations when developing regulatory proposals. The purpose of the RIAS is to understand the issue being regulated, the government's objectives, the regulation's costs and benefits, who was consulted and *who will be affected* (Treasury Board of Canada 2014). As such, RIASs are an ideal opportunity to identify potential vulnerabilities and ensure policies are aligned with constitutional obligations towards gender equality and non-discrimination.

All of the RIASs examined included a cost-benefit analysis explaining the regulations' predicted impacts on stakeholders. While costs and benefits to industry, consumers, government and health were identified, there was no analysis of costs and benefits disaggregated by gender. The analysis of implications for consumers would be an opportunity to identify particular groups that might be especially vulnerable to the changes, whether because they have lower incomes, have different transportation patterns, are employed in different sectors, or have different vulnerabilities, caregiving responsibilities or more.

Most of the RIASs (e.g. renewable fuels) are sensitive to the way in which the proposed regulations will impact consumers and industry players in different regions. The renewable fuel RIAS also documented increased prices of gasoline for consumers with regional variation. All of the RIASs indicate that stakeholders were consulted and describe the process in which this was achieved. Most indicate that they received comments from industry stakeholders as well as environmental groups, and occasionally a province. The RIASs did not report any consultations with women's groups or poverty advocates.

It is interesting to note that the RIAS for the renewable fuel regulation is the only one to note that a Strategic Environmental Screening was conducted. The *Cabinet Directive on the Environmental Assessment of Policy, Plan and Program Proposals* requires departments and agencies to evaluate the implications of policies, plans or programmes on the environment whenever the proposal may result in important environmental effects, whether positive or negative.

In sum, none of the RIASs for the federal climate regulations address the implications of the policies by gender. There is, similarly, no indication that women's groups were part of the consultation process. While several of the RIASs statements include some form of distributional analysis, the impacts considered were those on sectors and regions. Distributional impacts based on income levels, gender or other such factors were not considered.

Provincial climate policy

I also conducted a review of all the provincial and territorial climate-related legislation, policies and plans identified in Table 15.1 using a keyword search for: gender, women, men, low-income, vulnerable, senior and Aboriginal. My analysis found that only two provinces made any reference to gender in the climate policies and plans examined. Newfoundland and Labrador's funded a climate-related programme for girls (Newfoundland and Labrador Department of Environment and Conservation 2016) and Nova Scotia's Sustainable Transportation Strategy mentioned the importance of making transportation accessible for all, including gender (Government of Nova Scotia 2013).

The climate plans or policies of six provinces (Newfoundland and Labrador, Nova Scotia, New Brunswick, Québec, Manitoba, and British Columbia) included initiatives aimed at supporting low-income families. The policies or plans of the NWT, Alberta, Nova Scotia and Québec specifically reference the need to engage with Aboriginal communities. Québec's *Climate Change Action Plan* includes several initiatives aimed at supporting vulnerable groups, including seniors (Québec n.d.).

Carbon pricing

Much of the discussion about climate mitigation is focused on creating a transition to a low-carbon economy by establishing a price for carbon. Carbon pricing is important since it can facilitate a faster transition to a low-carbon economy, which will help everyone including the most vulnerable regions and people. However, care must be taken to design carbon-pricing policies in a way that does not exacerbate inequality, whether between nations or people. Gender considerations have generally not been integrated into carbon pricing discussions (Chalifour 2010; Women's Major Group 2012). Similarly, the crucial role of Indigenous peoples in a low-carbon economy has also not been adequately integrated into decision-making, in spite of international commitments such as the *United Nations Declaration on the Rights of Indigenous Peoples* (Dictaan-Bang-oa 2009).

Because of high levels of unpaid labour and lower levels of formal education, women may be at a disadvantage in some market-based systems. Carbon-offset systems, for example, have been criticized as potentially damaging for women in developing countries since offset projects may keep women out of areas traditionally used to collect food or fuel (Haigh & Vallely 2010). Women's stewardship role is often unrecognized (World Forest Coalition 2008). Relatedly, Haigh and Vallely (2010) argue that

> [c]arbon trading and other market-based systems emerge from a male-dominated system and masculine model of development and economic growth and primacy, in which certain work (typically conducted by men) is recognized and rewarded, while other tasks (those typically conducted by women) are not. (28)

It is well known that carbon prices are regressive (meaning that higher prices represent a larger proportion of income for low-income families) and that policies to attenuate those impacts are required. The BC carbon tax has a refundable low-income tax credit intended to address this. However, measures of distributional impacts of a tax policy are usually conducted at the household level rather than the individual level, meaning they are gender blind.

Carbon pricing policies generate enormous amounts of revenue. How that revenue is allocated has important gender implications. For instance, how will devoting revenue to corporate tax cuts versus public transit influence men as compared to women? If revenue is used to create incentives (e.g. to retro-fit homes or buy electric cars), who is accessing those incentives? In Canada, women spend an average of 1.5 more time than men on domestic unpaid work, and double the time caring for children (Milan, Keown & Urquijo 2011). What would be the effect on equality if some portion of carbon pricing revenues were invested in childcare? What about investments in job training targeted at helping women gain entry to high-paying jobs in the low-carbon economy?

Most evaluations of benefits from government spending highlight the amount of dollars spent rather than the actual benefits received (Yalnizyan 2005). As such, effective evaluations of the gender implications of tax policies (incidence and spending of revenues) may be incomplete (ibid.). Similarly, while recycling carbon tax revenues into personal and corporate tax cuts may achieve a goal of overall economic efficiency, it is important to analyse their gender impacts. Lahey's (2008) research has shown that women do not benefit as much as men from tax cuts. Taking into account not only income levels, but also other factors such as gender, race, age and region can also yield insights about how a given policy might impact particular groups (Bubna-Litic & Chalifour 2012; Chalifour 2010).

As the world moves towards putting a price on carbon, we should be asking ourselves how the design, implementation, monitoring and evaluation of carbon pricing policies will impact equality, and look for opportunities to shape climate policies in a way that also promotes equality.

Gender in climate adaptation policies

The ability of people to adapt to climate change depends upon many factors, including control over land, financial resources, health, mobility, food security and freedom from violence (Brodie, Demetriades & Esplen 2008; Lambrou & Piana 2006). Since women on average have lower incomes, less formal education and own less property than men, adapting to climate change may be more difficult. While there are regional differences, the gendered challenges in adapting to the impacts of climate change are also relevant in Canada. Some populations, such as Indigenous women, are particularly vulnerable to climate change due to high rates of poverty, pronounced health impacts and the general vulnerability of the Arctic to climate change (Bubna-Litic & Chalifour 2012; Wolski 2008).

The Canadian government has developed an adaptation policy framework (Environment and Climate Change Canada 2011) that identifies a number of priorities, including the natural environment (i.e. through protected areas), natural resources (e.g. forest sector), water resources and coastal zone management, municipal infrastructure, agricultural food production and human health (Environment and Climate Change Canada 2014). The human health component of the strategy includes information on vector-borne diseases and extreme heat events. It is noteworthy that while the National Report and the Adaptation Framework mention the importance of mainstreaming climate considerations into planning and policy development, there is no mention of gender or the need to conduct a GBA.

Most provinces have or are developing a climate adaptation policy. The four Atlantic provinces, for instance, released a joint adaptation strategy in 2008. Quebec's 2013–2020 adaptation strategy is a comprehensive plan with broad aims, including a $200 million investment. Ontario's strategy is also broad, and includes a focus on water resources and biodiversity.

While I have not conducted a comprehensive analysis of Canada's policies in support of adaptation, a quick review suggests that a gender lens has not been applied to these policies. Yet opportunities to ensure that adaptation policies are designed in order to promote gender equality abound. Research has shown that women are particularly vulnerable to the effects of extreme weather-related events. Adaptation policies should take gender into consideration to reduce vulnerabilities and create structures to reduce increases on caregiving burdens and health impacts. Establishing a process whereby women's perspectives are actively brought forward at every state of decision-making, including for adaptation policies, is critical.

Conclusion

Climate change has become a pivotal issue both internationally and within countries seeking economic prosperity and energy security in a carbon-constrained world. As jurisdictions reshape their policies, they have an opportunity to ensure that the resulting policies not only promote energy security and economic prosperity, but also equality and justice.

TABLE 15.1 Provincial/territorial climate mitigation policies

Jurisdiction	*Initiative*	*Implementation*
Newfoundland and Labrador	Lower Churchill Hydroelectric Project	2017
	Green Fund	2007
Prince Edward Island	PEI Climate Strategy	2008
	PEI Energy Accord	2010
Nova Scotia	*Environmental Goals and Sustainable Prosperity Act*	2009
	Sustainable Transportation Strategy	2013
	COMFIT-Community Feed-in Tariff Programme	2010
	Renewable Electricity Act	2007
New Brunswick	Blueprint Energy Plan	2011
	New Brunswick's Air Quality Regulations	1997
	Refurbished Point Lepreau Nuclear Generating Station	2012
Quebec	2013–20 Climate Change Action Plan	2013
	Emissions Cap and Trade System	2013
	Quebec Public Transit Policy	Completed
	Quebec's Energy Strategy	Completed
Ontario	Feed-in-Tariff	2009
	Cap-and-Trade System	2017
	Phase-out of Coal	2003–13
Manitoba	Manitoba Clean Energy Strategy	2012
	Clean Energy Transfer Initiative	2004
	Coal-Reduction Strategy	2012
	Cap-and-Trade	Announced in 2015
	The Biofuels Act	2003
	The Energy Savings Act	2012
Saskatchewan	*Management and Reduction of Greenhouse Gases Act*	2010
	Renewable Electricity Generation	2015
	Go Green Fund for Communities	2008
	Carbon Capture and Storage Initiatives	Ongoing
Alberta	The Climate Leadership Plan	2017
	Carbon Capture and Storage	2008
	Specified Gas Emitters Regulation	2007
British Columbia	Revenue Neutral Carbon Tax	2008
	BC Climate Action Charter	2007
	Renewable and Low Carbon Fuel Requirements Regulation	2008
Yukon	Climate Change Action Plan	2009
	Energy Strategy	2009
Northwest Territories	Biomass Energy Strategy	2012
	Energy Efficiency Technology Programme	
	Greenhouse Gas Strategy for NWT 2011–15	2012
Nunavut	Inuit Qaujimajatuqangit (Traditional Knowledge) of Climate Change in Nunavut	2005
	Nunavut Energy Strategy	2006

The analysis in this chapter has shown that while climate policies are being adopted at all levels across Canada, gender has been absent from the discussions. As a starting point, all governments should conduct gender-based analyses of existing and proposed climate policies. Doing so will reveal opportunities to shape those policies in a way that promotes human rights and equality, and fulfil our obligations under both domestic and international law. Policies developed with a gendered lens would help ensure, for instance, that investments in green infrastructure and community design support the needs of women, and that the transition to the low-carbon economy capitalizes on opportunities to reduce the wage gap and remove barriers for women to enter into traditionally male-dominated sectors.

All of humanity shares a common responsibility for reducing GHG emissions, but policy responses to climate change can and should be differentiated, if necessary, to ensure that gender perspectives are taken into consideration and that climate policies do not exacerbate existing inequalities. We cannot use a gender-indifferent approach to developing climate change policies. Rather, we need to be informed about how decisions differ by gender with respect to their contribution to GHG emissions, and we need to design policies that promote gender equality. Just as it would be inequitable to expect the same level of emissions reductions from countries that have contributed little to creating the climate change problem, especially at the outset, it would be inequitable to design policy responses to climate change that place a disproportionate burden on women – especially poor, racialized women.

Carbon emissions are fundamentally tied to the economy, which means that addressing climate change essentially requires an economic transformation. Integrating not only climate considerations but also a social justice perspective into our governance of energy, labour markets, the environment, health, agriculture, transportation, manufacturing, innovation and natural resource management will help ensure that Canada's policy response to climate change contributes not only to economic prosperity, but also to equality and social justice, including gender equality.

Acknowledgements

I would like to thank Adam Rochwerg, Jessica Earle, Kathleen Selkirk and Tyler Paquette for their research assistance. The author is grateful for the research funding provided by the Social Science and Humanities Research Council (SSHRC).

References

Aguilar, L 2013, "A path to implementation: Gender-responsive climate change strategies", in M Alston & K Whittenbur (eds), *Research, action and policy: Addressing the gendered impacts of climate change*, pp. 149–157, Springer, Netherlands.

Alston, M & Whittenbury, K (eds) 2013, *Research, action and policy: Addressing the gendered impacts of climate change*, Springer, Netherlands.

Bakker, I 2006, *Gender budget initiatives: Why they matter in Canada*, CCPA, Regina.

Brodie, A, Demetriades, J & Esplen, E 2008, *Gender and climate change: Mapping the linkages – a scoping study on knowledge and gaps*, Institution of Development Studies, Brighton.

Bubna-Litic, K & Chalifour, NJ 2012, "Are climate change policies fair to vulnerable communities? The impact of British Columbia's carbon tax and Australia's carbon pricing proposal on Indigenous communities", *Dalhousie Law Journal*, vol. 35, no. 1, pp. 127–178.

Canada 2007, *Gender budgets: An overview*, Library of Canada, No. PRB 07–25E, Ottawa, ON.

Chalifour, NJ 2010, "A feminist perspective on carbon taxes", *Canadian Western Law Journal*, vol. 21, no. 2, pp. 169–212.

Charlesworth, H 2005, "Not waving but drowning: Gender mainstreaming and human rights in the United Nations," *Harvard Human Rights Journal*, vol. 18, no. 1, pp. 1–18.

Cohen, MG 2014, "Gendered emissions: Counting greenhouse gas emissions by gender and why it matters", *Alternate Routes*, vol. 25, pp. 55–80.

Denton, F 2002, "Climate change vulnerability, impacts, and adaptation: Why does gender matter?" *Gender and Development*, vol. 10, no. 2, pp. 10–20.

Dictaan-Ban-oa, EP 2009, "Perishing past and pride: Indigenous women and climate change", *Women in Action*, vol. 2, pp. 49–52.

Environment and Climate Change Canada 2011, *Federal adaptation policy framework*, Environment Canada, Gatineau.

Environment and Climate Change Canada 2014, *Canada's sixth national report on climate change*, Environment Canada, Gatineau.

Ergas, C & York, Y 2012, "Women's status and carbon dioxide emissions: A quantitative cross-national analysis", *Social Science Research*, vol. 41, pp. 965–976.

Foreign Affairs and International Trade Canada 2013, *Mainstreaming of a gender perspective*, DFAIT, Ottawa, ON.

GenderCC Network 2007, "Women for climate: Justice position paper", United Nations Framework Convention on Climate Change, Thirteenth Conference of the Parties, Bali.

Government of Nova Scotia 2013, *Chose how you move sustainable transportation strategy*, (http://novascotia.ca/sustainabletransportation/docs/Sustainable-Transportation-Strategy.pdf).

Haigh, C & Vallely, B 2010, "Gender and the climate change agenda: The impacts of climate change on women and public policy", Women's Environmental Network Report, World Development Movement.

Intergovernmental Panel on Climate Change 2014, *Climate change 2014: Impacts, adaptation, and vulnerability, Part B – Regional aspects*, Working Group 11 Contribution to the Fifth Assessment Report of the IPCC, Geneva.

Johnsson-Latham, G 2007, "A study of gender equality as a prerequisite for sustainable development: What we know about the extent to which women globally live in a more sustainable way than men, leave a smaller ecological footprint and cause less climate change", Ministry of the Environment, Stockholm.

Lahey, KA 2008, *Critique of the Department of Finance gender analysis of tax measures in budgets 2006 and 2007*, Standing Committee on the Status of Women, Ottawa, ON.

Lambrou, Y & Piana, G 2006, *Gender: The missing component of the response to climate change*, United Nations Food and Agriculture Organization, Rome.

Masika, R (ed.) 2002, *Gender, development and climate change*, Oxfam, UK.

Milan, A, Keown, LA & Urquijo, CR 2011, *Families, living arrangements and unpaid work*, Statistics Canada, Ottawa, ON.

Newfoundland and Labrador Department of Environment and Conservation 2016, *GreenFund,* (http://www.env.gov.nl.ca/env/nlgf/).

Office of the Auditor General of Canada 2009, *Gender based analysis*, OAG, Ottawa, ON.

Québec n.d., "2013–2020 Climate Change Action Plan", (http://www.mddelcc.gouv.qc.ca/changementsclimatiques/plan-action-fonds-vert-en.asp).

Röhr, U 2001, "Gender and energy in the North, Expert Workshop: Gender perspectives for Earth Summit 2002, Energy, Transport, Information for Decision-making", 10–12 January, Berlin.

Sjolander, CT 2005, "Canadian foreign policy: Does gender matter?" *Canadian Foreign Policy*, vol. 12, no. 1, pp. 19–31.

Soares, R, Bartkiewicz, MJ, Mulligan-Ferry, L, Fendler, E & Kun, EWC 2013, *2013 Catalyst census: Fortune 500 women board director*, Catalyst.

Statistics Canada 2001, *Taking public transportation*, 2001 Census, (http://www12.statcan.ca/english/census01/products/analytic/companion/pow/publictrans.cfm).

Treasury Board of Canada 2014, *Guide to the federal regulatory development process*, TBS, Ottawa, ON.

Ugochukwu, B 2015, "Climate change and human rights: How? Where? When?" Centre for International Governance Innovation, CIGI Papers, no. 82, November, Waterloo, ON.

United Nations Framework on Climate Change 2015, *Report on gender composition*, UNFCCC, Paris.

US Bureau of Labor Statistics 2014, *Women in the labour force: A databook*, BLS Reports, Washington.

Warren, FJ & Lemmen, DS (eds) 2014, *Canada in a changing climate: Sector perspectives on impacts and adaptation*, Government of Canada, Ottawa.

Wolski, E 2008, "Culturally relevant gender-based analysis: A tool to promote equity", *Canadian Women's Health Network*, fall/winter 2008–9, vol. 11, no. 1, p. 26.

Women's Environmental Network & National Federation of Women's Institutes 2007, "Women's manifesto on climate change", (http://www.wen.org.uk/wp-content/uploads/manifesto.pdf).

Women's Major Group 2012, "A Gender perspective on the 'green economy': equitable, healthy and decent jobs and livelihoods", Women Rio+20 Steering Committee, Rio de Janeiro.

World Forest Coalition 2008, *Life as commerce: The impact of market-based conservation mechanisms on women*, World Rainforest Movement, Montevideo.

Yalnizyan, A 2005, *Canada's commitment to equality: A gender analysis of the last ten federal budgets (1995–2004)*, Canadian Feminist Alliance for International Action, Ottawa.

16

THE INTEGRATION OF GENDER IN CLIMATE CHANGE MITIGATION AND ADAPTATION IN QUÉBEC

Silos and possibilities

Annie Rochette[1]

Introduction

Climate change is a global threat. It is already impacting ecosystems and communities everywhere in the world, especially in developing countries and in the southern hemisphere. As summarized by the Intergovernmental Panel on Climate Change (IPCC), the atmosphere and ocean have warmed, snow and ice cover has diminished, sea levels have risen, and the concentrations of greenhouse gases have increased, undoubtedly linked to human activities (IPCC 2014, 2–4). Climate change action is required by all states, but developed states have a responsibility to do more than their fair share of mitigation efforts as they are responsible for the lion's share of greenhouse gas emissions, past and present. Climate change policies at the international and national levels address both mitigation and adaptation, mitigation measures designed to stabilize or reduce greenhouse gas emissions and adaptation strategies helping individuals and communities adapt to the changes (McNutt & Hawryluk 2009).

According to many international reports, women in developing countries will bear the greater burden of climate change impacts, even though they have not contributed much to the problem (Nellemann, Verma & Hislop 2011). Thanks to organizations like Women Environment Development Organization (WEDO) and the Global Gender and Climate Alliance, the gendered aspects of climate change are increasingly well known and have been considered in decisions of the international climate change regime since 2001. The Women and Gender Constituency was also created in 2009 to ensure that "women's voices and their rights are embedded in all processes and results of the UNFCCC framework, for a sustainable and just future, so that gender equality and women's human rights are central to the ongoing discussions" (Women and Gender Constituency 2009).

The gender dimension to climate change impacts and mitigation strategies has not been extensively examined in developed countries, even less so in Canada. This is surprising considering that the literature that does exist demonstrates the numerous gendered aspects of climate change mitigation and adaptation. Research into gender and climate change is largely focused on developing countries, although as MacGregor (2010) points out, this research has "focused almost exclusively on the material impacts of climate change on women in the Global South and has neglected to place equal emphasis on the gendered power relations and discursive framings that shape *climate politics*" (224, emphasis in original).

The purpose of this chapter is to summarize the findings of a small-scale qualitative research project[2] on the integration of gender in Québec policies and civil society. The chapter examines the obligations of the Québec government regarding the integration of gender in climate change mitigation and adaptation, based on international instruments and national laws, and how the government has fulfilled them. Second, drawing from the research project and from observations made during workshops with environmental and women's groups, the chapter also looks at the integration of gender in climate change within Québec civil society. Finally, the chapter explores some reasons for the absence of gender in climate change policy and action in Québec.

I start from the position that a gender-based and intersectional analysis of climate change mitigation and adaptation policies and strategies is necessary for two main reasons. First, studies have shown that perceptions and attitudes towards climate change are gendered, as are carbon footprints. Public education campaigns and mitigation efforts will thus be more effective in reducing carbon footprints if they reflect these gender differences (McCright 2010; Räty & Carlsson-Kanyama 2010; Terry 2009). Gender-specific mitigation campaigns will also avoid putting the burden of "saving the planet" disproportionately on women, for example, by overly focusing on actions that can be taken in the household (Buckingham & Külcür 2009; MacGregor 2006). Secondly, gender-sensitive mitigation and adaptation policies and strategies are essential to ensure that they do not exacerbate existing social inequalities (Terry 2009).

I adopt Sherilyn MacGregor's (2010) position on gender analysis:

> I would argue that gender analysis and the study of gender politics should involve the analysis of power relations between men and women and the discursive and social constructions of hegemonic masculinities and femininities that shape the way we interpret, debate, articulate and respond to social/natural/technological phenomena like war, economic crisis and climate change. (224)

A gender analysis thus not only points to the gendered perceptions, attitudes, carbon footprints and impacts of climate change, but also brings to light the fact that they are the product of social constructions of hegemonic masculinities and femininities that determine gender roles and relations, including the gendered division of labour, and continue to perpetuate social inequalities. For example, in 2008 women were still paid 83 per cent of men's salaries, education and experience levels no longer being a deciding factor explaining this gap. Rather, the gap can be explained by the type of work, women being mostly employed in the underpaid education and health sectors, whereas men are employed mostly in high paying primary (natural resources and energy) and secondary (construction and transportation) sectors (Vincent 2013). Women also have more often than men a precarious and/or part-time employment situation. They are more often than men the heads of single-parent households (Statistics Canada 2012). Moreover, in the household, women are still in the majority doing the grocery shopping, food preparation, household chores such as laundry and cleaning, childrearing, and caregiving for the elderly (Milan, Keown & Urquijo 2011).

Methodology

The methodology of the research project consisted first of a literature review on gender and climate change from the perspective of a developed country. The literature we reviewed came from a diversity of developed states and disciplines including sociology, political economy, communications, political ecology and law, at both the national and international levels. International treaties and decisions of the Conference of the Parties of the climate change regime were analysed in order to identify the obligations of Canada and Québec for the inclusion of gender in climate change policies. Then the Québec legislation, policies and strategies relating to both gender equality and climate change mitigation and adaptation were examined to identify the policy tools relating to gender equality and gender-based analysis, and to determine if and how gender had been integrated into climate change policies.

We also examined the integration of gender and climate change in civil society by looking at the websites of environmental groups working on climate change in Québec to see if gender was included in their campaigns and publications, and the sites of women's groups for action on climate change or other environmental issues. After this initial survey of groups, I contacted by email environmental groups working on climate change, as well as women's groups, to set up interviews with a representative in order to explore if and how they were working on gender and climate change, or, alternatively, the reasons for the lack of integration. Many of the women's groups contacted either did not reply to my request or refused to participate, for the simple reason that they did not work on climate change related issues. I finally managed to conduct semi-structured interviews with a few individuals working or volunteering within those two movements, five from the environmental movement (Greenpeace Québec, Équiterre, Réseau québécois des groupes écologistes, Réseau des femmes en environnement, Association québécoise de lutte contre la pollution atmosphérique) and five from the women's movement (Fédération des femmes du Québec, Conseil des Montréalaises, Association féminine d'éducation et d'action sociale, Fédération des agricultrices du Québec, Femmes en parcours innovateurs). I also met with five individuals we can generally put in the category of "decision-maker" and who are either elected (municipal councillors and mayors) or working for the provincial government. These interviews were not recorded or transcribed, but extensive notes were taken during and after the interviews.

As mentioned earlier, the findings summarized in this chapter also include observations made during a series of workshops I facilitated with environmental and women's groups following the research project. Since the main finding of the research was that gender was absent from both government and civil society actions on climate change, the main goals of the workshops were to encourage women's groups to think about how gender blind climate change policies have the potential to exacerbate social inequalities, and to push environmental groups to include gender, or at least social justice, in their climate change campaigns, in the hopes of encouraging strategic alliances between the two movements. Unless it was not possible, separate workshops for environmental groups on the one hand, and women's groups on the other, were therefore organized. Invitations to all of the workshops were sent via our partners in collaboration with regional Tables de concertation, which group together representatives from different women's groups in a given region, as well as through social media (Facebook events). Attendance at these workshops varied greatly from six people to more than twenty. The workshops for women's groups tended to be better attended than those for environmental groups.[3]

The three-hour workshops were designed to maximize interaction between participants and these interactions were a rich source of observation data. By asking participants to share what they knew about climate change, I wanted not only to gauge their base of knowledge but also to challenge the perception that climate change is the domain of experts and scientists (this was one of the findings of the research, as explained below) and to reframe climate change as an everyday social issue. The attendees' knowledge about climate change varied tremendously, but collectively, they knew more than they thought they did. Participants were also invited to brainstorm about mitigation strategies, both individual and collective. The notions of gender and gender-based analysis, the danger of reinforcing gender stereotypes, and gendered statistics on salary differences and unpaid labour were presented and discussed. After a short break, participants were then invited to apply a gender-based analysis to the mitigation strategies and impacts identified in the first part. The examples brainstormed by the participants were then completed with more concrete examples identified in studies from the literature. The pedagogical materials included a set of eleven fact sheets[4] about different aspects of gender and climate change, and a visual presentation (PowerPoint slides). Some of the findings discussed in this chapter include observations made during those workshops as well as comments made by the participants on the evaluation form, which asked participants about their previous knowledge, what they got out of the workshop and which actions they might take within their respective organizations following the workshop.

Findings and analysis

Obligations under international, national and provincial laws

There is no question that under international law Québec and Canada have the obligation to integrate gender into climate change mitigation and adaptation policies and actions. Both governments are parties to the 1979 *Convention on the Elimination of All Forms of Discrimination against Women* (CEDAW) and its additional protocol, which require governments to enact legislation that is free of discrimination. The *Beijing Platform for Action* (UN Women 1995) explicitly links together gender and the environment by encouraging governments to ensure women's participation in environmental decision-making, by integrating women's needs, preoccupations and opinions in sustainable development policies and programmes, and by reinforcing national mechanisms to measure the gendered impacts of development and environmental policies. Québec also reaffirmed its adherence to the Beijing Programme of Action at its tenth anniversary in 2005.

In international environmental law, a commitment to gender mainstreaming has been reiterated in many instruments since the 1992 Conference on Environment and Development in Rio. In the climate change regime, gender was not included in the 1992 *United Nations Framework Convention on Climate Change* (UNFCCC), but a decision of the Conference of the Parties (COP) in Marrakesh in 2001 recognized the importance of women's participation in UNFCCC instances and encouraged Parties to nominate women for these positions. Since the COP in Bali in 2007, and largely because of the work of the Global Gender and Climate Alliance, the international climate change regime minimally integrates a gender dimension to decisions and policies. For instance, a decision of the parties at the Cancun COP in 2010 recognizes that women's equality and active participation are significant for effective action on climate change, and that adaptation policies must be conscious of gender equality. In December 2014, at COP 20 in Lima, the international community adopted the *Lima Work Programme on Gender*, including a two-year plan of action on gender and climate change. However, *Paris Agreement* coming out of COP 21 is disappointing as women's equality and empowerment are relegated to the preamble of the agreement, with no new commitment by states to address the gender dimensions of climate change.

At the provincial level, Québec has a longstanding policy framework designed to encourage gender equality and gender-based analysis of laws and policies. First, the Québec *Charter of Human Rights and Freedoms* establishes some safeguards against discrimination and the preamble to the act recognizes the importance of equality between women and men. Québec has enacted gender equality policies since the 1970s. More recently, it adopted a government policy in 2006 and two plans of action (2007–10 and 2011–15). The more recent plan of action establishes 100 concrete actions in 26 different ministries and agencies. The Ministère du Développement Durable, Environnement et Lutte contre les Changements

Climatiques (MDDELCC) and the Ministère de l'Énergie et Ressources Naturelles, two ministries in charge of climate change and energy policies, are not part of these initiatives. The Québec government has also adopted an action plan for 2011–15, which identifies 35 government actions or projects that must carry out a gender-based analysis. Although the document "presents gender-based analysis (GBA), one of its governance tools, as a cross-cutting approach to gradually integrate gender equality principles into all government decisions as well as those of local and regional decision-making bodies" (Secrétariat à la Condition Féminine 2011), the ministries and agencies involved in these projects are those ministries with socially related missions (i.e. health, education, welfare) and do not include the two ministries responsible for climate change. The government is therefore still working in silos, integrating a gender-based analysis only for projects with a social purpose, and not seeing climate change as a social issue.

Gender-neutral climate change policies and strategies

It is not surprising to find that gender is completely absent from the Québec government climate change mitigation and adaptation action plans and strategies. On climate change mitigation, Québec has adopted two action plans to mitigate greenhouse gas emissions (2006–12 and 2013–20) that reflect the main sources of greenhouse gas emissions. Indeed, in Québec, the transportation sector is responsible for 43.5 per cent of greenhouse gas emissions, the industry sector for 28 per cent, the residential and tertiary sectors for 14 per cent, agriculture for 8 per cent, waste for 6 per cent and electricity for only 0.8 per cent of emissions (Gouvernement du Québec 2012a). In Québec, the production of electricity is relatively carbon neutral because the main source is hydroelectricity.

Québec has fixed a target of 20 per cent reduction of greenhouse gas emissions from 1990 levels by 2020. In order to attain this target, the Québec government in its latest action plan, *Québec in Action Greener by 2020: 2013–2020 Climate Change Action Plan*, sets 30 priorities for action, including sustainable land-use planning, citizen engagement, mainstreaming climate change into public administration, a carbon market, promoting public transit and alternative transportation, more fuel-efficient vehicles, reducing the carbon footprint of road freight transport by encouraging more efficient vehicles and driving, sustainable buildings, improving certain agricultural and livestock production practices and soil management, reducing the use of fossil fuels in agriculture, fostering bioenergy, and energy efficiency. In addition to these action plans, the Québec government has also adopted different strategies related to energy, electric transportation and sustainable mobility. Finally, the Québec government has introduced legislation to enable carbon trading and is also part of the Western Climate Initiative.[5] Many of the strategies and priorities of the Québec government for climate change mitigation have a gendered dimension. Yet the gender aspects of these measures are ignored.

For example, the Québec Climate Change Action Plan and the Québec Sustainable Mobility Strategy both provide for government investments in public transportation infrastructures to increase the offer by 30 per cent by 2020. Reducing greenhouse gas emissions related to transportation requires that people leave their cars at home to use active (e.g. walking and cycling) and public transportation. Women are already using public transportation more than men. This is explained in part by workplace location (men work more often in industries located outside city centres while women work more often downtown, readily accessible by public transportation), as well as life style and income (Johnsson-Latham 2007). Women and men also have different needs when it comes to public transportation. For example, a study in Montréal showed that in using public transportation, women are concerned about safety, but also about juggling family and professional obligations (Conseil des Montréalaises 2009). Many participants in the workshops pointed out that women often have to make many different stops before they get to their destination, including dropping off children at daycare and school; driving is thus more efficient than taking numerous bus and subway routes. The "school run" situation has received much attention in the UK where working mothers who are driving their children to school are being blamed for greenhouse gas emissions and traffic congestion (Terry 2009, 10). Terry notes that this easy scapegoating shows how easy it would be for gender-neutral government policies to disproportionately penalize gendered energy use (ibid.).

Carbon taxes are another way of inciting people to drive less and generally to reduce their energy consumption. However, as Chalifour (2010) has pointed out, these measures, if poorly designed, can have a regressive impact. Klein (2014) argues that carbon taxes should be used by governments to reduce greenhouse gas emissions and to fund the transition to a carbon-neutral economy, but she insists that these measures must be progressive and not regressive. Another strategy used by governments to mitigate greenhouse gas emissions related to transport is to subsidize the purchase of hybrid and electric cars. However, these incentives do little to decrease the consumption of goods (and thus the energy required to build these cars), and tend to benefit people with higher revenues who can afford to buy these vehicles. Because they generally have higher revenues than women, men are more likely to benefit from these programmes.

Domestic energy consumption is directly related to income (Clancy & Röhr 2003, 46; EIGE 2012, 23). Not surprisingly, then, one European study found that single men consume more energy than single women, most likely because they have a higher income (Räty & Carlsson-Kanyama 2010). Increased energy rates designed to reduce consumption and increase efficiency thus have gendered impacts. As Chalifour has pointed out, higher electricity rates can have a disproportionate impact on people living in poverty, racialized and Aboriginal women even more so, without directly affecting their energy consumption, because they cannot necessarily afford energy efficient appliances, and, as renters, they have little control over building and window insulation (Chalifour 2010, 199–200; Clancy & Röhr 2003, 45). Education campaigns designed to encourage households to reduce their carbon footprint by buying more environmentally friendly and less packaged goods, by recycling and composting, by buying local and organic produce, and by reducing or cutting meat consumption can also potentially put the burden of "saving the planet" disproportionately on women. Indeed, because of the gendered division of labour in the household, women are still doing most of the grocery shopping, cleaning and cooking, and these green practices are time-consuming, thus adding to their "care burden" (Buckingham & Külcür 2009, 668; MacGregor 2010, 69).

Because climate change is already affecting ecosystems and human activities and will continue to do so, the international community and governments are also planning for adaptation. In Québec, expected impacts of climate change include higher temperatures generally and a greater frequency and intensity of extreme climatic events such as heat waves, droughts, forest fires, torrential rains, storms and even tornados. Temperature rises in the north of the country will be greater than in the south and will result in permafrost degradation, the loss of ice cover, avalanches and violent winds. These will have huge impacts on northern communities, some of which might have to relocate (Gouvernement du Québec 2012c, 8–9). Climate change will also have impacts on human activities and infrastructures, including roads and municipal sewers, transportation systems, hydroelectric production, economic activities (agriculture, fishing, forestry, recreation and tourism), and human health (Gouvernement du Québec 2012b). Hotter and more frequent heat waves, especially in urban centres where heat islands ("îlots de chaleur") are common, can cause respiratory problems and in certain cases death. Increased temperatures also increase air pollution in urban centres (smog), intensifying respiratory problems. With increased temperatures and humidity, concentrations of pollens will also rise and the allergy season will also last longer (Gouvernement du Québec 2012b, 6).

Adaptation is considered an essential response to climate change and it involves making adjustments to ways of thinking and decision-making to reduce harm and to take advantage of potential opportunities. According to the Québec adaptation strategy, "climatic vulnerability is the product of three parameters: exposure to hazards (climatic events), sensitivity to them and adaptation capacity" (Gouvernement du Québec 2012b, 40). Individual vulnerability is influenced by factors such as the physical environment and socioeconomic status, as well as discrimination and social exclusion (Alber & Hemmati 2011). In the Québec adaptation strategy, vulnerable groups are identified as the elderly, children and infants, socially and economically disadvantaged groups, those with chronic illnesses and respiratory problems, Aboriginal people and residents of northern and remote communities. Women are in a majority in most of these groups or are often caring for children and the elderly, yet gender is not considered a factor of vulnerability in the Québec adaptation strategy.

The literature documenting the gendered impacts of climate change in developed countries gives us a few illustrations of why it is important to carry out a gender-based and intersectional analysis of adaptation strategies. For example, women are more vulnerable to illness and death during heat waves because they generally live longer than men and also more likely to live in precarious socioeconomic situations (especially racialized, immigrant and Aboriginal women) that would physically place them in heat islands in urban centres and make it difficult for them to adapt to these heat waves (for example, by being able to afford air conditioning). Women are also more vulnerable as they will more often than men take on the burden of caring for the elderly and children during heat waves (EIGE 2012).

Extreme climatic events, which will be more common and intense because of climate change, also have gendered impacts. Natural catastrophes have been shown to lower women's life expectancy, increase violence against women and women's burden as caregivers (Alston 2014; David & Enarson 2012; EIGE 2012; Enarson 1999; Enarson & Scanlon 1999, 104; Neumayer & Plümper 2007). Gender studies of natural disasters have also documented women's resilience and leadership in rebuilding their communities and mobilizing for environmental protection (Alston, Chapter 9 in this volume; David & Enarson 2012; Enarson 2013, 70).

Agriculture will also be impacted by climate change, whether these impacts are positive (longer harvest seasons) or negative (droughts, torrential rains). In Québec, the principal model for agriculture is still the family farm. A recent gender-based analysis of Québec farms shows that gender roles are still predominant, as women usually have a job outside the farm and take care of the house, while their male partners work on the land (Québec 2013). When the roles on the farm are gendered, the impacts of climate change on agriculture are gendered (Alston 2011).

The integration of gender and climate change in civil society

From the interviews conducted with individuals from the environmental and the women's movements, as well as the conversations we had during the gender and climate change workshops, it seems that even within civil society climate change and gender equality issues are treated separately. However, as we will see, this is starting to change. The environmental groups we talked to do not consider the gendered aspects of climate change, except for a few feminist activists within those groups, who told us they feel somewhat isolated. The women's groups we talked to were not directly involved in climate change activism although some had taken positions on certain environmental issues related to health and quality of life.

There are many reasons why the environmental and the women's movements in Québec are working in silos. First, climate change is perceived as a distant threat, both in time and in space; the impacts will be felt mostly in the developing world, and years from now. However, most participants did recognize that women in northern communities in Québec and Canada, especially Indigenous communities, will be impacted by climate change. Climate change also seems to be confused with other related environmental issues such as air pollution and the ozone layer.[6] Some of the interview participants who were from rural areas noticed an increase in extreme weather events such as torrential rains, but they did not seem to make the connection with climate change. For a few interview and workshop participants, water quality and quantity was a significant concern, but they did not necessarily see the relationship between this issue and climate change.

Another reason explaining the silos is that climate change is perceived as a scientific and highly technical issue, the discourse around climate change being the domain of experts. This perception is symbolized by the comment of one participant from a woman's group that climate change was "des Messieurs en sarraus blancs" (men in white coats). The technical and scientific discourse around climate change is thus very far from the daily life of lay people and women in particular. Most interview and workshop participants agreed that people living in Québec do not see how climate change will have an impact on their daily life or how they can reduce greenhouse gas emissions to make a difference. When the issue is framed in technological terms, people are less likely to feel responsible because researchers and engineers are seen as the primary actors (Wolf & Moser 2011, 559). This perception is also influenced by the discourse adopted by politicians and national environmental groups around climate change, which is largely based on the scientific debates, numbers games (per cent reduction, base year, etc.) and technical and economic solutions such as carbon capture and storage, cap and trade mechanisms, alternative renewable energies, and geoengineering; this discourse is not very accessible to the general public. Terry (2010) notes that the "main discourse is still a stereotypically 'masculine' one, of new technologies, large-scale economic instruments, and complex computer modelling" (6; see also Hultman, Chapter 2 in this volume and elsewhere). Climate change, however, is inherently a social, economic and political issue and it will have impacts not only on ecosystems, species and infrastructures, but also on social justice and equality (Skinner 2012, 2). The highly technical and scientific nature of this masculinist discourse has been criticized by many feminists because it makes women's concerns invisible and excludes them from the debate (Dankelman 2002; Hemmati 2008; MacGregor 2010), as eloquently stated by MacGregor (2010):

> However, since the 1990s and early 2000s, the growing attention to climate change has been accompanied by a relocation of the centre of environmental debate and action within (rather than outside) the scientific and policymaking institutions. This has brought men to the fore as policy experts, scientists, political advocates, entrepreneurs, commentators and celebrities. One could say that the rise of climate change to the top of the green agenda has brought about an apparent "masculinisation" of environmental politics. (230)

In Québec, the public debate around climate change is indeed dominated by men and characterized by a masculinist discourse. As one participant from a major environmental group pointed out, the absence of gender in climate change policy and action can be explained by the fact that climate change is closely related to energy, energy is closely related to power, and power is the domain of men ("une affaire de gars"). One participant who had worked in the climate change area said she was at the time the only woman in Québec working on this issue in the big environmental groups and that the Canadian delegation to the international negotiations was mostly composed of men. Indeed, at the 2010 UNFCCC Conference of the

Parties, women represented only 30 per cent of national delegations (MacGregor 2010, 230; Skinner, 4). When it comes to climate change, even the Québec mainstream media spaces seem to be almost entirely occupied by men from the environmental movement. Indeed, the spokespeople of the major environmental groups working on and speaking about climate change in Québec are all men. This fact was noticed by quite a few interview participants, but also in the workshops with women's groups. However, there seems to be a difference between the national environmental groups, which we interviewed, and the grassroots citizens' groups that have sprung up in the last few years around shale gas development, oil drilling and pipelines, as the latter seem to include more women at the forefront.

Buckingham and Külcür (2009) observe the same situation in the UK environmental NGOs, which involve a great number of women, but whose senior management positions are held mostly by men. These environmental NGOs do little to challenge the "gendering of experience and responsibility in the domestic sphere" and blindly reproduce the gender structures found in greater society (673–674). Not surprisingly, then, gender was also absent from most of their campaigns. Mandy adds that the lack of focus on the gendered impacts of climate change in many UK environmental groups might be explained in part by the explicit and non-explicit hierarchies that favour men, and the failure to adopt a gender analysis of the processes of their organizations and a "critique of the gender expectations of the wider culture" (Mandy, Chapter 6 in this volume).

Finally, in Québec, women's groups do not take concerted action on climate change for the main reason that they have limited resources. Indeed, even though the women's movement in Québec is well structured and organized, and has historically benefited from both provincial and federal funding (contrary to the environmental movement, still underfunded in this province as compared to other provinces), decreasing funding to women's programmes by the federal and now the provincial government has meant that women's groups have to do more with less resources. With these cutbacks, they are struggling to meet their primary mission, let alone take on the issue of climate change.

The findings from the research project and workshops thus suggest that environmental and women's groups are working largely in silos, although they also show a willingness to consider gender and climate change. In 2010, MacGregor commented on these silos:

> The lack of attention by Western/Northern feminists to climate change politics and the regressive solutions that it may be ushering in, is worrying. Feminists have been critical of environmental scholars for their blindness to gender. It is now time to be critical of feminist scholars who are blind to the environmental crisis. (236)

The silos are problematic for many reasons. First, gender neutral climate change mitigation and adaptation strategies run the risk of exacerbating social inequalities and of putting the burden of saving the planet disproportionately onto women.

Secondly, the failure to see the relationship between gender equality and climate change is a missed opportunity for the strategic alliances that are much needed in these times of pro-oil development provincial and federal governments. Indeed, by framing climate change mitigation and adaptation as a feminist issue, that is as an issue of social justice and equality, the women's movement in Québec is more likely to get involved, thus lending a strong voice and hand to the environmental movement in the fight against climate change, oil development and pipelines.

The workshops seem to have been effective in informing both women's groups and environmental groups about the gender dimension of climate change mitigation and adaptation. The evaluation forms from the gender and climate change workshops confirm that the connection between climate change and gender and social equality is not well known among environmental groups and women's groups. Surprisingly, many participants, including those from the women's movement, reported not knowing or not understanding the notion of gender before the workshop. Most, if not all participants at these workshops report that after attending the workshop, they see the gendered aspects of climate change as an important issue, whether they were from an environmental group, a group working on women's issues, or individual citizens. Many of the participants from women's groups say they will go back to their groups and think of concrete actions to mitigate climate change and many of the participants from environmental groups said they would try to include a gender dimension to their activities in order to reach more people and to be more effective. The idea that strategic alliances are possible if climate change mitigation and adaptation are framed as a social justice issue also seems to have been heard by many of the workshop participants.

The situation has changed a bit since we conducted our interviews. The Québec Federation of Women (Fédération des femmes du Québec, FFQ[7]) has recently adopted the position that it must address the environmental, as well as the gendered, impacts of neoliberal economic policies. Adoption of this sphere of action sets the stage for the most significant women's organization in Québec to take position and action on environmental issues, including climate change.

The move of the Québec women's movement to address environmental issues and climate change has been influenced by the global women's movement, more specifically the World March for Women (WMW). In 2015, the theme for the March was "Free our bodies, our earth and our territories", making connections between women's bodies, labour, violence against women, Aboriginal women and environmental protection. The connection between the WMW theme and gender and climate change was mentioned by one participant during the research interviews, but was a recurring theme in the gender and climate change workshops conducted with women's groups in different regions. In the context of the TransCanada pipeline debate, a grassroots women's group – Le Mur de femmes contre les oléoducs et les sables bitumineux – has also emerged as a separate voice, staging demonstrations on International Women's Day, the World March for Women and other events, sometimes in a joint action with Femmes Autochtones Québec to protest against pipelines and the exploitation of the Alberta oil sands.[8]

Contrary to expectations, there was not much resistance to the feminist movement tackling environmental issues from either the women participating in our workshops, or from those participating in the FFQ discussions. There has historically been a reticence from feminists to address environmental issues, mostly for the problematic position of being associated with nature. The essentialist view that women are in a better position to protect the environment because of their inherent "ethic of care" is thus an important issue to address (internally) for the women's movement's involvement in climate change and other environmental issues to ensure that feminist objectives are not undermined.

Linda Alcoff speaks of a third way (i.e. between the feminist poststructuralist view of deconstructing gender and the cultural feminist view of reappropriating and revaluing the feminine), which is to "explore the possibility of a gendered subject that does not slide into essentialism" (Alcoff 1988, 422; MacGregor 2006, 50). According to Alcoff, "the concept of positionality allows for a determinate though fluid identity of woman that does not fall into essentialism: woman is a position from which a feminist politics can emerge rather than attributes that are objectively identifiable" (Alcoff 1988, 435; MacGregor 2006, 50). By keeping in mind that gender differences are socially constructed, we can work to deconstruct them at the same time. MacGregor (2006) proposes a feminist ecological citizenship. As she says, "[f]eminist citizenship has the potential to be a positive political identity that allows women to express their gender-related concerns for environmental quality but that does not forever tie women (in general) to the private sphere of care and maternal virtue" (6). There is therefore a non-essentialist and fundamentally political basis for feminists to be involved in climate change mitigation and adaptation.

Conclusion

This chapter has shown that Québec government policies on climate change mitigation and adaptation are gender-blind and that civil society actors are still largely working in silos on these issues. However, the interviews and workshops conducted also show that once participants saw some concrete examples of how climate change can be gendered, these actors are willing to learn more and to act. The next step will be to see what strategies and alliances the environmental and women's movements can build with each other and with more grassroots citizens' groups, in order to build a more equal, ecological and carbon-neutral society. Governments must also play their part and conduct gender-based and intersectional analyses of climate change mitigation and adaptation strategies before these strategies are implemented.

On a theoretical level, environmental and feminist theories and critiques of the root causes of climate change and environmental degradation can contribute to this goal. For this to happen, however, the environmental movement needs to challenge and change its internal masculinist structures and implicit and explicit hierarchies and pay attention to gender. Feminists, on the other hand, must go beyond pointing out how climate change will hurt women to adopt, as MacGregor (2010) rightly argues, a "normative position on the human exploitation of the planet or the intrinsic value of the non-human world" (228), as the FFQ has recently done. Ecofeminist theories can help to build bridges because they challenge neoliberalism and a capitalist market for their dualistic exploitation of nature, women and "others"; question the binaries at work in this exploitation and the exclusionary and alienating power of the Western scientific discourse; and offer ways in which to reimagine our relationship to nature.

Notes

1 This chapter is based on a small-scale research project carried out in partnership with the Réseau des Femmes en Environnement (RFE) and funded by the Services aux collectivités of the Université du Québec à Montréal. The chapter is also based on observations made during a series of workshops with environmental and women's groups that were funded by the Québec Ministère de l'Éducation, de l'Enseignement Supérieur et de la Recherche and in partnership with the RFE, the Réseau québécois des groupes écologistes (RQGE) and Relais-Femmes. I also want to thank the two research assistants who worked on the research project, Sophie Gramme and Florence Lavigne Le Buis and highlight the excellent work of Marie-Claude Plessis-Bélair on the educational workshop materials.
2 The report (in French) can be found online (Rochette 2013).
3 This might be partly explained by the theme of the 2015 World March of Women, which is "Free our bodies, our earth, our territories", increasing the links between women and the environment.
4 These fact sheets can be found at Réseau québécois des groups écologistes, and L'intégration du genre dans la lutte et l'adaptation aux changements climatiques au Québec.
5 The Western Climate Initiative (WCI) is a cap and trade programme involving four Canadian provinces (Québec, BC, Manitoba and Ontario) and seven American states, including California.
6 These findings on perceptions of climate change as being distant and confused with other environmental issues concord with those of numerous other studies reviewed by Wolf and Moser (2011, 548).
7 The Québec Federation of Women is the largest feminist organization in Québec. Its membership consists of women's groups and individual members from all over the province.
8 See their Facebook page for the different events.

References

Alber, G & Hemmati, M 2011, "Gender perspectives: Debunking climate policy myths", Commonwealth Ministers Reference Book, London, (http://www.commonwealth ministers.com/articles/gender_perspectives_debunking_climate_policy_myths/).

Alcoff, L 1988, "Cultural feminism versus post-structuralism: The identity crisis in feminist theory", *Signs: Journal of Women in Culture and Society*, vol. 13, pp. 405–436.

Alston, M 2011, "Gender and climate change in Australia", *Journal of Sociology*, vol. 47, no. 1, pp. 53–70.

Alston, M 2014, "Gender mainstreaming and climate change", in L Nare & P Akhtar (eds), *Women's Studies International Forum, Special Issue on Gender, Mobility and Social Change*, vol. 47, Part B, November, pp. 287–924.

Buckingham, S & Külcür, R 2009, "Gendered geographies of environmental injustice", *Antipode*, vol. 41, no. 4, pp. 659–683.

Chalifour, NJ 2010, "A feminist perspective on carbon taxes", *Canadian Journal of Women and the Law*, vol. 22, no. 1, pp. 169–212.

Clancy, J & Röhr, U 2003, "Gender and energy: Is there a northern perspective?" *Energy for Sustainable Development*, vol. 7, no. 3, pp. 16–22.

Conseil des Montréalaises 2009, "Pour qu'elles embarquent! Avis du Conseil des Montréalaises sur l'accessibilité du transport collectif et son impact sur la qualité de vie des Montréalaises", Montréal.

Dankelman, I, 2002, "Climate change: Learning from gender analysis and women's experiences of organizing for sustainable development", *Gender and Development*, vol.10, no. 2, pp. 21–29.

David, E & Enarson, E 2012, *The women of Katrina: How gender, race, and class matter in an American disaster*, Vanderbilt University Press, Nashville, TN.

EIGE 2012, "Review of the implementation in the EU of area K of the Beijing Platform for Action: Women and the environment – Gender equality and climate change", European Institute for Gender Equality, Luxembourg.

Enarson, E 1999, "Violence against women in disasters: A study of domestic violence programs in the United States and Canada", *Violence Against Women*, vol. 5, no. 7, p. 742.

Enarson, E 2013. "Two solitudes, many bridges, big tent: Women's leadership in climate and disaster risk reduction", in M Alston, & K Whittenbury (eds), *Research, action and policy: Addressing the gendered impacts of climate change*, pp. 63–74, Springer Science+Business Dordrecht.

Enarson, E & Scanlon, J 1999, "Gender patterns in flood evacuation: A case study in Canada's Red River Valley", *Applied Behavioral Science Review*, vol. 7, no. 2, pp. 102–124.

Gouvernement du Québec 2012a, "Québec in action greener by 2020: 2013–2020 climate change action plan", Bibliothèque et archives nationales.

Gouvernement du Québec 2012b, "Québec in action greener by 2020: 2013–2020 government strategy for climate change adaptation", Bibliothèque et archives nationales.

Gouvernement du Québec 2012c, "Québec in action greener by 2020: 2013–2020 government strategy for climate change adaptation", Bibliothèque et archives nationales.

Gouvernement du Québec 2013, "Relève agricole féminine: Des parcours qui se distinguent de ceux des hommes – analyse différenciées selon les sexes des caractéristiques de la relève agricole du Québec", Bibliothèque et archives nationales.

Hemmati, M. 2008, "Gender perspectives on climate change: Background paper to the interactive expert panel at the UN Commission on the Status of Women", United Nations 52nd session, p. 9.

IPCC 2014, "Climate change 2014: Synthesis report, Contribution of Working Groups I, II and III to the Fifth Assessment Report of the Intergovernmental Panel on Climate Change", Geneva, Switzerland.

Johnsson-Latham, G 2007, "A study on gender equality as a prerequisite for sustainable development", Report to the Environment Advisory Council, Ministry of the Environment, Sweden.

Klein, N 2014, *This changes everything: Capitalism vs the climate*, Penguin Random House, Toronto, ON.

L'intégration du genre dans la lutte et l'adaptation aux changements climatiques au Québec n.d., (http://www.rqfe.org/GenreChangementClimatique).

MacGregor, S 2006, *Beyond mothering earth: Ecological citizenship and the politics of care*, University of British Columbia Press, Vancouver, BC.

MacGregor, S 2010, "'Gender and climate change': From impacts to discourses", *Journal of the Indian Ocean Region*, vol. 6. no. 2, pp. 223–238.

McCright, AM 2010, "The effects of gender on climate change knowledge and concern in the American public", *Population and Environment*, vol. 32, no. 1, pp. 66–87.

McNutt, K & Hawryluk, S 2009, "Women and climate change policy: Integrating gender into the agenda", in AZ Dobrowolsky (ed.), *Women and Public Policy in Canada: Neoliberalism and After?* pp. 107–124, Oxford University Press, Oxford, (https://www.academia.edu/223610/Women_and_Climate_Change_Policy_Integrating_Gender_into_the_Agenda).

Milan, A, Keown, L-A & Urquijo, CR 2011, "Families, living arrangements and unpaid work: Women in Canada – A gender-based statistical report 2010–2011", Government of Canada, Ottawa.

Nellemann, C, Verma, R & Hislop, L (eds) 2011, "Women at the frontline of climate change–gender risks and hopes: A rapid response assessment", United Nations Environment Programme, Birkeland Trykkeri AS, Norway, (http://www.grida.no/publications/rr/women-and-climate-change/).

Neumayer, E & Plümper, T 2007, "The gendered nature of natural disasters: The impact of catastrophic events on the gender gap in life expectancy, 1981–2002", *Annals of the Association of American Geographers*, vol. 97, no. 3, pp. 551–66.

Räty, C & Carlsson-Kanyama, A 2010, "Energy consumption by gender in some European countries", *Energy Policy*, vol. 38, pp. 646–649.

Reseau quebecois des groups ecologists n.d., (http://rqge.qc.ca/genre-et-changement-climatique/).

Rochette, A 2013, "Québec: Croiser genre et changement climatique", (http://www.mediaterre.org/genre/actu,20130527141414.html).

Secrétariat à la Condition Féminine 2011, "Turning equality in law into equality in fact: 2011–2015 – Government action plan on gender equality", Ministère de la Culture, Communication et Condition Féminine, Bibliothèque et archives nationales.

Skinner, E 2012, "Genre et changement climatique: Panorama", BRIDGE, Institute of Development Studies, UK.

Statistics Canada 2012, "Portrait of families and living arrangements in Canada: Families, households and marital status", 2011 census population, (http://www12.statcan.gc.ca/census-recensement/2011/as-sa/98-312-x/98-312-x2011001-eng.cfm).

Terry, G 2009, "No climate justice without gender justice: An overview of the issues", *Gender & Development*, vol. 17, no. 1, pp. 5–18.

UN Women 1995, "Beijing Platform for Action: Report of the Fourth World Conference on Women, Beijing, 4–15 September, (http://www.un.org/womenwatch/daw/beijing/platform/index.html).

Vincent, C 2013, "Pourquoi les femmes gagnent-elles moins que les hommes?" *Canadian Public Policy*, vol. 39, no. 3, p. 473.

Western Climate Initiative (WCI) n.d., (http://www.wci-inc.org/).

Wolf, J & Moser, CS 2011, "Individual understandings, perceptions, and engagement with climate change: Insights from in-depth studies across the world", *WIREs Climate Change*, vol. 2, pp. 547–569.

Women and Gender Constituency 2009, "United Nations Framework Convention on Climate Change", (http://womengenderclimate.org/).

17

URBAN FORM THROUGH THE LENS OF GENDER RELATIONS AND CLIMATE CHANGE

Cases from North America and Europe

Penny Gurstein and Sara Ortiz Escalante

Introduction

While cities are only 2 per cent of the world's landmass, they consume 78 per cent of the world's energy and produce more than 60 per cent of all carbon dioxide and other GHG emissions, through energy consumption and transportation use (UN Habitat n.d.). As climate change becomes increasingly more evident in our environment, greenhouse gas (GHG) emissions have been linked to unsustainable urban forms. Recent studies reveal that dense, compact, mixed-use communities served by public transit allow for greater opportunities for economic integration, and lower transportation use and impact on the environment. They also produce lower GHG emissions than low-density single detached housing (Wheeler 2012).

What has not been made so evident is the link between gender, urban form and climate change. Research has been done on the differentiated impacts of climate change by gender, the absence of women in climate policy and the role women could play if fully involved (Masika 2002). In addition, the intersectionality of social identities within communities and nations creates inequities based on income, race, age and sexual orientation, among other factors (D'Agostino & Levine 2011). Given the increasing vulnerability of women to climate change impacts why is there an absence of gender considerations in urban climate change policies at local levels?

Urban form, the structural and infrastructure elements such as building heights and density, natural and open space features, transportation corridors, public facilities and activity centres, defines a city physically. As will be shown below, urban form impacts consumption patterns. Shwom and Lorenzen (2012) make a compelling argument that reducing consumption can be an important route to reducing the GHG emissions that cause climate change. The largest GHG-intensive consumption is in housing and transportation. Changing consumption patterns is not just an individual choice but a societal one. Creating urban models that reduce consumption not only reduces GHG emissions but also change gender relations by providing easier access to economic and social opportunities. For example, providing a comprehensive public transport system allows for the reduction of private automobile use but also benefits women's access to employment. In this way, gender, climate change and urban form are inextricably linked.

This chapter uses a gender and climate change lens to analyse urban form with a particular focus on the lived experience of women in their capacity as wage earners and caregivers in their homes and communities, and correspondingly the impact of urban form on increasing or reducing climate change. Case studies of policies and programmes in North America and Europe illustrate the impact of differing urban forms on the gendered experience and provide insights into how these urban policies and programmes can address both gender and climate change concerns. In particular, case studies in Austria, Spain and Canada are used to illuminate the various jurisdictional contexts in which gendered policies, programmes and projects can be embedded. The European cases demonstrate opportunities for government intervention at the city and neighbourhood levels, while the Canadian case illustrates how community-based initiatives can address differing gender experiences.

Gender, climate change and urban form

The forces that drive urban land use include income, population, transportation costs and the opportunity cost of urban land, as well as social and economic forces such as gender roles.

In the planning field, there have been two main strategies to facing climate change: "(1) fostering more compact forms of development with improved transit land-use characteristics; and (2) encouraging energy efficient, 'green' new construction and retrofits of existing housing" (Mueller & Steiner 2011, 93). Research on climate change has focused on the technical aspects of making housing more energy efficient and reducing GHG emissions (Killip 2013; Makantasi & Mavrogianni 2015). This type of research examines how the design of the building, house or unit could help mitigate GHG emissions through an increase in domestic energy efficiency, renewable energy and construction modernization; but there is little analysis that goes beyond the building and does not look at how the type of housing (from single-unit houses to multi-unit buildings), its density and its location contributes to GHG emissions.

Williams et al. (2013) conducted a study of how climate change can be mitigated in English suburbs and argue that "to adapt successfully suburbs require a socio-technical approach" (518). While buildings and their environment (streets, public spaces etc.) have to be adapted to reduce climate change emissions, a change in the socio-economic behaviour of residents is also required. Ewing and Rong (2008) address the issue of mitigating climate change through the promotion of a compact model of living: "Energy conservation, and the associated reduction in greenhouse gases, can be thought of as just one more reason to encourage compact development and discourage sprawl. Compact development provides a double benefit, typically reducing transportation energy use and emissions by 20 to 40 per cent relative to sprawl" (22). Feminist scholars have also advocated for mixed-use compact cities as an urban model that better serves women's everyday lives and can help reduce gender inequalities (Gilroy & Booth 1999; UN Habitat 2014).

A compact city promotes relatively high residential density with mixed land uses, an efficient public transport system and an urban plan that encourages non-motorized transportation (walking and cycling), low energy consumption and reduced pollution. It is seen as a more sustainable form than urban sprawl because it is less car dependent, resulting in less infrastructure provision. The resulting dense resident population increases opportunities for social interaction (Dempsey 2010). A compact city is not achieved by building high-rise developments in all parts of a city per se. Rather, it is planning for an overall compact urban form that promotes mixed use and public transport. For example, case studies illustrate that by shifting from large, single-detached homes to a mixture of large and small single-detached homes, townhouses and small apartment buildings, a suburban community can realize a 22 per cent reduction in emissions (Senbel et al. 2014). However, the capacity of compact development to reduce vehicle kilometres travelled (VKTs) and GHG emissions are tempered by other factors such as income and household consumption (Eluru et al. 2010), and the tendency for households that are more amenable to using alternative modes of transportation than the automobile self-selecting to live in neighbourhoods that offer that choice (Chatman 2009). In particular, transportation patterns differ by gender.

Research on women's daily mobility acknowledges the diversity and complexity of trips people make in all the spheres of everyday life, and that the majority of journeys are trip-chaining and multi-tasking (two features of women's travel) because they combine different activities from paid work, to care and domestic work as well as community work (Greed 2008; Hanson 2010, Law 1999). These trips are shorter and more complex, usually carried out in a polygonal spatial pattern (Lynch & Atkins 1988; Sánchez de Madariaga 2013).

Studies demonstrate that women's main modes of mobility are walking or public transportation (Sánchez de Madariaga 2013; Whiztman 2013), while men have greater access to and dependency on private transportation. While women in North America have higher rates of private transportation use, globally, women still have less access to private cars and less driver's licenses (Lynch & Atkins 1988; Miralles Guasch, Martínez Melo & Marquet 2016). These authors conclude that women's transportation use generates less GHG emissions than men, and that transportation policies should promote transportation modes used by women such as public transport.

Traditional models of households, where men work in the productive paid sphere, and women are responsible for domestic work have not only reproduced the sexual division of labour, but also not responded to the needs of those in charge of reproductive and care work making women's everyday lives more complicated.

Women continue to be primarily responsible for maintaining the home and in childcare, even if they work in the paid and productive sphere (Gilroy & Booth 1999; Healey 1997; UN Habitat). In the case of wealthier households, help may be hired, but 83 per cent of domestic workers are still women (UN Women 2015). There is a diversity of households that exist ranging from lone parent-headed households, to extended families, among others. In addition, the needs and size of a nuclear family household evolve over time as children grow up and leave the home. For example, in Spain in 2013, 54 per cent of the total households consisted of couples with or without children, while 25 per cent of households were of only one person, and 8 per cent were female lone parent households (INE 2014).

This perpetuation of gender roles, the sexual division of labour and the public–private divide has been reproduced through urban sprawl and the suburban housing model. Suburbia in early Anglo-American planning was inspired by the nineteenth-century vision of women's place; sending women to the suburbs would not expose them to the temptations and distractions of the city and would make them focus on their reproductive and domestic role in the private space of the home (Fainstein 2005). Hayden (1980) critiques this model of "suburban sprawl" as it does not recognize the needs of employed women and their families; the energy and time involved in combining paid work with domestic responsibilities does not allow women to advance socially and professionally. This North American suburban model has spread to other regions of the world, including European countries such as Spain, where suburbs started to appear in the 1960s as vacation homes for people in the city, and later in the 1990s and early 2000s would become an alternative to unaffordable housing in the city.

Within these communities, Angel, Sheppard and Civco (2005) suggest the participation of women in the labour market has resulted in an increase in household income and increased household urban land use. This greater buying power has allowed for larger homes to be bought and multiple vehicles within a household making possible the travel distance between home and work to become longer, thus increasing a household's carbon footprint. The proliferation of suburban communities disconnected from work opportunities has been attributed to this phenomenon. While the resulting sprawl is occurring, a countertrend is creating the intensification of low-density communities that can also be attributed to women's participation in the labour market. Single use suburban communities are being transformed into mixed-use locales with opportunities for work such as service industries that can be taken up by women in large numbers (Garreau 1992).

While there is a broad consensus in the urban planning literature of the benefits of compact communities, rapid urbanization has challenged efforts to encourage compact, integrated and inclusive cities increasing both climate change and gender disparities. Housing affordability has become a challenge worldwide but is particularly acute in cities that are seen as magnets for employment opportunities. In these cities housing prices have risen faster than incomes, creating financial stress on households. Coupled with this is that in the last several decades there has been a steady and significant erosion of public sector involvement in housing provision throughout the world and a trend towards privatization of housing delivery. In cities such as Vancouver housing prices have forced people to move out to the periphery. Low income as well as middle-class households have moved to the suburbs and small urban areas in search of housing affordability, taking with them the problems associated with poverty (Kneebone & Berube 2014).

The failure to address gender equality into urban planning and economic development is preventing women and girls from benefiting from urban life. How women navigate urban life reflect their options and limitations. Women are particularly vulnerable if their housing limits their access to employment and their ability to perform their reproductive work. Low-income, migrant, Indigenous, young and senior women, as well as female single-head households, are often more impacted by the increase of housing prices, and the result is that they may live in inadequate, unhealthy housing conditions. Women also encounter discrimination in the rental market, reflecting women's disproportionate poverty.

Climate change exacerbates inequalities in key aspects of urban life including access to resources. A gender-sensitive response requires an understanding of existing inequalities between women and men, and the ways in which climate change can intensify these inequalities (Masika 2002). It also requires recognition of the capacity of individuals to contribute knowledge on their lived experiences and to participate effectively in creating positive change. To make this change also requires measures that address consumption patterns to encourage what Shwom and Lorenzen term the "socially organized consumer"; a consumer whose preferences, values and beliefs are not prefixed but change over time in an ongoing process. Sustainable lifestyles are created through daily routines that are impacted by the environment they are in. An urban form that provides access to resources for all will be more sustainable than one that creates barriers and imposes restrictions.

Policies to address gender, climate change and urban form

While the literature links urban form outcomes within gender relations and climate change, the current situation in two countries, Canada and Spain, illustrate how policies to reduce GHG emissions generally lack this perspective.

Canada

In Canada, mitigation of climate change is being addressed more by the provinces than the federal government, though the new 2015 federal government has committed to more action. Before the 2015 election, Canada had the worst climate change policy of all wealthy nations, and the fourth-worst among all nations. Denmark ranked as best, followed by Sweden and Portugal. The US ranked forty-third (Tencer 2012). A quarter of Canada's emissions are in the transportation sector, mainly automobiles, and another quarter is in residential, commercial and industrial fuel consumption. Canada was particularly signaled out for very poor performance on current emissions, renewable energy and climate policy. In contrast to this assessment, the Government of Canada's (2016) position is that they are taking action in four main arenas: actions to reduce Canada's GHG emissions, investments to help Canadians adapt to a changing climate, research to inform decision-making and the development of policies and programmes, and leadership in international climate change efforts.

While the federal government's primary focus is on adaptation not mitigation, many communities in Canada have policies and programmes to address mitigation through encouraging compact communities. At the federal level, specific measures that relate to the urban realm focus on the use of building materials that are more energy efficient in the construction of housing but these measures found in the National Building Code of Canada are only enforceable if adopted by a jurisdiction (the provinces as well as some cities) that regulates construction. There are no specific policies that link gender and climate change.

The provinces of British Columbia, Alberta, Manitoba, Ontario and Quebec have established programmes to reduce emissions (*Climate Change and Emissions Management Act*: Specified Gas Emitters Regulation 2007, 27, *Greenhouse Gas Reduction Targets Act SBC 2007*, Chapter 42, Ministry of the Environment and Climate Change 2016). Within these provinces, municipalities are enacting measures to address climate change, in particular by encouraging more non-motorized transportation especially biking. However, these measures are counterbalanced by the continuing increase in urban sprawl in Canadian cities attributed in part to a lack of affordable housing and supply that does not address the demand for appropriate and adequate housing in inner cities. In the Metro Vancouver region, for instance, the energy and material consumption on areas such water, food, transportation and buildings has resulted in an ecological footprint in 2006 of an area approximately 36 times larger than the region itself (Moore, Kissinger & Rees 2013).

While a number of governments have responded with policies to reduce emissions through planning, auto emissions standards and aggressive municipal targets, implementing high-density development continues to be challenging. The City of Vancouver, Canada, for example, has had a climate change action plan since 2005, and recently introduced an EcoDensity Charter and a Greenest City 2020 Implementation Plan, both intended to contribute to lower housing and transportation energy consumption. Public resistance to the model, however, has galvanized to create significant impediments to the implementation of the plan due to resistance to densification in detached housing neighbourhoods.

Spain

Spain ranked 27th in all developed countries for effective policies on climate change (Germanwatch 2013). EU directives and mandates guide most Spanish climate change policies. However, under the current conservative government, the country has not met many EU standards, showing the government's lack of commitment in environmental issues. Climate change has been addressed in the urban form and housing arena mostly through legislation on energy efficiency and renewable energies. In particular, two Royal Acts were passed in 2006 and 2007 to incorporate energy efficiency criteria and the use of solar, thermal and photovoltaic energy in new buildings or those buildings under renovation. In the region of Catalonia, where Barcelona is located, the regional government developed a Framework Plan for the Mitigation of Climate Change in Catalonia 2008–12 that has been renewed. However, as in Canada, urban sprawl has spread in major cities in Spain. The case of Spain also demonstrates the great emphasis on technical aspects in urban form and housing to reduce GHG emissions, and the lack of social and economic policies, which makes gender issues invisible in climate change policies.

Nevertheless, in 2007, Spain enacted new legislation, *Ley Orgánica 3/2007*, to advance gender equality. This law has specifically focused on improving women's conditions in the economy, promoting affirmative action in the labor market and family-work reconciliation policies. However, women have been particularly impacted by the economic recession of 2008, and a central conservative government has been oblivious to women's needs and rights. Using the excuse of the economic downturn, the government developed austerity policies focused on maintaining employment in male dominated industries – especially construction and banking. Simultaneously, the government eliminated policies that gave support to care work, such as programmes for the elderly, the sick or other dependents. These services have been privatized in the sense that responsibilities have been returned to the "family", and therefore women. In response to these policies and the devastating economic crisis, grassroots feminist organizing has increased, advocating for systemic changes to eliminate the patriarchal structures engrained in all aspects of everyday life.

Comparing Canada and Spain

When comparing the two countries, we can see similarities, differences and opportunities to learn from each other. The models of urban form are different: while in Canada the suburban model is predominant, as a response to affordability and space needs (though densification is occurring in both inner cities and suburbs), in Spain the compact city model predominates (though suburban communities are increasing).

Spain seems to have been more proactive in incorporating a gender lens into legislation at the regional and national level. However, the application of the current legislation has been very dependent on the political party in government, and recently, the ruling parties have prevented the implementation of this legislation. Better employment and childcare policies are needed in addition to housing policies that particularly address gender and climate change.

Both countries need to be more proactive in incorporating a gender lens into climate change policies, as well as relating these policies to housing issues, and urban form and mobility systems. Gendered policies would provide a pluralistic approach that recognizes the differential implications for women and men, and within these gender categories, and the need for intersectional considerations.

Major cities in Canada and Spain, such as Vancouver and Barcelona have good public transportation systems complemented by other mobility policies, such as cycling infrastructures that make good examples of sustainable mobility. But both cities need to continue investing in sustainable modes of mobility that focus on improving people's everyday lives at the same time as reducing GHG emissions. Mobility policies still could benefit from using a gender lens to plan mobility networks, not only based on the journey-to-work, but on the needs of the broad spectrum of the population and the diversity of hours and trips that people make.

These two countries' records demonstrate that while climate change is being addressed to varying degrees at different jurisdictional levels, gender and climate change is not. As well, there is no specific link between gender, climate change and urban form. There is a need for further research and debate about how higher densities and housing construction using a model of compact cities with less car dependency and more mixed-uses can benefit both reduction of GHG emissions and gender equality. There is also a need to understand how consumption patterns can be altered through the use of gender and climate change policies.

Research on this is being conducted on the rehabilitation of low-density residential suburbs in Catalonia, Spain.[1] The studied residential suburbs were built on steep slopes isolated from commercial, residential and cultural centres. In addition to the physical and infrastructure costs, local governments also learnt that there were social costs to this development strategy. The goal of the research was to analyse the needs and problems that inhabitants faced in these communities to create an action plan for rehabilitation that would improve people's everyday lives in all their spheres of work (productive, reproductive, personal and political) and increase environmental sustainability.

The research found that in low-density suburbs, households with limited resources, and especially the women within them, have impediments to access job, leisure and educational opportunities. A mother living in a low-density single-use residential area spends on average between five and seven hours daily going from home to work, to school and to the diverse activities of children of different ages (Muxí & Gallart 2012). The suburbs examined did not provide the conditions, services as well as public spaces needed. In response to these needs, the project generated ideas for land use, urban planning and urban improvement that provide "solutions that create infrastructures for everyday life" (ibid.). The study proposed rehabilitation actions in five strategic areas: increase connectivity and proximity through more sustainable modes of transportation; promote physical and social mix through densification and mixed-uses; improve public transportation and pedestrian infrastructures; foster a sense of community through the promotion of participation; and promote environmental sustainability by reducing energy consumption and promoting local biodiversity.

While research such as above is critical in identifying issues and proposing solutions, without government action very little can be accomplished. Outlined below are cases from Canada and Europe that tease out a more fine-grained relationship between gender, climate change and urban form. These cases point to the need for policies that more explicitly address this relationship at a variety of scales from the site specific and neighbourhood levels to the regional and for the importance of an intersectionality perspective on the linkages. They also point to the importance of government regulations, programmes and policies that ensure compliance.

Cases that promote the intersectional linkage between gender, climate change and urban form

The cases below provide brief snapshots of what is possible if planning for urban form is viewed through a gender and climate change lens. The first case addresses gender mainstreaming at a citywide level. The second case refocuses neighbourhoods through government intervention. The third case is of a form of housing that addresses the need to rebalance gender experiences and consumption patterns by sharing resources.

Gender mainstreaming in Vienna, Austria

In Vienna, there is an administrative authority responsible for gender-specific aspects of urban planning called the "Co-ordination Office for Planning and Construction Geared to the Requirements of Daily Life and the Specific Needs of Women" (Kail & Irschik 2007). This office has been tasked with data collection and the gender mainstreaming of any plans and programmes within the city starting with the 2003 master plan, "Transport Vienna". The office's research found that in Vienna, 59 per cent of all car journeys are made by men, 60 per cent of all pedestrian trips by women, and that women do more multitasking on those trips than men.

Following from the master plan was the proposal for a gender mainstreaming pilot district to test how to reduce transportation use and improve quality of life. The proposal to focus on a district to implement changes was seen as advantageous as Vienna's urban policy development is decentralized. It is at the district level where traffic and road network policies and development decisions get made and district councils have financial responsibility for the public realm.

Mariahilf was selected in 2002 as a pilot district to test gender mainstreaming as a methodological approach to transport and traffic planning. Mariahilf was favoured because of the district's existing focus on the public realm and pedestrian traffic, and it had an active district women's commission. The municipal departments involved were responsible for road building, public lighting and traffic, and were chosen as core departments because of the importance of their work for the district's public realm. Performance measures to ascertain that gender mainstreaming was systematically taken into account were established. In order to strengthen awareness of the social dimensions of transportation and traffic planning, a record was made of anticipated user groups (school-aged children, older people, small children with escorts etc.) and their routes, as well as the impact on vulnerable users with greater demands (mobility-impaired persons, children travelling alone etc.). To create a continuous pedestrian route network, the improvements to encourage pedestrian traffic introduced in Mariahilf included the widening of 1,000 metres of pavement, the construction of 40 street crossings, improved lighting and the redesign of three public squares. All of these measures improved access to non-motorized transportation and reduced the reliance on automobile use, addressing both gender and climate change priorities. Nevertheless, the project had limitations. Motorized traffic in the primary street networks still took precedent over pedestrians.

This district pilot project is one of more than 60 other pilot projects on gender mainstreaming that has been carried out in Vienna that is reshaping urban planning in the city. Another project, "Frauen-Werk-Stadt I" (Women-Work-City), conceived in 1993 was designed to support the care work of women (Foran 2013). The project consists of 359 housing units in a series of apartment buildings surrounded by courtyards with childcare facilities to minimize the distance parents travel to take their children to daycare. It also includes commercial space for shops within the housing block, medical facilities and a police station, with public transit in close proximity.

Vienna's example is being adopted by other European cities, although Vienna is now opting to call what it is doing a "Fair Shared City" rather than gender mainstreaming so as to address the intersectionality of gender and inequality (ibid.). The Central European Urban Spaces project is using gender analysis to improve the urban environments of eight Central European countries and the British Royal Town Planning Institute has created a toolkit to promote gender analysis in information gathering and planning (Gender News 2014).

Neighbourhoods that support walking and social interaction: The case of the Superilles in Barcelona, Spain

The Superilles (super-blocks) is an urban planning strategy, currently implemented by the city government of Barcelona, that has the potential to mitigate climate change and better respond to women's everyday lives. Superilles are compact and efficient neighbourhoods (a polygon of existing blocks) created by the closure of main intersections to passing traffic. In this way, the closed intersection becomes a large public space that supports walkability and community involvement. In the Superilla, the network of interior streets is returned to the main users of the public space: pedestrians and bicyclists, as well as transportation for distribution, emergencies and residents' vehicles with a speed limit of no more than 10 kph. Passing vehicles and citywide public transportation circulate on the edges. These edges are connected to the rest of the city through a network of public transportation.

Within the Superilla street priorities are changed, and non-motorized transportation occupies the space traditionally used by automobiles and other motorized vehicles. In Barcelona, standing automobiles occupy between 65 and 70 per cent of the public space, while residents' mobility by automobile does not reach 25 per cent of total trips. The Superilles respond to the disproportion between the amount of space that automobiles occupy in relation to the amount of people who move by them, and return the public space to those that use it in a more intensive and sustainable way: pedestrians. In addition, transportation research in Barcelona shows that mobility patterns differ by gender. The main mode of transportation of women of different ages is walking and public transportation, while men more intensively use automobiles. Therefore, the Superilles project will have a positive impact on women's everyday lives, because it prioritizes their mode of mobility. The Superilles project also promotes social interaction through widening sidewalks, increasing street furniture such as benches, as well as playground and socialization spaces, and improving the connectivity between facilities and services, such as schools, health centres and local retail stores; and promotes social cohesion through the use of public participation in its design and development.

This strategy, originally promoted by ecologist Salvador Rueda from Barcelona's Agency of Urban Ecology (Rueda n.d.) was first implemented in the District of Gràcia in Barcelona, but now the current government of the city is developing a plan to expand to five pilot areas in four neighbourhoods of the city (Ajuntament de Barcelona 2016). To promote equity and address climate change, the current city government's plans to include gender mainstreaming in all their departments.

Neighbourhoods with shared resources: Cohousing

Cohousing neighbourhoods combine the autonomy of private dwellings with the advantages of shared resources and community living (Canadian Cohousing Network n.d.). The concept originated in Denmark in the 1960s and came to the US in 1988. One hundred and nineteen cohousing communities have been completed since then internationally, including in Canada and many countries in Europe. Residents own their own homes, which are clustered around a "common house" with shared amenities. The amenities may include a kitchen, dining room, workshops, offices, guest rooms, childrens' play areas, garden, laundry and other community facilities. While each home has its own kitchen, the shared kitchen allows for communal meals, which are seen as integral to community building. There is resident involvement in the planning, design, ongoing management and maintenance of their community.

Cohousing is designed to address the changing nature of families, resource limitations and environmental concerns. Given the participation of women in the workforce it provides an urban form where work-life balance can be accommodated through shared facilities such as shared childcare and spaces for work to be conducted close to home. Shared resources also reduce consumption and energy use. For example, a laundry room shared by residents is more energy efficient than laundries in each unit. In this way, residents change their consumption habits.

Cohousing also provides opportunities for innovations to occur. A new cohousing project, the first in Vancouver, incorporates solar energy and electric vehicle features (Renewable Cities 2016). The 31-unit development was built on three double lots and maximizes shared space to reduce energy expenditure. It has solar panels on the roofs that will generate electricity for the building's common areas, and excess energy will be fed into the citywide grid and the grid will supply energy when the panels can't. The solar project is cooperatively owned and has been designed to accommodate electric vehicles. One solar panel is dedicated to electric vehicle charging and 30 per cent of the parking stalls are wired for charging.

The cases above illustrate how a gender and climate change perspective can be incorporated into urban planning and design, and by so doing can result in better planned communities for all citizens that lessen transportation reliance, energy expenditure and resource consumption. They also illustrate that for societal change that promotes equity and sustainability to occur system-wide, multi-sectoral approaches are needed.

Conclusion and moving forward

An analysis that links gender and climate change will go far in tackling urban form problems. If the gendered experiences of women in their productive and reproductive lives are addressed by lessening consumption and energy expenditure then climate change imperatives will also be addressed. Societal patterns of consumption needed to be changed. However, without government policies and programmes encouraging this transformation, very little change will occur.

That so little public policy addresses gender and climate change points to the need for more women and men with a feminist perspective to be elected to municipal offices and for more specific policies focused on these interrelated issues. The allocation of funding priorities in cities need to prioritize women and their families including the development of affordable housing and food security strategies, childcare facilities, better public transit and safer communities. One method that has been effective in other countries such as Brazil is a participatory budget process that provides opportunities for public input into decisions around a city's budgetary allocation.

Effective participatory planning processes that engage diverse communities of women in identifying their needs and ways to address them is necessary. Women's involvement has to go beyond generic citywide advisory committees, which only involve them at a cursory level. Instead, having women involved in project and community-specific processes as work progresses provides more real involvement. In Barcelona, Col·lectiu Punt 6, a feminist planning organization, has advocated for a decade for the inclusion of a gender perspective in urban planning, in particular through the use of collective and bottom-up participatory processes centred on women' s everyday lives. Col·lectiu Punt 6 proposes gender-transformative urban planning that "acknowledge[s] and make[s] visible women's experiences, and activities, needs, and responsibilities associated with domestic and care work, . . . respond[s] to the consequences of having a female sexualised body in public space, and the temporal dimension of everyday life, that looks beyond the productive life and responds to the different times when domestic and care work are developed" (Ortiz Escalante & Gutiérrez Valdivia 2015, 116).

It is also necessary to ensure that a gender lens is used when developing community plans and evaluating their effectiveness. Such a lens would ensure that the lived experience of women's lives and their needs for community engagement, safety and accessibility, and health and well-being is addressed in these plans. There is also a need to ensure that a common ground can be reached between the diverse communities to identify changes, which could help many or all of these groups. For example, a Vancouver City policy a number of years ago was changed to make corner curb cuts into ramps rather than actual curbs. It took years for this change to become widespread because there were so many corners to change, but it gradually happened. And this seemingly small change was not only very good for people with disabilities, but also improved accessibility for people pushing children in strollers and seniors, thereby increasing walking as a transportation mode.

Moving forward will require public support inspired by a new vision that conjures up a compelling future for women and girls in our city. Planning departments and municipal governments are often open to the idea of supporting women's empowerment and increasing women's participation in planning processes. However, allowing women to intervene directly in actual changes is not always welcomed by formally trained planners and architects, who might perceive themselves as the only experts (Ortiz Escalante & Gutiérrez Valdivia 2015). Feminist planners can push changes in this respect. For example, Col·lectiu Punt 6 conduct training workshops on urban planning from a gender perspective at the municipal and regional levels that provide tools and strategies to be applied in local contexts and open a dialogue among those less convinced, which can move to later implementation of gender policies (ibid.).

Finally, policies need to be more proactive in incorporating a gender lens in climate change policies, and relating policies to housing, urban form and mobility at local, regional and national levels. Policies need to address differential vulnerabilities, capacity building to provide tools and resources to adapt or mitigate climate change, and resilience strategies. Addressing the myriad of issues encompassing gender, climate change and urban form requires a concerted effort by all levels of government and a perspective that recognizes the important role that gender plays in shaping experiences and the environment.

Note

1 The study began in 2012 with the participation of Col·lectiu Punt 6, a cooperative of architects, urban planners and sociologists that work on urban planning projects, research and participatory processes from an intersectional gender perspective (Muxí 2013). And the study has continued in 2015 in the county of Garraf, located south of Barcelona, with funds of the Government of the Province of Barcelona.

References

Ajuntament de Barcelona 2016, *Superilles*, (http://ajuntament.barcelona.cat/superilles/ca).

Angel, S, Sheppard, S. & Civco, D 2005, *The dynamics of global urban expansion*, The World Bank, Washington, (http://www.citiesalliance.org/publications/homepage-features/feb-06/urban-expansion.html).

Canadian Cohousing Network n.d., (http://cohousing.ca).

Chatman, DG 2009, "Residential choice, the built environment, and nonwork travel: Evidence using new data and methods", *Environment and Planning A*, vol. 41, no. 5, pp. 1072–1089.

Climate Change and Emissions Management Act: Specified Gas Emitters Regulation 139, 2007, Edmonton, Alberta, p. 27, (http://www.qp.alberta.ca/1266.cfm?page=2007_139.cfm&leg_type=Regs&isbncln=9780779738151).

D'Agostino, M & Levine, H 2011, "Feminist theories and their application to public administration", in M D'Agostino & H Levine (eds), *Women in public administration: Theory and practice*, pp. 3–13, Jones & Bartlett Learning, Sudbury, MA.

Dempsey, N 2010, "Revisiting the Compact City?", *Built Environment*, vol. 36, no. 1, pp. 5–8.

Ewing, R & Rong, F 2008, "The impact of urban form on US residential energy use", *Housing Policy Debate*, vol. 19, no. 1, pp. 1–30.

Fainstein, SS 2005, "Feminism and planning: Theoretical issues", in SS Fainstein & LJ Servon (eds), *Gender and planning: A reader*, pp. 120–140, Rutgers University Press, New Brunswick, NJ.

Foran, C 2013, "How to design a city for women: A fascinating experiment in 'gender mainstreaming'", *The Atlantic Citylab*, 16 September, (http://www.citylab.com/commute/2013/09/how-design-city-women/6739/).

Garreau, J 1992, *Edge city: Life on the new frontier*, Anchor Books, New York.

Gender News 2014, "Gendered innovations transform housing and neighborhood design: Sustainable gender-aware housing and neighborhood design cuts down on transportation and improves the quality of people's lives", The Clayman Institute for Gender Research, 25 March, (http://gender.stanford.edu/news/2014/gendered-innovations-transform-housing-and-neighborhood-design).

Germanwatch 2013, "The Climate Change Performance Index: Results 2013", Climate Action Network, Europe, (http://germanwatch.org/en/download/7158.pdf).

Gilroy, R & Booth, C 1999, "Building an infrastructure for everyday lives", *European Planning Studies*, vol. 7, no. 3, pp. 307–324.

Government of Canada 2016, "Canada's way forward on climate change", Climatechange.ga.ca, (http://www.climatechange.gc.ca/default.asp?lang=En&n=72F16A84-1#X-201409191350046).

Greed, C 2008, "Are we there yet? Women and transport revisited", in T Uteng & T Cresswell (eds), *Gendered mobilities*, pp. 243–256, Ashgate, Aldershot.

Greenhouse Gas Reduction Targets Act, SBC 2007, Chapter 42, Victoria, BC.

Hanson, S 2010, "Gender and mobility: New approaches for informing sustainability", *Gender, Place and Culture*, vol. 17, no. 1, pp. 5–23.

Hayden, D 1980, "What would a non-sexist city be like? Speculations on housing, urban design, and human work", *Signs*, vol. 5, no. 3, pp. S170–S187.

Healey, P 1997, *Collaborative planning: Shaping places in fragmented societies*, Macmillan Press, London.

INE 2014, "Las formas de convivencia", Instituto Nacional de Estadística, (http://www.ine.es/ss/Satellite?L=es_ES&c=INECifrasINE_C&cid=1259944407896&p=1254735116567&pagename=ProductosYServicios%2FINECifrasINE_C%2FPYSDetalleCifrasINE).

Kail, E & Irschik, E 2007, "Strategies for action in neighbourhood mobility design in Vienna: Gender mainstreaming pilot district Mariahilf", *German Journal of Urban Studies*, vol. 46, no. 2, (http://www.difu.de/node/5949).

Killip, G 2013, "Transition management using a market transformation approach: Lessons for theory, research, and practice from the case of low-carbon housing refurbishment in the UK", *Environment and Planning C: Government and Policy*, vol. 31, no. 5, pp. 876–892.

Kneebone, E & Berube, A 2014, *Confronting suburban poverty in America*, Brookings Institution Press, Washington, DC.

Law, R 1999, "Beyond 'women and transport': Towards new geographies of gender and daily mobility", *Progress in Human Geography*, vol. 23, no. 4, pp. 567–588.

Ley Orgánica 3/2007, de 22 de marzo para la igualdad efectiva de mujeres y hombres, Boletín Oficial del Estado, núm. 71, Jefatura del Estado, p. 12611–12645, (https://www.boe.es/boe/dias/2007/03/23/pdfs/A12611-12645.pdf).

Lynch, G & Atkins, S 1988, "The influence of personal security fears on women's travel patterns", *Transportation*, vol. 15, no. 3, pp. 257–277.

Makantasi, AM & Mavrogianni, A 2015, "Adaptation of London's social housing to climate change through retrofit: A holistic evaluation approach", *Advances in Building Energy Research, forthcoming*, pp. 1–26.

Masika, R 2002, "Gender and climate change: Editorial", *Gender and Development*, vol. 10, no. 2, pp. 2–9.

Ministry of the Environment and Climate Change 2016, "Ontario", (http://www.ene.gov.on.ca/environment/en/resources/STD01_076573.html).

Miralles Guasch, C, Martinez Melo, M & Marquet, O 2016, "A gender analysis of everyday mobility in urban and rural territories: From challenges to sustainability", *Gender, Place and Culture*, vol. 23, no. 3, March, pp. 398–417.

Moore, J, Kissinger, M & Rees, W 2013, "An urban metabolism and ecological footprint assessment of Metro Vancouver", *Journal of Environmental Management*, vol. 124, pp. 51–61.

Mueller, EJ & Steiner, F 2011, "Integrating equity and environmental goals in local housing policy", *Housing Policy Debate*, vol. 21, no. 1, pp. 93–98.

Muxí Martínez, Z. 2013, *Postsuburbia*, Comanegra, Barcelona.

Muxí Martínez, Z. & Gallart, AA 2012, "Millorar la vida quotidiana a les urbanitzacions de baixa densitat/Improving everyday life in sprawling urban areas", *POSTsuburbia TE'TSAB12*, p. 16–21, (https://issuu.com/carlesbaigescamprubi/docs/postsuburbia).

Ortiz Escalante, S & Gutiérrez Valdivia, B 2015, "Planning from below: Using feminist participatory methods to increase women's participation in urban planning", *Gender & Development*, vol. 23, no. 1, pp. 113–126.

Renewable Cities 2016, "On location at BC's first community-owned solar energy project", 4 April, (http://www.renewablecities.ca/news-updates/east-vancouver-community-solar-project).

Rueda, S n.d., "Las supermanzanas: Reinventando el espacio público, reinventando la ciudad", (http://transit.gencat.cat/web/.content/articles/arxius_ponencies_v_congres/las_supermanzanas.pdf).

Sánchez de Madariaga, I 2013, "Mobility of care: Introducing new concepts in urban transport", in I Sánchez de Madariaga & M Roberts (eds), *Fair shared cities: The impact of gender planning in Europe*, pp. 33–48, Ashgate Publishing Limited, Surrey.

Senbel, M, Giratalla, W, Zhang, K & Kissinger, M 2014, "Compact development without transit: Life-cycle GHG emissions from four variations of residential density in Vancouver", *Environment and Planning A*, vol. 46, no. 5, pp. 1226–1243.

Shwom, R & Lorenzen J 2012, "Changing household consumption to address climate change: Social scientific insights and challenges", *Wiley Interdisciplinary Reviews: Climate Change*, September/October, vol. 3, no. 5, pp. 379–395.

Tencer, D 2012, "Canada climate change policy ranks worst in wealthy world: Climate Action Network", *The Huffington Post Canada*, 5 December, (http://www.huffingtonpost.ca/2012/12/05/canada-worst-climate-policy_n_2246238.html).

UN Habitat n.d., "Climate change", (http://unhabitat.org/urban-themes/climate-change/).
UN Habitat 2014, "Women and housing: Towards inclusive cities", UN Habitat, Nairobi.
UN Women 2015, "Progress of the world's women 2015–2016: Transforming economies, realizing rights", UN Women, New York.
Wheeler, S 2012, *Climate change and social ecology: A new perspective on the climate challenge*, Routledge, London.
Whitzman, C 2013, "Women's safety and everyday mobility", in C Whitzman, C Legacy, C Andrew, F Klodawsky, M Shaw & K Viswanath (eds), *Building inclusive cities: Women's safety and the right to the city*, pp. 35–52, Routledge, London.
Williams, K, Gupta, R, Hopkins, D, Gregg, M, Payne, C, Joynt, JLR, Smith, I & Bates-Brkljas, N 2013, "Retrofitting England's suburbs to adapt to climate change", *Building Research & Information*, vol. 41, no. 5, pp. 517–531.

18

CANADIAN INDIGENOUS FEMALE LEADERSHIP AND POLITICAL AGENCY ON CLIMATE CHANGE

Patricia E. Perkins

Introduction

The Canadian federal election of 2015 was a watershed moment for women's political agency, Indigenous activism and climate justice in Canada. Since 1990, skyrocketing fossil fuel extraction, especially in the Alberta tar sands, had generated escalating environmental crises on First Nations territories. Extreme weather events due to climate change were impacting communities across the country, with particular implications for women's caring and other unpaid work. Ten years of attacks on women's organizations and priorities by the conservative government of Prime Minister Stephen Harper had angered female voters. In response, Indigenous and settler women's organizing on climate and environmental justice, fossil fuel extraction and voting rights was an important factor in Harper's October 2015 defeat. Justin Trudeau, elected on promises to address climate change, Indigenous rights and gender equity, now faces the challenge of delivering on both distributive and procedural climate justice.

This story of extraction, climate change, weather, unequal impacts, gender and political agency in a fossil fuel-producing country in the Global North has implications for gender and climate justice globally. Canada contains within its borders many examples of environmental racism stemming from fossil fuel extraction and climate change that parallel global injustices. The politics of addressing these inequities is key to a successfully managed energy transition away from fossil fuels. In the Canadian case at least, women's leadership – especially *Indigenous* women's leadership – is emerging as crucial.

Extraction, environmental justice and Indigenous women's leadership

Canada, the world's fifth-largest fossil fuel producing country, more than doubled its production of crude oil and equivalents between 1990 and 2014, from

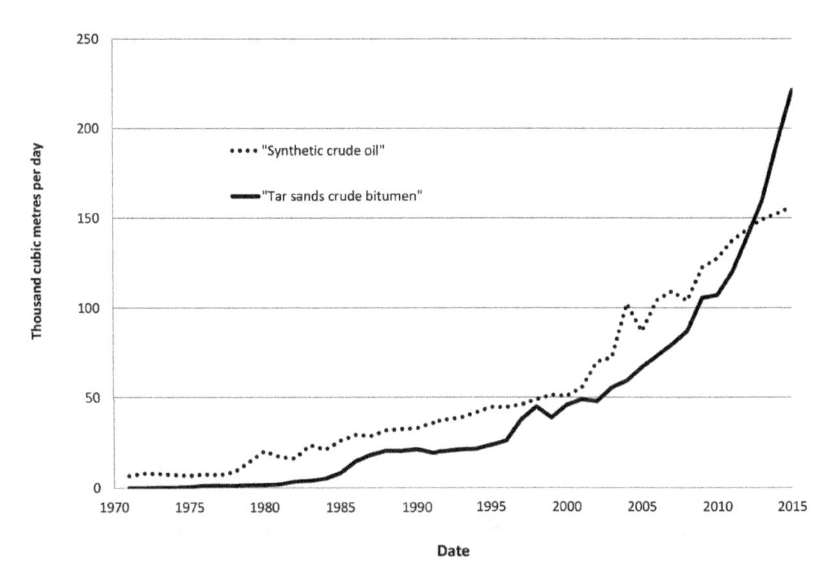

FIGURE 18.1 Canadian crude oil and tar sands production, 1971–2015

Source: Canadian Association of Petroleum Producers (2016, Table 3.16a).

264.9 to 611.4 thousand cubic metres per day (Canadian Association of Petroleum Producers 2016). As shown in Figure 18.1, the extraction of bitumen from the Alberta tar sands north of Fort McMurray increased tenfold during this period.

The environmental impacts of tar sands extraction are massive: boreal deforestation, toxic tailings ponds covering more than 75 square km, hazardous air emissions, vast use of surface and underground water, and toxic seepage into the Athabasca River upstream from First Nations settlements at Fort McKay and Fort Chipewyan and into Lake Athabasca, where fish and other wildlife are now showing abnormalities (Gillespie 2008; McDiarmid 2013). The social and human health effects are gendered, with elevated women's cancer rates downstream from the tar sands (Young 2014), Indigenous people shifting to non-traditional foods over fears of contamination (McCarthy & Cryderman 2014), and prostitution, sexual violence and harassment of women linked to industries associated with fossil fuel extraction and pipeline construction (Zuckerman 2012).

Alberta bitumen producers are pressing for pipeline construction to get their product to the coasts or Great Lakes for export, but all routes cross First Nations territories and face strong opposition from Indigenous and environmentalist groups concerned about spills and impacts on forests and wildlife, watersheds and coastal ecosystems. Fracking for natural gas has also sparked strong opposition from First Nations and environmental groups, frequently led by women (Troian 2013).

The Harper government (2006–15) pushed fossil fuel development, increasing energy subsidies to more than $34 billion per year by 2010. Canada was providing more subsidies to petroleum as a proportion of government revenue than any other country besides the US and Luxembourg (Anderson 2014, Climate Action Network 2012). Canada repeatedly won the Climate Action Network's Fossil awards at Conference of the Parties (COP) climate change negotiations and the Rio+20 meeting in Rio de Janeiro, and the Lifetime Unachievement Fossil Award at the Warsaw COP in 2013. Christian Holz, executive director of Climate Action Network Canada, stated,

> Canada is in a league of its own for its total lack of credibility on climate action. The utter lack of a credible climate policy plan on the part of the Harper government has gone a long way towards undermining Canada's standing in the world, even as a clear majority of Canadian citizens seek action and leadership on climate change. (Climate Action Network 2013, 1)

In opposition to the Canadian government's stance on climate and the tar sands, Canadian women have raised their voices and organized a range of responses, grounded in their own communities and experiences. Their actions address both the differential *impacts* of climate change and fossil fuel extraction – distributive justice – and the *political agency* necessary to change those impacts – procedural justice (Paavola, Adger & Huq 2006). Besides world-famous activists such as Elizabeth May, Naomi Klein and Maude Barlow, there are dozens of Indigenous women activists whose organizing laid the groundwork by 2015 for a huge political shift in Canada.

In November 2012, at a Saskatoon teach-in in opposition to federal Bill C-45 (which weakened environmental laws and water protection), four young women launched a national movement, Idle No More, by calling for nationwide protests, rallies and demonstrations against resource exploitation, especially on First Nations territory, and for respectful consultation with First Nations on land claims, treaties, water protection and resource-sharing. These women were Nina Wilson (Lakota and Plains Cree from Kahkewistahaw First Nation, Treaty 4 territory), Sheelah McLean, Sylvia McAdam (Nêhiyaw Cree First Nation, Treaty 6 territory) and Jessica Gordon (Pasqua First Nation, Treaty 4 territory). Idle No More included a surge of protests in the winter of 2012–13, including round dances and flash mobs at shopping malls and plazas in Regina, Edmonton, Winnipeg, Saskatoon, Vancouver, Toronto and Ottawa; blockades of railway lines, roads and border crossings in Ontario, Quebec, Manitoba and British Columbia; and solidarity protests in the US, Europe, New Zealand, Egypt and elsewhere. Idle No More was fundamentally about Indigenous rights, protecting the earth and protecting women (Ross 2013). Foreign Policy magazine named Gordon, McAdam, McLean and Wilson "Leading Global Thinkers of 2013".

Idle No More built on a long history of Indigenous women's advocacy on environmental justice issues in Canada. For example, Dorothy McDonald-Hyde, the first woman to be elected chief of the Fort McKay First Nation (1980–88), campaigned for water, housing and infrastructure on the reserve, led a roadblock against a logging highway in 1981, and in 1983 brought charges against Suncor's

Fort McMurray plant for spills into the Athabasca River, setting a precedent for oil industry accountability with Indigenous communities (Donnelly 2012, 4).

Sheila Watt-Cloutier, former President and International Chair of the Inuit Circumpolar Council (ICC), is a spokesperson for Arctic Indigenous peoples and in 2005 launched the world's first international legal action on climate change, a petition to the Inter-American Commission on Human Rights. Her book, *The Right to be Cold*, about the effects of climate change in Inuit communities, was published in 2015. She is the recipient of many honorary doctorates and global environmental awards, including the Right Livelihood Award in 2015.

The former chief of the Xeni Gwet'in First Nation, Marilyn Baptiste, led a battle against the Prosperity copper and gold mine which would have destroyed Fish Lake, the source of spiritual identity and livelihood for Indigenous peoples in the south Chilcotin area of British Columbia. Both in federal environmental hearings and via a one-woman blockade in 2011 to prevent construction crews from reaching the proposed mine site, Baptiste's leadership catalysed opposition to the mine. The Supreme Court of Canada's Tsilhqot'in decision in 2014 granted First Nations land title to more than 1,700 square kilometres in the Nemiah Valley where Baptiste lives. Baptiste was awarded the Goldman Prize for grassroots environmental activism in April 2015.

In 2009, Eriel Tchekwie Deranger (Athabasca Chipewyan First Nation, Dene, Treaty 8 territory), the Tar Sands Campaigner for the Rainforest Action Network, climbed to the top of a Toronto flagpole and dropped a banner saying "Please Help Us Mrs. Nixon.com", urging Royal Bank of Canada CEO Gordon Nixon's wife Janet Nixon to help the activists pull RBC's investments in the Alberta tar sands (Hunter 2009). Deranger later became the communications manager of the Athabasca Chipewyan First Nation, which is suing Shell Canada regarding unmet Impact Benefits Agreements obligations, and she continues to campaign against tar sands projects and publicize their impacts. In a 2014 speech at Harvard University, she said "Our people and our Mother Earth can no longer afford to be economic hostages in the race to industrialize our homelands. It's time for our people to rise up and take back our role as caretakers and stewards of the land" (Winton 2014). Deranger's mother and father met at the American Indian Movement protests at Wounded Knee, and campaigned against Indigenous land takeovers by Dorado and other mining companies (Ball 2012).

Melina Laboucan-Massimo (Lubicon Cree First Nation, treaty 8 territory), an activist, documentary filmmaker, and tar sands campaigner with Greenpeace Canada and the Indigenous Environmental Network, has publicized the health and environmental impacts of tar sands exploitation in her northern Alberta homeland, including the effects of the 28,000-barrel Little Buffalo pipeline spill in April 2011 (Laboucan-Massimo 2015).

Theresa Spence, Chief of the Attawapiskat First Nation from 2010 to 2015, called attention in 2011 to appalling living conditions on the reserve, where many families were living in tents, more than 200 homes needed replacement or major repairs, and spring flooding endangers both infrastructure and the drinking water supply. In the winter of 2012, she carried out a six week hunger strike during the Idle No More protests while living in a teepee on an island in the Ottawa River, asking for the Prime Minister and Governor General to meet with chiefs to discuss living conditions on reserves. A group of Manitoba women chiefs, after meeting with her, stated, "We share Chief Spence's deep concern for the future of our nations and echo Chief Spence's call for restoring our relationship with the Crown to reflect the original spirit and intent of the treaties" (CBC 2013).

Crystal Lameman, Treaty Coordinator and Communications Manager for the Beaver Lake Cree Treaty 6 territory) is leading their legal action against the Alberta and Canada governments because of the damage tar sands mining has caused on their traditional lands, covered by an 1876 treaty. This is the first legal consideration of First Nations treaty rights in relation to the tar sands (DeSmog Canada 2013).

Along with other young Indigenous women, Helen Knott (Prophet River First Nation, Treaty 8 territory), a social worker and community activist, set up the Rocky Mountain Fort protest camp on 13 December 2015 to block land clearing and construction of the "Site C" dam on the Peace River in northeastern British Columbia, which was approved by a joint federal-provincial review in 2014. The dam would flood 107 kilometers of the Peace Valley, destroying farmland and sacred burial grounds. Although challenges are making their way through the courts, BC Hydro obtained an injunction to remove the protesters and plans to continue construction (McSheffrey 2015; Smith 2015).

Led by women tar sands activists, First Nations representatives met in January 2016 in Edmonton to develop an *Indigenous Climate Change Action Plan* based in the *UN Declaration on the Rights of Indigenous Peoples* (Connors 2016; Indigenous Climate Action 2016; Perry 2016).

There are many additional examples from across Canada of extraction, pipeline and climate justice protests and campaigns led by women activists, fighting for both distributive and procedural justice (Gray-Donald 2015).

Climate change and women's activism in Canada

In Canada, the effects of climate change are gendered, as they are everywhere. Extreme weather events, warming cities, melting sea ice and permafrost, ice storms, floods, droughts and forest fires cause economic and social impacts that affect women differently from men, as a function of their gendered economic and social positions. For example, cleaning up after a flood, moving house during a fire, finding alternate sources of food and caring for depressed partners and sick or upset children are usually women's responsibilities. Women and men also have differential access to redress and to political processes shaping public responses.

A growing body of literature, documented elsewhere in this book, shows how women often bear the brunt of extreme weather events, at both the household and the community levels. Women's gendered social roles, economic positions and expertise derived from activism related to inequality are logical reasons for their climate activism. As a result, environmental and climate justice movements often employ organizing and activist techniques developed within the feminist movement, such as consciousness raising, unmasking patriarchy and contextual reasoning – the grounding of the movement's theorizing in women's lived experiences rather than abstractions (Weiss 2012). At the same time, environmental and climate justice activism changes the lives of the women involved and, by extension, other women, forcing them to confront the constraints they face – time, work and other opportunities, political agency and so on – and thereby creating the conditions and potential for more radical change (Weiss 2012, 6). This seems to describe the process that has been playing out in Canada since the late 1990s, with a huge push from Indigenous women's grounded, culturally-embodied activism (Nixon 2015).

Indigenous chief and activist Arthur Manuel comments in his book *Unsettling Canada: A National Wake-Up Call* that women have long held leading roles in Indigenous activism on land, rights and the environment, and the majority of young Indigenous activists today are women (2015, 211). Clayton Thomas-Muller, tar sands campaign organizer for the Indigenous Environment Network and Defenders of the Land, also emphasizes the importance of Indigenous women's leadership (Thomas-Muller 2014).

Says *The Feminist Wire* website,

> Indigenous women activists and academics have shown how the foundation of contemporary capitalism was contingent on industrial resource extraction of Indigenous people's land, which was also simultaneously fully reliant on disempowering any positive ethic towards nature and women. This was achieved by installing European forms of gender relations and dismantling women's power, aided by the appropriation of Indigenous women's bodies. Residential schools were perhaps the strongest tools in reinscribing balanced gender relations of North American Indigenous matrilocal societies into the unequal ones of patriarchal models imposed by European colonizers and settlers. For the women's contingency (at the September 2014 Peoples' Climate March in in New York City), the centrality of resisting the colonization of Mother Earth, Terra Madre, and Pachamama is paramount. (Gorecki 2014)

Indigenous women identify very clearly the connection between environmental and gender justice. The 2014 climate march was led by a group of Indigenous people including Melina Laboucan-Massimo, who commented,

> Violence against the earth begets violence against women. I think when we don't deal with both of them we're not ever really going to resolve the issue of the colonial mind and the colonial mentality and the values of patriarchy and the values of capitalism that essentially exploit the land and exploit our women. (Gorecki 2014)

Kanehsatà:ke Mohawk activist Ellen Gabriel stated,

> Indigenous women were targets of the *Indian Act* because they (European colonizers) knew that the power rested with the women. And right now it's a man's world. In fact, it's a rape culture because in Canada, rape of Indigenous women has gone on with impunity and the government of Canada refuses to . . . have an inquiry because it profits them to continue to oppress Indigenous People . . . and it's another form of genocide as far as I'm concerned. (ibid.)

At least 1,200 Indigenous women, and perhaps far more, have been murdered or reported missing since 1980 in Canada. Bella Laboucan-Massimo, Melina's sister, who died July 20, 2013, is one of them. Indigenous women are eight times more likely to be killed than non-Indigenous women in Canada (Inter-American Commission on Human Rights 2014, 11; Kirkup 2016; Narine 2015). Calls for a federal inquiry into this problem long went unheeded by the Harper government; Indigenous women started their own lists (It Starts With Us; Walk4Justice) and continue to call for official investigations.

Outcome: Canada's 2015 election

By 2015, the connections among fossil fuel extraction, climate change, extreme weather events, violence against women and women's activism were manifest. Polls prior to the 2015 Canadian federal election showed a big split between women and men on climate change and other environmental issues. A nationwide random-sample poll conducted for the Climate Action Network in April 2015 found that while a majority of all Canadians felt that protecting the climate is more important than building pipelines or developing the Alberta tar sands (61%), the gender breakdown was also significant: 74 per cent of women agreed, compared to 52 per cent of men; 68 per cent of women said that building the Energy East pipeline to take tar sands oil to Atlantic ports was unethical because it is harmful to the environment, while only 43 per cent of men agreed; 71 per cent of women said that a federal government promise to legally limit carbon pollution was important or very important, compared to 49 per cent of men; 72 per cent of women wanted to see a federal commitment to phase out oil, gas and coal and replace them with renewable energy; 47 per cent of men felt the same way. And according to this poll, women constituted 67 per cent of those Canadians who had not yet decided who to vote for in the federal election; they were apparently waiting to see whether any of the (male) candidates would represent their views. Louise Comeau, Executive Director of Climate Action Network, commented,

> Women have the power to move Canada forward on climate protection. If Canada is going to step up to the plate, women need to be at the forefront of what is clearly an ethical issue with serious implications for our children. (Climate Action Network 2015)

A pre-election poll in September 2015 showed that the environment and health were about twice as important to women as to men, while "the economy" was much less important for women than for men (CBC News 2015a). Another 2015 poll found that women were more likely than men to consider climate change a serious problem (90 per cent of Canadian women compared to 77 per cent of Canadian men), to be concerned it will harm them personally (71 per cent to 51 per cent) and to say that major lifestyle changes are needed to solve the problem (81 per cent to 66 per cent) (CBC News).

During the 2015 campaign, Justin Trudeau promised to launch a national public inquiry into missing and murdered Indigenous women and girls (MMIW). He also said that he would include an equal number of men and women in his cabinet; ensure gender-based impact analysis in cabinet decision-making; develop a national childcare framework to ensure affordable fully inclusive childcare is available to all families who need it; implement a federal gender violence strategy; strengthen gun control; phase out subsidies for the fossil fuel industry; cancel the Northern Gateway pipeline; re-do the expansion review process for the Kinder Morgan Trans Mountain pipeline; and provide new funding for Indigenous languages and culture, education and schools, post-secondary student support, water supply, roads, Métis self-government and economic development. He promised to enact the recommendations of the Truth and Reconciliation Commission on the legacy of residential schools; to review, repeal and amend all existing laws that do not respect Indigenous rights or that were passed without proper consultation; to ensure every new policy and law meets with the principles of the *UN Declaration on the Rights of Indigenous Peoples* (UNDRIP); and to guarantee that First Nations communities have a veto over natural resource development in their territories (Trudeaumetre 2016).

While Harper made it clear that climate change was not a priority for him (Scrimshaw 2015), NDP candidate Thomas Mulcair promised affordable childcare, women's shelter enhancement programs and a strategy to end violence against women, as well as an inquiry into MMIW (Onstad 2015). A campaign debate focusing on "women's issues" was cancelled after Harper refused to participate and then Mulcair backed out; only Trudeau and candidates of the minor parties were willing to come (Armstrong 2015; Hansen 2015; Up for Debate 2015). Although the New Democratic Party, traditionally Canada's most progressive party, fielded more women candidates in the 2015 election (40% of candidates, compared to 33% for the Liberals and below 20% for the Conservatives) (Murphy 2015), the NDP downplayed its gender and climate positions, in an attempt to appeal to more centrist voters, while the Liberals emphasized theirs.

Voter turnout in the 2015 Canadian federal election reversed a 30-year trend of declining participation. There was a 7.39% increase in voter turnout in the 2015 election, compared to the previous federal election in 2011 when 2.736 million more Canadians showed up at the polls, "a clear indication of the degree to which the 2015 election galvanized the Canadian public", and the vast majority of these new voters voted for Trudeau (Majka 2015).

Elections Canada does not provide an immediate breakdown of voters by gender, so the precise role of women's votes in deciding the result is not yet known. In the 2011 election, however, 2 per cent more women than men voted, which meant that half a million more women than men turned out to vote; this led one commentator to note that "women vote bigger than their demographic" (McInturff 2015).

Also a big factor in the 2015 election was the impressive First Nations voter turnout. In 2014, Harper's Conservative government – perhaps threatened by the success of Idle No More and wanting to disenfranchise Indigenous voters – passed Bill C-50, which he called the "Fair Elections Act", making it harder to vote without approved identification. This backfired, however, when Indigenous activists led by women conducted a massive voter registration and election mobilization campaign. Some reserves saw voter turnout spike by up to 270 per cent in the 2015 election. In the riding of Kenora, Ontario, where 40 Northern Ontario First Nations are located, some First Nations polling stations ran out of ballots and either photocopies were used or voters waited patiently for more ballots to be brought in (Puxley 2015).

Tania Cameron, a band councilor for Dalles First Nation who started "First Nations Rock the Vote" on Facebook as part of her organizing efforts (which also included "ID clinics" where people could find out if they were registered or had the required documents to cast a ballot), commented,

> It was so heartening to see . . . Harper's intent was to suppress the Indigenous vote and that motivated me. It just caught on. I think the excitement of getting rid of the Harper government, showing Harper that his oppression tactics weren't going to work – I think that was a huge motivator for many people who decided to step up . . . I was thinking were're going to see a turnout that Harper never expected. (ibid.)

According to Manitoba First Nations activist Leah Gazan, the Indigenous voter turnout was also related to the Conservative government's Bill C-51, passed in 2015 (which increased police powers, allowed domestic spying and could criminalize Indigenous activists), and to Harper's weakening of federal environmental protection and support for Aboriginal organizations.

> He was quite violent with Indigenous people through aggressive cuts and aggressive legislation that aimed to silence Indigenous people. (But) as much as he attempted to divide, he really brought people on Turtle Island together, she said. (ibid.)

The voter-registration drives apparently brought a high percentage of Indigenous voters to the polls, meaning that although they make up just 4.3 per cent of the population in Canada, Indigenous people voted much bigger than their demographic – and this largely benefited the Liberals (Grenier 2015).

Trudeau's Liberals won 184 of 338 seats in the election, with 39.5 per cent of the vote, to 31.9 per cent (99 seats) for the Conservatives, and 19.7 per cent (44 seats) for the NDP. A record ten Indigenous Members of Parliament were elected in October 2015, and in Kenora, Conservative Natural Resources Minister Greg Rickford went down to defeat (Puxley 2015). Jody Wilson-Raybould, a former crown prosecutor, treaty commissioner and regional chief with the British Columbia Assembly of First Nations, which represents the interests of 203 First Nations in BC, won her Vancouver riding and was then appointed Trudeau's Justice Minister (Stueck 2015).[1]

Immediately following his electoral victory, Trudeau announced his half-female cabinet: "Because it's 2015!" Besides Wilson-Raybould, the female cabinet ministers appointed included Catherine McKenna (Environment and Climate Change), Chrystia Freeland (International Trade), Carolyn Bennett (Indigenous and Northern Affairs), Jane Philpott (Health), Patricia Hajdu (Status of Women), Marie-Claude Bibeau (International Development and La Francophonie), Diane Lebouthillier (National Revenue), Kirsty Duncan (Science), MaryAnn Mihychuk (Employment, Workforce Development and Labour), Judy Foote (Public Services and Procurement), Mélanie Joly (Canadian Heritage), Diane LeBouthillier (National Revenue), Maryam Monsef (Democratic Institutions), Carla Qualtrough (Sport and Persons with Disabilities) and Bardish Chagger (Small Business and Tourism).

Trudeau and a delegation of ten federal and provincial ministers and other politicians attended the 2015 Paris climate change conference, and Trudeau supported the inclusion of Indigenous rights in the resulting agreement. First Nations leader, Grand Chief Edward John, also at the First Nations Summit there, said:

> Canada has taken a very supportive role in Paris which is absolutely welcome given where we have been over the last decade on this issue. The only recourse we have had is to the courts. (Prystupa 2015a)

Environment and Climate Change Minister Catherine McKenna met with environmental activists at COP21, endorsed the goal of limiting global warming to 1.5 degrees C, and advocated a national carbon tax (McSheffrey 2016, Prystupa 2015b).

Trudeau's government may not take particularly radical action on climate change, but at least the discourse has changed: politicians can no longer sweep climate justice under the rug and pretend it's not an issue for Canada. Clearly the level of policy discussion on gender and climate justice in Canada was transformed by the 2015 election outcome – driven largely by women and Indigenous leaders and voters.[2]

Conclusion

Returning to the question of distributive and procedural climate justice for women in Canada, the story traced in this chapter about women's leadership and organizing on environment, Indigenous and intersectional issues illustrates a convergence of gendered impacts, awareness and action.

As noted in climate justice theory, it is those on the front lines of climate change – both extreme weather events and extraction – who are most aware of its impacts and most knowledgeable about how they should be addressed; this puts women at the forefront of climate justice struggles. It is no surprise that Indigenous women, facing health and livelihood crises due to fossil fuel extraction on their territories, are leading movements to address this issue at its source. Besides the gendered economic and social roles that all women face in a patriarchal society, cultural factors also lead Indigenous women to assert their voices and leadership on matters related to water, health, education and livelihoods.

It is important to underscore that the climate justice struggle in Canada, led mainly by Indigenous women, is not subject to claims or control by Western environmental and/or feminist movements (Nixon 2015). These women's activism highlights a key distinction in how gender justice and climate justice are linked in Canada (and likely in other countries that both produce and consume fossil fuels). It is the toxic effects of fossil fuel *production* itself – water and air pollution, ecosystem impacts on fish, wildlife, soils and forests, and particularly in Alberta the huge scale of government-subsidized tar sands operations, trampling on local governance processes and Indigenous land rights – that first and most clearly demonstrate the deathly problematic nature of the economic system that produces climate change. The impacts of fossil fuel *consumption* – greenhouse gas emissions leading to extreme weather events and weather variability, and their myriad health, social and economic outcomes – while global in their implications, are longer-incubating; their gendered effects also have great political significance. Women's leadership to address all of these gendered climate-related injustices is a powerful political force in both fossil fuel producing and consuming countries.

Notes

1 Harper's Environment Minister Leona Aglukkak, former Chair of the Arctic Council who had served in the Nunavut government as Minister of Health and Social Services and the Minister Responsible for the Status of Women, was defeated in the 2015 election. She was the first Inuk to serve in the federal Cabinet (CBC News 2015b). Earlier, her statements on climate change had ignited controversy when, during a 2013 television interview, she avoided agreeing that the climate is changing. Asked whether ice was melting in the Arctic, she sighed and said it was "debatable." The important thing, she said is that

> people that live in the Arctic become experts and are engaged in that . . . Because we live in that environment every day. We are seeing the changes every day or no changes – what have you – and we have valuable information to contribute to research. (True North Smart and Free 2013)

She also seemed reluctant to use the term 'climate change' during the interview.

> I was in Oslo, just recently at the climate ch-ah climate conference, ah environment ministers conference, sorry, she said. (ibid.)

At the time Aglukkaq was appointed Environment Minister in 2013, the conservative National Post commented,

> Insiders say Aglukkaq is hard-working, bright, looks after her constituency and is highly managed from the top. She seldom goes off script and almost never scrums with reporters. (Cheadle 2013)

Just as elsewhere in Canada, there is a broad political spectrum in Nunavut with significant support for all political parties (Ducharme 2015); Aglukkaq lost her seat in the 2015 election to Hunter Tootoo, a Liberal.

2 However, a 2 March 2016 meeting on climate change in Vancouver between Trudeau and provincial premiers "fell to shambles", according to Athabasca Chipewyan Chief Allan Adam, who said,

> I think Canada's in a crisis and it ain't going to get any better now. Canada failed terribly, the provinces failed terribly in regards to addressing this issue. (Moran 2016)
>
> Trudeau's failure to include the Congress of Aboriginal Peoples or the Native Women's Association of Canada in the Vancouver meeting led to criticism and disappointment. (Weber 2016)

In Ontario, Indigenous leaders including Ontario Regional Chief Isadore Day called for more formal recognition and engagement with government on climate change strategies (Garlow 2016).

References

Anderson, M 2014, "IMF pegs Canada's fossil fuel subsidies at $34 billion", *The Tyee*, 15 May, (https://thetyee.ca/Opinion/2014/05/15/Canadas-34-Billion-Fossil-Fuel-Subsidies/).

Armstrong, J 2015, "Mulcair, Harper not attending debate on women's issues", *Global News*, 24 August, (http://globalnews.ca/news/2182494/mulcair-harper-not-attending-debate-on-womens-issues/).

Burn, Donald H 1999, "Perceptions of flood risk: A case study of the Red River flood of 1997", *Water Resources Research*, vol. 35, no. 11, pp. 3451–3458.

Canadian Association of Petroleum Producers 2016, "Statistical Handbook, Table 3.16a".

CBC News 2013, "Chief Theresa Spence to end hunger strike today", 23 January, (http://www.cbc.ca/news/politics/chief-theresa-spence-to-end-hunger-strike-today-1.1341571).

CBC News 2015a, "Vote Compass: Economy and environment rate as top issues", 10 September, (http://www.cbc.ca/news/politics/vote-compass-canada-election-2015-issues-canadians-1.3222945).

CBC News 2015b, "Conservative's Leona Aglukkaq out as Liberals sweep the North", 19 October, (http://www.cbc.ca/news/canada/north/conservative-s-leona-aglukkaq-out-as-liberals-sweep-the-north-1.3279293).

Cheadle, B 2013, "Jury out on whether Leona Aglukkaq will be help or hindrance as environment minister", *National Post*, 21 July, (http://news.nationalpost.com/news/canada/canadian-politics/jury-out-on-whether-leona-aglukkaq-will-be-help-or-hindrance-as-environment-minister).

Climate Action Network 2012, "End tax breaks to Big Oil", *Climate Action Network*, (http://climateactionnetwork.ca/issues/getting-off-fossil-fuels/end-tax-breaks-to-big-oil-2/).

Climate Action Network 2013, Canada wins "Lifetime Unachievement" Fossil award at Warsaw climate talks, *Climate Action Network*, 22 November, (http://climateactionnetwork.ca/2013/11/22/canada-wins-lifetime-unachievement-fossil-award-at-warsaw-climate-talks/).

Climate Action Network 2015, "National poll shows Canadians want leadership on climate protection", *Climate Action Network*, 7 April, (http://climateactionnetwork.ca/2015/04/07/61-of-canadians-say-protecting-the-climate-more-important-than-pipelines-and-tarsands/).

Connors, D 2016, "Groundbreaking meeting sets course for Indigenous climate action", *Rabble.ca*, 11 February, (http://rabble.ca/blogs/bloggers/council-canadians/2016/02/groundbreaking-meeting-sets-course-indigenous-climate-actio).

DeSmog Canada 2013, "The most important tar sands case you've never heard of", *Huffington Post*, 28 May, (http://www.huffingtonpost.ca/desmog-canada/beaver-lake-cree-tar-sands_b_3334160.html).

Donnelly, G 2012, "Indigenous women in community leadership case studies: Fort McKay First Nation, Alberta, International Centre for Women's Leadership", Coady International Institute, Antigonish, NS.

Ducharme, S 2015, "Aglukkaq comes under fire at CBC Nunavut all-candidates forum: Conservative incumbent trades jibes with Liberal Tootoo", *Nunatsiaq Online*, 14 October, (http://www.nunatsiaqonline.ca/stories/article/65674aglukkaq_comes_under_fire_at_nunavut_candidates_forum/).

Garlow, N 2016, "First Nations have a key role in climate change action", *Two Row Times*, 29 February, (https://tworowtimes.com/news/first-nations-need-to-take-a-key-role-in-climate-change-action/).

Gillespie, C 2008, "Scar sands", *Canadian Geographic*, vol. 128, no. 3, June, p. 64–78.

Gorecki, J 2014, "'No climate justice without gender justice': Women at the forefront of the People's Climate March", *The Feminist Wire*, 29 September, (http://www.thefeministwire.com/2014/09/climate-justice-without-gender-justice-women-fore-front-peoples-climate-march/).

Government of Canada 2015, "Background on the inquiry", 4 December, (http://www.aadnc-aandc.gc.ca/eng/1449240606362/1449240634871).

Gray-Donald, D 2015, "Protesters keep shutting down the Line Nine oil pipeline", *Vice*, 20 December, (https://www.vice.com/en_au/article/protesters-keep-shutting-down-the-line-9-oil-pipeline).

Grenier, E 2015, "Indigenous voter turnout was up – and Liberals may have benefited most", *CBC*, 16 December, (http://www.cbc.ca/news/politics/grenier-indigenous-turnout-1.3365926).

Hansen, J 2015, "The Up for Debate campaign helped bring meaningful change", Amnesty International, 29 October, (http://www.amnesty.ca/blog/debate-campaign-helped-bring-meaningful-change).

Hunter, E 2009, "EcoChamber #15: Meet the woman at ground zero of the rat-sands fight", *THIS*, 5 August.

Indigenous Climate Action 2016, "Indigenous peoples meeting on climate change", (https://www.indigenousclimateaction.com/indigenous-peoples-climate-meeting).

Inter-American Commission on Human Rights 2014, "Missing and murdered Indigenous women in British Columbia", Canada, Organization of American States.

Kirkup, K 2016, "Carolyn Bennett says there are more than 1,200 missing or murdered Indigenous women", *Huffington Post*, 16 February (http://www.huffingtonpost.ca/2016/02/15/carolyn-bennett-missing-m_n_9238846.html).

Laboucan-Massimo, M 2015, "Awaiting justice: Indigenous resistance in the tar sands of Canada", *openDemocracy*, 13 May, (https://www.opendemocracy.net/5050/melina-loubicanmassimo/awaiting-justice-%E2%80%93-indigenous-resistance-to-tar-sand-development-in-cana).

Majka, C 2015, "Election 2015: The triumph of voter engagement", *Rabble.ca*, 30 October, (http://rabble.ca/blogs/bloggers/christophermajka/2015/10/election-2015-triumph-voter-engagement).

Manuel, A 2015, *Unsettling Canada: A national wake-up call*, Between the Lines, Toronto, ON.

McCarthy, S & Cryderman K 2014, "Oil sands pollutants contaminate traditional First Nations' foods: Report", *The Globe and Mail*, 6 July, (http://www.theglobeandmail.com/news/national/oil-sands-pollutants-affect-first-nations-diets-according-to-study/article19484551/).

McDiarmid, M 2013, "Alberta lakes show chemical effects of oilsands, study finds", *CBC News*, 7 January, (http://www.cbc.ca/news/politics/alberta-lakes-show-chemical-effects-of-oilsands-study-finds-1.1331696).

McInturff, K 2015, "Filling in the blanks: What do the polls say about women voters?" *Behind the Numbers*, 18 September, (http://behindthenumbers.ca/2015/09/18/filling-in-the-blanks-what-do-the-polls-say-about-women-voters/).

McSheffrey, E 2015, "Judge grants injunction to BC Hydro against Site C protesters", *National Observer*, 29 February, (http://www.nationalobserver.com/2016/02/29/news/judge-grants-injunction-bc-hydro-against-site-c-protesters).

McSheffrey, E 2016, "McKenna energized by Canada's climate leadership opportunity", *National Observer*, 2 March, (http://www.nationalobserver.com/2016/03/02/news/mckenna-energized-canadas-climate-leadership-opportunity).

Michelin, O 2013, "Women activists from Elsipogtog First Nation", *Aboriginal Peoples Television Network*, 17 October.

Moran, B 2016, "'Canada failed terribly, the provinces failed terribly', Chiefs disappointed after climate talks with PM, Premiers", *APTN National News*, 3 March, (http://aptnnews.ca/2016/03/03/canada-failed-terribly-the-provinces-failed-terribly-chiefs-disappointed-after-climate-talks-with-pm-premiers/).

Murphy 2015, "Canadian elections hinge on women – at the polls and behind the scenes", *The Guardian*, 1 July.

Narine, S 2015, "Violence against aboriginal women not an aboriginal-only issue", *Windspeaker*, 9 December.

Nixon, L 2015, "Eco-feminist appropriations of Indigenous feminisms and environmental violence", *The Feminist Wire*, 30 April, (http://www.thefeministwire.com/2015/04/eco-feminist-appropriations-of-indigenous-feminisms-and-environmental-violence/).

Onstad, K 2015, "The Chatelaine Q+A: NDP leader Thomas Mulcair", *Chatelaine*, 14 September, (http://www.chatelaine.com/living/politics/election-2015-the-chatelaine-qa-ndp-leader-thomas-mulcair/).

Paavola, J, Adger, WN & Huq, S 2006, "Multifaceted justice in adaptation to climate change", in WN Adger, J Paavola, S Huq & JJ Mace (eds), *Fairness in adaptation to climate change*, pp. 263–278, MIT Press, Cambridge, MA.

Perry, E 2016, "First Nations crafting an Indigenous climate action plan", Work and Climate Change Report, 22 February.

Prystupa, M 2015a, "Trudeau fights to keep Indigenous rights in Paris climate deal", *National Observer*, COP21, 7 December, (http://www.nationalobserver.com/2015/12/07/news/trudeau-fights-keep-indigenous-rights-paris-climate-deal).

Prystupa, M 2015b, "Climate Action Network nudged Canada's best at COP21", *National Observer*, 13 December, (http://www.nationalobserver.com/2015/12/13/video/climate-action-network-nudged-canadas-best-cop21).

Puxley, C 2015, "Anger, disenfranchisement behind surge in aboriginal voter turnout", *The Globe and Mail*, 25 October, (http://www.theglobeandmail.com/news/national/anger-disenfranchisement-behind-surge-in-aboriginal-voter-turnout/article26966275/).

Ross, G 2013, "The idle no more movement for dummies", *Indian Country Today Media Network*, 16 January, (https://indiancountrymedianetwork.com/culture/thing-about-skins/the-idle-no-more-movement-for-dummies-or-what-the-heck-are-all-these-indians-acting-all-indian-ey-about/).

Scrimshaw, Mackenzie 2015, "Harper addresses climate change on campaign trail: It's not a priority", *IPolitics*, 4 September, (http://ipolitics.ca/2015/09/04/harper-addresses-climate-change-on-campaign-trail-its-not-a-priority/).

Smith, J 2015, "Activist fights the BC dam project", *The Toronto Star*, 24 February.

Stueck, W 2015, "Aboriginal leader Jody Wilson-Raybould elected as Liberal in Vancouver Granville", *The Globe and Mail*, 20 October, (http://www.theglobeandmail.com/news/british-columbia/aboriginal-leader-jody-wilson-raybould-elected-as-liberal-in-vancouver-granville/article26882821/).

Thomas-Muller, C 2014, "The rise of the native rights-based strategic framework," in S D'Arcy, T Black, T Weis & JK Russell (eds), *A line in the tar sands: Struggles for environmental justice*, pp. 240–252, Between the Lines, Toronto, ON.

Troian, M 2013, "Mi'kmaq anti-fracking protest brings women to the front lines to fight for water", *Indian Country Media Network*, 11 October, (https://indiancountrymedianetwork.com/news/first-nations/mikmaq-anti-fracking-protest-brings-women-to-the-front-lines-to-fight-for-water/).

Trudeaumetre 2016, Trudeaumetre.ca (https://www.trudeaumetre.ca).

True North Smart and Free 2013, "Case #42: Environment Minister Aglukkaq unclear on climate change", *True North Smart and Free*, 2 October, (http://www.truenorthsmartandfree.ca/incident/environment-minister-aglukkaq-unclear-climate-change).

Up for Debate 2015, "Ending global poverty begins with women's rights", Oxfam, Canada, (http://www.oxfam.ca/upfordebate).

Weber, B 2016, "Climate change: Aboriginal leaders tell Trudeau they want seat at the table", *Winnipeg Free Press*, 3 February, (http://www.winnipegfreepress.com/arts-and-life/life/greenpage/trudeau-meeting-in-vancouver-with-aboriginals-on-climate-370857841.html).

Weiss, C 2012, "Women and environmental justice: A literature review", Women's Health in the North (WHIN), Australia, (http://www.whealth.com.au/documents/environmentaljustice/EJ-literature-review-HR-DP.pdf).

Winton, J 2014, "Eriel Deranger: Fighting the world's largest industrial project, the Alberta tar sands", *Cultural Survival*, 5 February, (https://www.culturalsurvival.org/news/eriel-deranger-fighting-worlds-largest-industrial-project-alberta-tar-sands).

Young, L 2014, "Alberta report finds Fort Chipewyan has higher rates of three kinds of cancer", *Global News*, 24 March, (http://globalnews.ca/news/1227635/alberta-report-finds-fort-chipewyan-has-higher-rates-of-three-kinds-of-cancer/).

Zuckerman, E 2012, "Are tar sands pipelines positive or negative for women?", *Huffington Post*, 21 May, (http://www.huffingtonpost.com/elaine-zuckerman/pipelines-damage-women_b_1525569.html).

19

USING INFORMATION ABOUT GENDER AND CLIMATE CHANGE TO INFORM GREEN ECONOMIC POLICIES

Marjorie Griffin Cohen

Introduction

In this century and especially since the economic crisis of 2008 many have had a sense that we are on the brink of seismic societal changes. The experience of the *Great Economic Recession* appeared to spike the idea that a revolutionary vision was essential to rethink the usual organization of capitalist societies. Since then enhanced media and activists' focus on the upward spiral of inequality, the devotion of most governments to "austerity" policies, the timid acceptance of shoddy corporate behaviour, and the ratcheting down of the collective provision of social supports that people really need has continued the strong sense that a different vision of the future is needed.

For some the pending apocalypse of very rapid climate change appears to be the possible catalyst for change. In a fairly short period of time environmentalists' concerns about climate change pivoted from being important to very few to being widely recognized as a major problem. Not continuing in the same direction that has brought about climate change needs decisive action, and the good part for many is that it might be the issue that could unite people from a variety of different groups and different ways of thinking.

This chapter examines the ways that the inclusion of gender in the analysis of a green economy and concepts of green jobs could be transformational in thinking about the nature of the economy and climate change initiatives. The discussion of green jobs and the green economy currently covers a very wide spectrum of analyses of what needs to be fixed, how to do this and what the impact of any solution may be on the economy. This ranges from the "light green" approach that has a fairly upbeat view of the ways that personal responsibility for climate change can make enormous differences, to the "dark green" approach that sees any solution only in radical political change in capitalism (Steffen 2009). Within this range is not only the "bright green" approach of the convergence of technological change with social innovation for positive and effective results, but also the no-growth green prescription with implications for permanent austerity.

The argument in this chapter is two-fold. One argument is that the most prominent ideas about green jobs and a green economy take the social organization, including the gendered division of labour, as given and assume that the issues of climate change are able to be appropriately dealt with within a system that accommodates market-oriented solutions. I see these objectives and solutions to climate change as a denial of the scale of the magnitude of changes needed and argue that the basis of social organization also needs to change. The second main argument in this paper relates to the social framework of those on all sides of the political spectrum regarding a green economy: there is a common position held by economists on both the left and the right that informs the framework for policy discussions on a green future. The problem is that this framework tends to see only certain sectors of the economy as the basis for both stability and growth. In virtually all ideas about a green future the significance of social reproduction, and the gender implications of its role in creating a green economy, is not a crucial part of change.[1] These approaches have in common similar ideas about what is productive and unproductive work and this, then, inhibits a shift in thinking about what kinds of economic activities could be foundational for economic success in the future. These solutions are obscured because of the gendered construction of our economy and the longstanding theoretical priorization of economic activities based on gender.

The arguments will be developed by first examining the divergent views on green jobs and what constitutes a green economy. It will show how the understanding of these concepts and their definitions matter with regard to solutions. The second part will discuss the economic biases against work related to "social reproduction" and set this in the context of both "green capital" and more progressive and inclusive thinking about climate change. The third section will examine the possibilities for a different kind of change with the inclusion of gender analysis in the discussion, as well as point to the major structural obstacles to creating a rational, inclusive economy to meet the needs of both the environment and people.

Concepts of green jobs and a green economy

The concept of green jobs has been an attempt to reconcile the differences between trade unions and environmentalists about policy to deal with climate change. The environmentalists' attempts in resource-extracting provinces in Canada to downsize or eliminate this work brought significant conflict with trade unions in these sectors. The most dramatic occurred in British Columbia where tensions escalated to a situation referred to as "the war in the woods". The premier at the time called environmentalists "eco-terrorists" and enemies of the province (Berman & Leiren-Young 2011).[2] The premier was not a right-wing climate change denier, but leader of the New Democratic Party of British Columbia, and was simply supporting his constituency of trade unionists in a province heavily dominated by resource extraction.[3] Ultimately it was essential that the trade union movement reconcile its position on climate change with the increasingly obvious necessity for governments to do something about it. This has not been easy because these industries, where men's employment dominates, are the "dirty" industries that are most likely to be affected negatively by any policies to reduce greenhouse gas emissions (GHG), something that would lead to increasing unemployment in some sectors that in Canada had already been battered by the effects of free trade agreements.

Like trade unions, the corporate sector initially challenged any carbon reduction initiatives as something that would be bad for it and for the economy in general. But as with almost every change that sweeps through society, in a fairly short period of time corporations found ways to create markets for new initiatives related to climate change. Most significantly, domestic businesses and transnational corporations over time actively positioned themselves as the primary agents for government policy considerations for action on climate change. Companies like Morgan Stanley and Goldman Sachs have clean technology divisions and most new clean energy initiatives in Canada involve constricting the public sector and expanding the private. Even companies in the Canadian tar sands (the most polluting resource extraction on earth) are also engaged in "green" technology, especially wind and bio-fuels. These, of course, are minuscule parts of their operations, but goals towards sustainability and "cleaner" production feature heavily in their public communications (Goldman Sachs 2011; Suncor 2011). Some environmentalists applaud the shift of capital towards wearing a "green" coat and see it as a "step in the right direction", but altogether the corporate vision of good green policy is contentious. Many groups attempting to represent social forces outside the corporate sphere, such as the People's Summit that was organized as a counterpoint to the business presence at Rio+20, called for mobilization against the corporate capture of the green economy discussions being held at the UN.

Trade unions were converted to the idea of green jobs through the spectre of considerable job creation that was expected to arise from new green investment. The understanding is that jobs will be lost in the short and medium term, but new types of jobs would offset these losses over time. There is no common definition of a green job, although the US Bureau of Labor Statistics finds the most consistent use of the term as referring to "jobs related to preserving or restoring the environment", a fairly fluid definition that could be either narrowly or widely interpreted. The categories of activities nearly universally cited are jobs that relate to "renewable energy, energy efficiency, pollution prevention and cleanup, and natural resource conservation" (Sommers 2013, 5). This has translated into a heavy focus on creating jobs related to the generation of green energy (wind, solar, biomass), public transportation, green building (primarily retro-fits and green construction), waste management, and agricultural and forest management. These include jobs in the manufacturing, construction and distribution aspects of the sectors. Sometimes tourism is included as a site for increasing green jobs as well.

As a proportion of the labour force of any advanced industrial nation, the numbers with green jobs is small. Counting green jobs is not consistent between nations and environmental or labour groups with specific political objectives tend to count green jobs in positive ways that accentuate the numbers. So, for example, in Canada it is claimed by a major environmental group that green jobs in the energy sector now outnumber those in the tar sands (Clean Energy Canada 2014). The impression is that green jobs are growing at a fast past and are overtaking those in conventional or dirty energy production: this needs to be taken with a bit of scepticism because of the inclusion of conventional hydro sources in this count. These are indeed green jobs, in that they have no GHG emissions, but since large hydro has always dominated in Canadian electricity generation, the count does not indicate a large change to something really new.

According to one analysis of trade union climate change action in Canada, policy development by labour has been confined almost entirely to the two largest private sector unions: the United Steelworkers (USW) and UNIFOR (Nugent 2011). Considering the public and governments' focus on GHG reductions it is not surprising that blue-collar identified unions are the most active of all trade unions on climate change and work and also that the discussions on both green jobs and the green economy tend to focus almost exclusively on a certain sub-set of industries. Government policy on climate change and most decidedly government support is in the interest of both business and labour groups in this group of industries. The public sector trade unions, whose members are predominately female, are mostly silent on how green jobs could be created in their sectors, although most do support climate change initiatives in general. The result of confining green jobs to male dominated sectors means that any discussion at all of gender inclusion is basically just a nod to the idea of desegregation of these sectors. In some countries there is, at times, a concerted attempt to include women in green jobs initiatives, such as during the stimulus spending during the 2008–10 recession, but the common practice, at least in the US, was for any new programmes to be withdrawn when the stimulus ended (Cohen 2015). The attempted integration of these traditional male dominant sectors over the years has had virtually no success in increasing female representation permanently in the industry in North America. For example, in Canada females have accounted for between 4 and 6 per cent of the building trades' total labour force for the past 40 years. The proportion is about the same in the US.

Few countries have gendered counts of green jobs, but one French study showed the results of programmes related to stimulus packages associated with the 2008–10 recession. This study made a distinction between green jobs, and greening jobs, with green jobs defined as jobs whose purpose is to measure, prevent, control or correct negative impacts on the environment. Greening jobs are not directly linked to environmental cleanup, but integrate new competences that take into account the "environmental dimension" to a significant and measurable extent. As Table 19.1 shows, the proportion of females in each sector with green jobs is small, although it is somewhat better with greening jobs. Most notable is that it is the sectors requiring relatively high levels of education where women are better represented.

TABLE 19.1 France: Gender composition of green jobs and greening jobs

Green jobs	*Male*	*Female*
Wastewater and pollution treatment	94%	6%
Energy and water production and distribution	85%	15%
Environmental protection	89%	11%
Other occupations*	72%	28%
Greening jobs		
Construction	95%	5%
Transport and logistics	88%	22%
Agriculture and forestry	81%	19%
Green space	94%	6%
Other*	83%	27%

* Covers mainly technicians and engineers; overall skill level in this category is higher than average.

Source: DARES Report (2012).

Green economy

The term "green economy" is recognized as being first used in a report's title in 1989 in the UK with regard to "sustainable development", although the concept was not developed within the paper itself (Pearce, Markandya & Barbier 1989). The term only became widespread during the discussions about the multiple crises associated with the economic recession of 2008 (UN n.d.) and is often referred to as a Green New Deal. The main focus for some governments and international organizations, like the UN, is how the two problems relating to the environment and the worst worldwide economic crisis since the Great Depression of the 1930s could be alleviated with stimulus spending to deal with job losses and low effective demand. Most rich countries instituted stimulus packages, with a small proportion of the spending going to green initiatives (Table 19.2). This varied from government to government, with most rich countries having fairly small commitments. In Canada there was no attempt in government work-related measures focusing on green initiatives to include women in actual jobs or job training and although most other rich countries had some programmes designed for marginalized groups, including women, the focus was primarily on multiple objectives with gender inclusion being only one of many.[4] If these programmes had been strongly supported with ample resources this might have made a difference for women, but that was not the case (Cohen 2015). Nor did any of those programmes that did focus on women have a lasting impact on the gender composition of the green jobs sectors.

The green economy has as many different types of definitions as does green jobs. In general, definitions of the green economy can be structured along a continuum of increasing specificity. On one end are formulations that highlight the progressive socio-economic possibilities latent in the green economy project, and on the other are ideas of how the green economy can coincide with capitalist

TABLE 19.2 Green stimulus spending

Country	*Green spending as % of total stimulus spending*
China	38%
Korea	81%
France	21%
Germany	13%
United States	12%
Canada	8%
Australia	7%

Source: Green Economy Post (2009).

market relations in a fairly narrow set of industries. The most influential and often cited statement on the subject is the United Nations definition. It is also potentially the most inclusive and would not necessarily be confined to a capitalist market economy: UNEP defines a green economy as one that results in "improved human well-being and social equity, while significantly reducing environmental risks and ecological scarcities" (UNEP 2011, 16). In practice, the use of the term tends to deal more with what is possible within the constraints of the current system and sees incremental improvements through new business opportunities as appropriate progress. For example, ECO Canada (2010) cites as its working point the definition in the *Stern Review* that sees that,

> action on climate change will also create significant business opportunities, as new markets are created in low-carbon energy technologies and other low-carbon goods and services. These markets could grow to be worth hundreds of billions of dollars each year, and employment in these sectors will expand accordingly. The world does not need to choose between averting climate change and promoting growth and development. Changes in energy technologies and in the structure of economies have created opportunities to decouple growth from greenhouse gas emissions. Indeed, ignoring climate change will eventually damage economic growth. (11)

The main attractiveness of the concept of a green economy is that there does not appear to have to be a choice between economic growth and environmental sustainability and that new industry to support the shift away from fossil fuel use could provide economic growth. Policies to support these solutions normally rely heavily on both government support of new private-sector industries, but most importantly, on the adequate pricing of the ecosystem. If this occurs the market mechanism is expected to discourage the use of fossil fuels and encourage green substitutes. The path to a sustainable economy would, then, be one that could tolerate expanded growth based on new technologies developing to deliver cleaner energy, and developing new competitiveness in green business. The role of innovation is critical in the reconfiguration of the capital stock to expand profitability, with a great deal of reliance on technological solutions that will make clean energy more efficient and cheaper (Eaton 2013). But also important is government support of the private sector initiatives through heavy subsidies to producers of green technologies and significant tax breaks for green industries.

Green taxes are the primary method through which pricing carbon occurs. The major debate among carbon pricing proponents in recent years is whether a carbon tax (on users of carbon) is preferable to cap and trade systems. So far, these are the main public policy approaches to dealing with GHG emissions, although some governments have been more proactive in banning the use of coal, an approach that uses regulation more forcefully to change behaviour.[5] With carbon pricing methods the major changes will be in the type of energy resources that are used and their distribution. It is clear that unless very carefully designed carbon pricing is a regressive tax and cap and trade so far has had problems where it has been most advanced in Europe and not been as effective as anticipated, primarily because it runs into serious difficulties during an economic downturn. But mainly it is a trading system for nothing tangible and is potentially subject to all of the manipulation that is rife in this type of trading. The main objections to carbon pricing as the primary policy for reducing GHG emissions are that any changes that occur will not meet the GHG reduction requirements, and take the existing economic system as given. This "green capitalism" relies heavily on continued economic growth and assumes the compatibility of economic and environmental objectives (Tienharra 2014).

Market-based solutions do not usually deal with the increased inequalities and consequent social justice issues that arise from the distributional aspects of green pricing, except for saying that the policy needs to be designed carefully to ensure social justice, something that rarely happens in governments' designs. The gender discussion associated with the market-solutions approach is usually confined to the ways that the incidence of the tax falls most heavily on the groups that have low incomes and how this might be rectified. So if gender is taken into consideration at all, it is as a subset of class. Any other discussion of market-based solutions related to gender issues are absent, a feature that has been widely noticed by feminist environmentalists.

The main counterpoint to the rosy scenario of green growth through the normal working of the market economy with appropriate carbon pricing is the no-growth model. In this approach a green economy would attempt to decouple consumption from growth, although most ideas about a no-growth economy tend not to deal with the real issues of economic hardship for masses of people. But some ideas are innovative could become more significant when coupled with ideas about how the economy could be sustained without material growth. Most ideas about a green economy deal with the problem of meeting people's needs primarily as a distribution issue in relation to a green economy. For those who envisage a no-growth economy this would be a serious issue in making decisions about sharing reduced resources and material goods, something requiring rationing and a strong political will. But for the most part the ideas about a green economy and how to reduce GHG emissions treat environmental equity issues as an add-on to the more pressing problem of combating climate: "climate justice" is low on the scale of policy priorities.

Social reproduction

Very few advocates for a green economy have a vision that encompasses in a serious way the kinds of work that women have historically done in all kinds of different economic contexts.[6] This is not accidental. Both the left and right analysis of economic and social policy have a blind spot with regard to what is considered productive labour and, therefore, what is most important for support through economic policy. Repeatedly governments in Canada point to the engines of growth as emanating from the manufacturing and resource sectors. Meeting people's needs in the public services sectors, even though these activities are a substantial part of the market economy, is almost consistently described as something that is paid for by the real engines of growth. This is not just a problem of semantics and uninformed politicians, but is at the very heart of ideas of what constitutes value and growth in any economy. This bias stems from a very long-standing failure to understand the basis of social reproduction as a significant economic element within any economy.

When feminists focus on social reproduction, it tends to be on the significance of unpaid housework as the overlooked element of the economy, and stress the ways the household itself supports the labour in the paid economy (Folbre 2004; Waring 1989). The observation that the very nature of capitalism is dependent on this type of work is crucial because of its implications for the ways governments treat the household: government resources to support the household and labour within the household are rarely treated as an economic investment. Also, support of household labour is usually considered in economic policy decisions as a social services expenditure that is only possible because other sectors of the economy have earned enough money to be taxed.[7] Any expenditure in the household sector is treated as consumption, while, in contrast, anything spent on supporting the privatized corporate sector is treated as supporting investment.

There is no doubt that unpaid labour is regarded as unproductive as a result of a long process of power negotiations that involved not simply class relations, but also gender relations. But what is or is not productive labour is also something that is shaped by ideas of what counts as the foundation of wealth and value at a given period of time. So, for example, the Physiocrats in the eighteenth century saw one sector, the agricultural sector, as driving all others, and work in this sector was the only productive work. Because this was the sector that produced true wealth the Physiocrats felt economic policy needed to reflect this by giving priority to the agricultural sector. In a similar way the resource and manufacturing sectors are considered, currently, to be the true engines of growth. As one well-known progressive economist in Canada says "real investment [in the form of] buildings, structures, equipment, tools and machinery [constitutes] the most important leading edge of growth and job creation in the economy". Like most economists, he does recognize that investment in human capital is important, but wants it made clear that "the benefits of human capital should not take away from a focus on tangible capital investment as a central source of economic growth and job creation" (Stanford 1999, 94).

This kind of thinking is common on both the left and the right, and economic policy tends to be shaped with this thinking almost subconsciously as its context. But this idea of manufacturing and resources as the leading edge of the economy is not borne out by the actual structure of the workforce, because the service sector of the economy of wealthy nations dominates economic activity. This includes financial services, business services, government services and care services, which together account for three-quarters of all employment and national income in most rich countries.

The point that I want to stress is that what "counts" in a capitalist economy is whatever gets produced and sold and there is not a requirement that anything bought or sold have a physical value. Capitalism does not specify what is to be produced, nor how it should be distributed, nor, most importantly, that production necessarily meet basic human needs. Selling anything will make the system itself work, whether it is something useful like food, wasteful like cars, or something useless like paper stocks or mutual funds. Any kind of selling puts money into the economy and people to work and it could function with any sector as a lead sector (as seems to be occurring now with the financialization of the economy). Many economists over the ages have understood this, as John Maynard Keynes' recommendation for a way to cure unemployment shows:

> If the Treasury were to fill old bottles with banknotes, bury them at suitable depths in disused coalmines which are then filled up to the surface with town rubbish, and leave it to private enterprise on well-tried principles of laissez-faire to dig the notes up again . . . there need be no more unemployment and, with the help of the repercussions, the real income of the community, and its capital wealth also, would probably become a good deal greater than it actually is. It would, indeed, be more sensible to build houses and the like; but if there are political and practical difficulties in the way of this, the above would be better than nothing. (1936, Book 3, Chapter 10, Section 6)

This flexibility of what constitutes value and wealth within capitalism does not mean that what is produced is irrelevant to people, as Keynes understood, rather, the actual nature of what sets off the activity is not all that important to the economy itself.

The real issue is the logic of our economic system and whether there is any way to better match real needs with production and employment. More employment is important, but our analysis cannot stop there and any discussion of economic policy ultimately has to deal with the economy as it is presently structured, even if we want to change it. The calls for no growth by some environmentalists, feminists and community economic development promoters, in the name of material sanity, usually pay no attention to the dreadful consequences a no-growth policy could have for real people. But, ultimately anyone dealing with climate change needs to struggle with the issue. It is irrational to perpetuate the proliferation of more and more irrelevant things that destroys our environment, simply to make the economy function. Those non-economists who question the narrow focus of economists are correct – it is the glut of things and the maldistribution of income that lead to overproduction crises in the world economy. Within the constraints of the existing economy and with the objective of having more good jobs and a rising material standard of living, economic growth is essential.

The huge growth in the services sectors of advanced industrial nations indicates that real prosperity may not be tied, ultimately, to manufacturing and resource extraction. The bias against certain aspects of the services sector (particularly those services directly meeting people's needs) has a long-standing logic that types this type of work as unproductive labour, very much along the lines of analyses that understand unpaid work associated with the household as unproductive work. That bias against the social reproduction dimensions of the economy, which is rooted in gendered differences, is reflected in the downward trend of government spending as a proportion of the national income on social programmes. In what follows I will show how the richest countries on earth have progressively reduced the proportion of government spending towards meeting peoples' needs and how this has affected the economy. This will also give an indication that there is a strong relationship between the way that social reproduction is treated in government policy and economic growth. Following this I will then discuss possibilities of growth itself and how this could occur to both meet people's needs and meet the demands for action on climate change.

Austerity policies

Rich countries in Europe and North America have been guided in recent years by neoliberal economic policies that encompass a whole range of policy directions that are assumed to be good for the economy. These "austerity" policies are informed by a theory called "expansionary austerity" (sometimes known as the German view) that believes that by aggressively cutting taxes, cutting government spending on social programmes and downsizing the public sector the private sector of the economy will expand considerably.[8]

In order to understand how significant public support of "social reproduction" is to the existing economic order I have tried to see what happens when there are serious reductions to social spending over a long period of time. In particular I examine whether the theory of "expansionary austerity" did in fact improve economic performance or whether it had a negative correlation through low growth, increased inequality and lower wages. What follows deals primarily with the reductions in social spending and show how this relates to the growth rate of one country, Canada.[9]

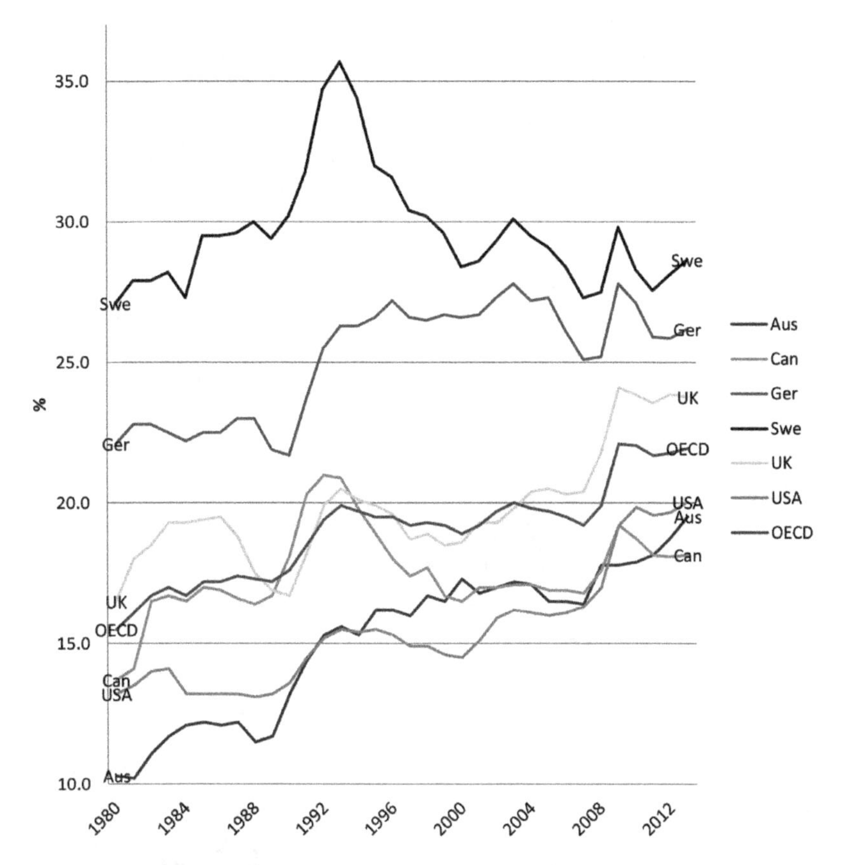

FIGURE 19.1 Public social expenditures as percentage of GDP (select countries)

Source: Calculated from OECD Stat (n.d.).

Figure 19.1 shows the relative standing of the six OECD countries that are the focus of this book and the OECD average from 1980–2012 in public social expenditures relative to national income.[10] There are some interesting differences between countries. Sweden, of course, is at the top, but its proportion of public social spending to GDP has dropped steadily since the early 1990s.

In contrast, Germany increased its share in the early 1990s and reached a fairly steady proportion until the 2008 recession. In the early 1990s Canada and the UK were above the OECD average, but Canada's decline has been remarkable and it is now last in this group and, with the US and Australia, is below the OECD average. The increase in public social spending to GDP improved temporarily in all countries as a result of the stimulus packages to counter the 2008 recession. This makes it very clear that governments do understand that spending on social reproduction can be an economic stimulus. Canada, in its reports on the success of stimulus packages, gave expenditure and tax multiplier information about its stimulus spending. This showed that spending money for low-income household and the unemployed was one of the highest stimulus actions, while reductions in business and other taxes actually withdrew money from the economy (Canada 2011).

Figure 19.2 looks at total government spending as a proportion of GDP for these rich countries and the OECD average. The high point for all of the countries was 1991, a time just before austerity policies were becoming more entrenched and effective, and when the rules of major free trade agreements were being put into place. While there was an increase in social spending among all governments to counter the 2008 deep recession, most have resumed the downward trend in the significance of government spending. Canada's position here is somewhat surprising. While Canada is ranked last in all of these countries in its public social spending, its total government spending ranks third, and is only below Sweden and

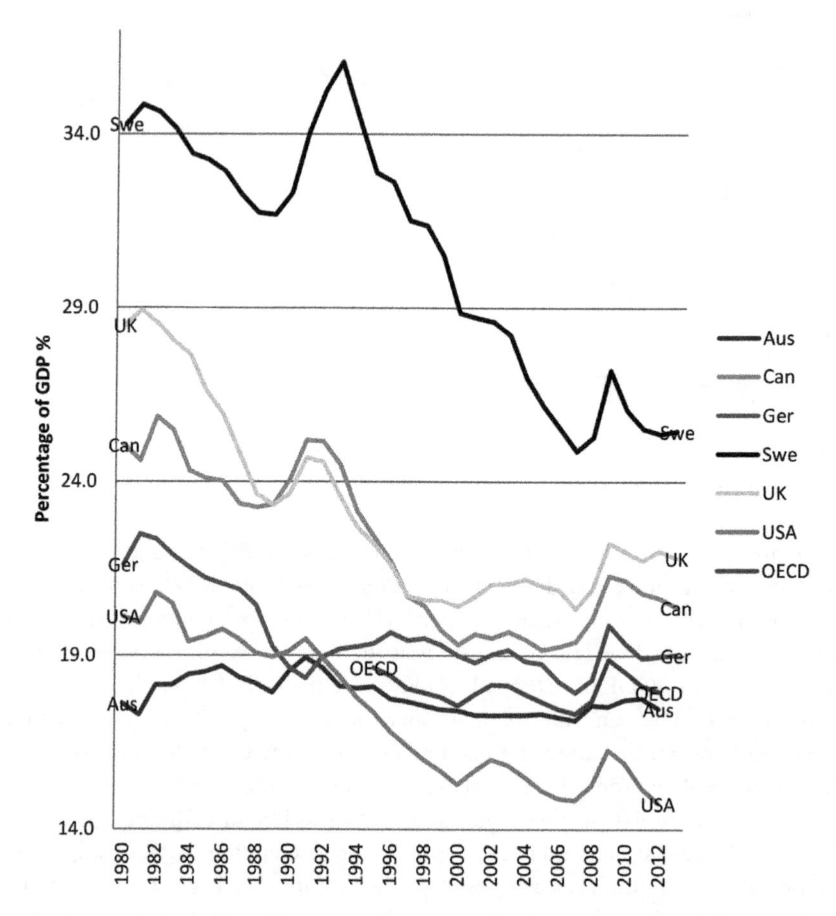

FIGURE 19.2 Government spending as percentage of GDP (select countries)

Source: Calculated from OECD Stat (n.d.).

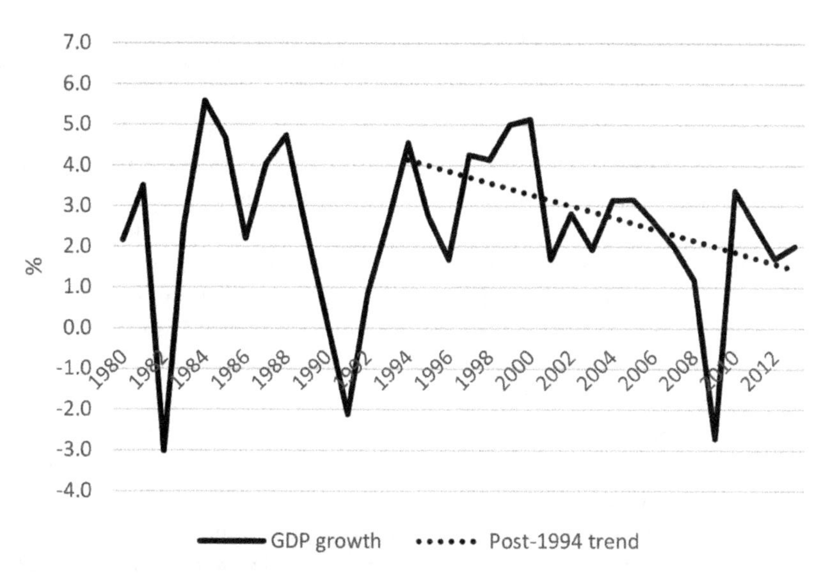

FIGURE 19.3 GDP growth rate: Canada

Source: Calculated from OECD Stat (n.d.), National Accounts, Main Aggregates, Gross Domestic Product.

Germany. This means Canada's governments do spend, but the spending on social reproduction is not keeping pace with other rich countries.

My interest is in knowing if the nature of public spending (i.e. what sectors are the focus of spending) is significant for its economic performance and whether the proportion spent on social programmes is a drain on the economy, as the austerity approach assumes. The results for Canada indicate that there is a strong likelihood that reduction in government social spending as a proportion of GDP puts a drag on the economy. Figure 19.3 shows the trend line for GDP growth, indicating that the steady downward rate of the growth trend is evident since Canada began its austerity programmes.[11] (In Canada the major changes occurred in the 1994 budget when transfers to the provinces were cut and no longer directed towards specific areas of social spending. This is also when there were major reductions to benefits for the unemployed and a confiscation of the Employment Insurance surplus in this budget.)

Why gender matters

In another study I calculated the GHG emissions by gender for Canada. The focus was on the distinctions between males' and females' work patterns. Because the labour force is highly segregated by gender, with the resource and manufacturing sectors dominated by males and the services sectors dominated by females, the GHG emissions from work are heavily dominated by males. As Table 19.3 shows, males account for 76 per cent of the GHG emissions through work, transportation, and the household while females account for 24 per cent of the total GHG emissions.[12] The point of this is not to point to male reponsibility for GHG emissions, per se, but to understand and reinforce which sectors of the economy are the problem sectors. So far, cleaning up these sectors has been the focus for public policy on climate change issues. While this is a necessary action, it still leaves the basic economic structures intact, both in terms of the kinds of sectors that remain dominant and the gendered division of labour. While integrating male dominated sectors is one strategy for dealing with future investments in a green labour force, it is a limited strategy, in that it deals with only a small segment of the economy.

A different vision for an economy that focuses on meeting rational needs of people would need a value system that understood the critical nature of social reproduction to the successful functioning of the system. As was mentioned earlier, sustaining an economy without growth is extremely difficult in any economic system, but is particularly problematic within a capitalist system. This difficulty needs to be factored into any shifts in the sectors that are promoted within the society. We now know that the exploitation and exports of resources, and the maniacal manufacturing of more and more things that are irrelevant to a good life are highly damaging to the environment. The superstructure of global free trade systems encourages and promotes this kind of economic activity and insists that all countries pursue it. Using a gendered perspective on economic organization and an understanding of the possibilities of social reproduction activities contributing to growth could point to the beginnings of an economic vision that could counter the current dominant ideas about a green economy.

One possibility would be to reconsider ideas about the "stationary state" as a possibility for the material aspects of life, much in the way that John Stuart Mill (1848) talked about the stationary state, while growth sectors could

TABLE 19.3 GHG emissions by gender in Canada (total MT by sector and gender % of GHG by sector)

Categories	*Total GHG emissions*	*Female % GHG*	*Male % GHG*
Industrial	455,050 (MT)	24.0%	76.0%
Transportation			
Road	133,701	11.0%	89.0%
Air	6,200	54.0%	45.0%
Residential	41,000	50.0%	
Share*	635,951	23.5%	76.5%

Source: Cohen (2014).

* The total GHG emissions in Canada = 692,000. This table includes only categories where it was possible to ascertain gender differences. So, for example, it does not include transportation other than road and air.

involve increases in the use of work in providing services that people need. Mill asked the crucial question about economic activity in an industrial capitalist economy:

> To what goal? Towards what ultimate point is society tending by its industrial progress? When the progress ceases, in what condition are we to expect that it will leave mankind? It must always have been seen, more or less distinctly, by political economists, that the increase of wealth is not boundless: that at the end of what they term the progressive state lies the stationary state . . . (IV. 6. 2)

Mill understood that the relentless pursuit of profit need not continue forever in order for the economy to function well, but that shifting what is the focus of the economy is possible:

> It is scarcely necessary to remark that a stationary condition of capital and population implies no stationary state of human improvement. There would be as much scope as ever for all kinds of mental culture, and moral and social progress. (IV. 6. 9)

A true and successful stationary state would require no growth in population (as Mill envisions), but there is another possibility. With a stationary state in material goods, and a growth in labour directed towards care services, even population could grow. It is in the areas of care services that countries like Canada are deficient and where the state's contribution, relative to the national income, has been declining steadily.

Material needs do need to be met and the redistribution of wealth within and among nations will certainly need to happen in order to arrive at a stationary material state. In the meantime, crucial shifts in government policy towards support for social reproduction could increase economic activity without increasing GHG emissions. This would mean recognizing the economic contributions of social reproduction to the strength of the system. But all of these approaches would require the strong assertion on the part of the state and its willingness to shift economic priorities away from the focus on material goods and financial services, towards working in the interests of social development and welfare.

Conclusions

Restoring an environment to one where there is a future that is sustainable for human beings will require a dramatic shift in the economic structures that currently condition the way we live, what we buy and how we work. The current economic thinking of both the left and the right political perspectives tends to subconsciously undervalue the kinds of work associated with women, both on the market and within the home. As a result, public support for this kind of work has not only declined in rich countries, but has been accompanied by lower rates of growth and greater inequalities. It is possible, however, for economies to prosper and still cut GHG emissions through less production in traditional areas of growth – namely material goods. This could be possible with a greater emphasis on the expansion of investment and work in the services that people need.

Notes

1 Social reproduction includes the activities of both males and females, and the ways that the market, the state, the community, the household and the individual are involved in meeting the direct needs of people. The state's role includes activities that directly and universally support the household (medical care, education, pensions, labour regulation and support), as well as specific programmes that are more targeted to meet the needs of specific populations (social assistance, disability aid, employment insurance, pensions, child care, housing). At various capitalistic stages each share undertaken by the actors in this process is different, with the state assuming a larger or smaller influence on the social security systems designed to support social reproduction, depending on the time, state of development and political ideology in ascendance (Cohen 2013, 235).
2 The "war in the woods" was most dramatically obvious when environmentalists stopped logging in old-growth forests. The result was more than 800 arrests of environmentalists and a mass arrest of more than 300 on one day, 9 August 1993. At the time the population of British Columbia was about 3.5 million, so the proportion willing to support the environmentalists by being arrested was relatively large. This "war" continued until the end of the decade. The danger for loggers was more than confronting protestors, since on many occasions they encountered nail-spiked trees and other serious hazards as they worked (Nursall 2013).
3 The NDP is a centrist social democratic party in Canada.
4 In the US these objectives included poverty reduction and training minorities that normally have difficulty finding employment, such as former prisoners, and those with multiple barriers to employment, such as poor, single mothers on welfare (Cohen 2015).
5 Ontario was the first jurisdiction in Canada that relied heavily on coal in electricity generation to outlaw its use (Ontario 2015).
6 Two significant exceptions to this are Mara Kuhl's study commissioned by the Greens/EFA Group in the European Parliament (2012), and Stefan Kesting (2011).
7 The premier of British Columbia, Canada, for example, sees that "resource development, which includes exporting liquefied natural gas to Asia, developing new mines and building the proposed Site C hydroelectric dam, represents an opportunity to ensure B's place as an economic powerhouse in Canada, while providing continued support for education, health care and social services and public infrastructure" (Canadian Press 2014).
8 The theory of the ability of "austerity" as a way to stimulate the economy is known as "expansionary austerity", or "expansionary fiscal contraction", and was given credibility by one of the first papers on this topic written by Francesco Giavazzi and Marco Pagano (1990).
9 The full study is available in a book entitled *State and austerity* (Cohen forthcoming).
10 Cross-country comparisons are always difficult. The OECD figures have been used because they are the most reliable. These figures include all levels of government, but for some reason exclude expenditures on education except through federal transfers.
11 Changes in the growth rate have a wide variety of causes, and are clearly not attributable only to changes in government spending on social programmes. In an export-dependent country like Canada that relies heavily on world prices, global changes have a big impact on the economy.
12 A detailed breakdown of the industries by gender and GHG emissions is available in Cohen 2014.

References

Berman, T & Leiren-Young, M 2011, *This crazy time: Living our environmental challenge*, Knopf, Canada.

Canada 2011, "Canada's economic action plan year 2: A seventh report to Canadians", Department of Finance Canada, (http://www.fin.gc.ca/pub/report-rapport/2011-7/index-eng.asp).

Canadian Press 2014, "Christy Clark firm on 5 pipeline conditions, balanced budget", 23 January, *CBC*, (http://www.cbc.ca/news/canada/british-columbia/christy-clark-firm-on-5-pipeline-conditions-balanced-budget-1.2507950).

Clean Energy Canada 2014, "Tracking the energy revolution", (http://cleanenergycanada.org/wp-content/uploads/2014/12/Tracking-the-Energy-Revolution-Canada-.pdf).

Cohen, MG 2013, "Neo-Liberal crisis/social reproduction/gender implications", *University of New Brunswick Law Journal*, vol. 64, pp. 234–252.

Cohen, MG 2014, "Gendered emissions: Counting greenhouse emissions by gender and why it matters", *Alternate Routes*, vol. 25, pp. 55–80.

Cohen, MG 2015, "Gender and government action on climate change: The US example", *Women & Environments International*, vol. 94/95, fall/winter, pp. 11–16.

Cohen, MG forthcoming, "Austerity's role in economic performance: The relationship between social reproduction spending, the economy, and people", in S McBride & B Evans (eds), *State and austerity*, University of Toronto Press, Toronto.

DARES Report 2012, "Jobs in the green economy: Typology and characteristics", European Working Conditions Observatory (EWCO), (http://www.eurofound.europa.eu/observatories/eurwork/articles/jobs-in-the-green-economy-typology-and-characteristics).

Eaton, D 2013, "Technology and innovation for a green economy", *Review of European Community & International Environmental Law*, vol. 22, no. 2, pp. 62–67.

ECO Canada 2010, "Labour Market Research Survey, Environmental Careers Organization", (http://www.eco.ca/pdf/Defining-the-Green-Economy-2010.pdf).

Folbre, N 1994, *Who pays for the kids? Gender and the structures of constraint*, Routledge, New York.

Giavazzi, F & Pagano, M 1990, "Can severe fiscal contractions be expansionary? Tales of two small European countries", *NBER Macroeconomics Annual*, vol. 5, pp. 75–122.

Goldman Sachs 2011, "Clean technology and renewables", (http://www.goldmansachs.com/what-we-do/investment-banking/industry-sectors/clean-tech-and-renewables.html).

Green Economy Post 2009, "Which country has the greenest stimulus package?" *The Green Economy Post*, (http://greeneconomypost.com/country-greenest-stimulus-package-674.htm).

Kestins, S 2011, "What is 'green' in the green new deal: Criteria from ecofeminist and post-Keynesian economics", *International Journal of Green Economics*, vol. 5, no. 1, pp. 49–64.

Keynes, JM 1936, *The general theory of employment, interest, and money*, Polygraphic Company of America, New York, Book 3, Chapter 10, Section 6, (www.marxists.org/reference/subject/economics/keynes/general-theory/).

Kuhl, M 2012, "The gender dimensions of the green new deal: An analysis of policy papers of the Greens/EFA New Deal Working Group", Berlin, (http://www.dr-mara-kuhl.de/fileadmin/user_upload/GND_Kuhl_ENGL.pdf).

Mill, JS 1848, *Principles of political economy*, Longmans, Green and Co, London, (http://www.econlib.org/library/Mill/mlP.html).

Nugent, JP 2011, "Changing the climate: Ecoliberalism, green new dealism, and the struggle over green jobs in Canada", *Labour Studies Journal*, vol. 36, no. 1, pp. 58–82.

Nursall, K 2013, "War in the woods mass arrests 20 years ago prompted lasting change", *Canadian Press*, 11 August, (http://bc.ctvnews.ca/war-in-the-woods-mass-arrests-20-years-ago-prompted-lasting-change-1.1406602).

OECD Stat n.d., "Social expenditure: Aggregated data", (http://stats.oecd.org/Index.aspx?datasetcode=SOCX_AGG#).

Ontario 2015, *Bill 9 An Act to amend the Environmental Protection Act to require the cessation of coal use to generate electricity at generation facilities*, Legislative Assembly of Ontario, (http://ontla.on.ca/web/bills/bills_detail.do?locale=en&Intranet=&BillID=3001).

Pearce, D, Markandya, A & Barbier, E 1989, *Blueprint for a green economy*, Earthscan, London.

Sommers, D 2013, "BLS green jobs overview", *Monthly Labor Review*, January, pp. 3–16.

Stanford, J 1999, *Paper boom: Why real prosperity requires a new approach to Canada's economy*, CCPA-Ottawa, J Loriner, Toronto, ON.

Steffen, A 2009, "Bright green, light green, dark green, gray: The new environmental spectrum", *Worldchanging*, 27 Feb, (http://www.worldchanging.com/archives/009499.html).

Stern, N 2006, "Stern review: The economics of climate change", HM Treasury, London, (http://mudancasclimaticas.cptec.inpe.br/~rmclima/pdfs/destaques/sternreview_report_complete.pdf).

Suncor Energy 2011, "Report on sustainability", (http://sustainability.suncor.com/2011/en/responsible/1829.aspx).

Tienharra, K 2014, "Varieties of green capitalism: Economy and environment in the wake of the global financial crisis", *Environmental Politics*, vol. 23, no. 2, pp. 187–204.

UN n.d., "Green economy", Department of Economic and Social Affairs, (https://sustainabledevelopment.un.org/index.php?menu=1446).

UNEP 2011, "Towards a green economy: Pathways to sustainable development and poverty eradication", United Nations Environmental Program, (http://www.unep.org/publications/contents/pub_details_search.asp?ID=4188).

Waring, M 1989, *If women counted*, Macmillan, London.

INDEX